AF352795

Bioprocess Monitoring and Control

Bioprocess Monitoring and Control

Editor

Bernd Hitzmann

MDPI • Basel • Beijing • Wuhan • Barcelona • Belgrade • Manchester • Tokyo • Cluj • Tianjin

Editor
Bernd Hitzmann
University of Hohenheim
Germany

Editorial Office
MDPI
St. Alban-Anlage 66
4052 Basel, Switzerland

This is a reprint of articles from the Special Issue published online in the open access journal *Processes* (ISSN 2227-9717) (available at: https://www.mdpi.com/journal/processes/special_issues/bioprocess_control).

For citation purposes, cite each article independently as indicated on the article page online and as indicated below:

LastName, A.A.; LastName, B.B.; LastName, C.C. Article Title. *Journal Name* **Year**, *Article Number*, Page Range.

ISBN 978-3-03936-932-4 (Hbk)
ISBN 978-3-03936-933-1 (PDF)

Contents

About the Editor

Bernd Hitzmann, Dr. rer. nat., is a professor at the University of Hohenheim. After his studies and doctorate at the University of Hannover, he worked at the California Institute of Technology, Pasadena, USA, and at ABB's corporate research center, in Heidelberg, Germany. He habilitated at the University of Hannover. Since 2011, he has been Head of the Department of Process Analytics and Cereal Science, Institute of Food Science and Biotechnology, at the University of Hohenheim, Germany. His expertise lies in bioprocess monitoring and control, spectroscopy, modelling, and chemometrics.

Preface to "Bioprocess Monitoring and Control"

Bioprocesses have been carried out for thousands of years. When they were first used, based on experience, one knew how to carry out these processes so that the required product, mainly food, was obtained in the desired quality. In the time of Antoni van Leeuwenhoek (1632–1723), the important protagonists of bioprocesses (microorganisms) became accessible. In particular, the production of beer, i.e. ethanol, was one major player, which gave rise to the improvement of our knowledge of bioprocesses significantly. That temperature is one of the most important process variable was recognized very early. However, during the 20th century, a lot of improvements in process development were gained. At the beginning of that century, examples were the development of a fed-batch process for baker yeast production and the patent of the production process of acetone and butanol (1915). In 1949, Monod published his model for the specific growth rate of microorganism. Not only did process development evolve, but process measurement systems such as spectroscopy also did, at BASF in 1938. These are important for the collection of on-line process information. From 1941 to 1944, industrial penicillin, i.e. medication drugs, began to be used, and glutamic acid production began in 1957. Around that time (1960), Rudolf E. Kalman published a new approach to linear filtering and prediction problems, i.e. the Kalman filter.

That all these themes are still of high importance is obvious. In particular, due to the initiative "A European Green Deal" of the European Commission, the further development and optimization of the bioprocesses involved in the bioeconomy are of fundamental importance. For a sustainable economy, bioprocesses must be central. They must be carried out in an optimal way, so that resources are saved.

In this book, methods and techniques are provided for the monitoring and control of bioprocesses. From new developments for sensors, the application of spectroscopy and modelling approaches, the estimation and observer implementation for ethanol production and the development and scale-up of various bioprocesses and their closed loop control information is presented. The processes discussed here are very diverse. The major applications are cultivation processes, where microorganism were grown, but also, the incubation process of birds' eggs, as well as an indoor climate control for humans, will be discussed. Altogether, in 12 chapters, nine original research papers and three reviews are presented. In the first chapter, a model-based soft sensor is presented, which is based on the adjustable structure geometric observer, and can be used to estimate important variables in a bioreactor. The performance of this approach was compared to the performance of an extended Kalman filter. In the second chapter, a new control method for a fed-batch yeast cultivation is discussed, based on dielectric spectroscopy to control the specific growth rate at different set points. Siderophores are important in areas where the availability of iron is low, because iron is one of the critical growth limiting factors for mainly all aerobic microorganisms. How the production of siderophores can be carried out is demonstrated in the third chapter. The production of an antimicrobial agent is discussed in the fourth chapter, as well as its inhibitory properties against different microorganisms. In the fifth chapter, the effect of nitrate and perchlorate on selenate reduction in a batch reactor is presented. A study of a bioleaching process for the extraction of metals from a flotation concentrate is discussed in the sixth chapter. In this contribution, results regarding the influence of two typical flotation frothers on the sensitivity of bacteria in the mesophilic mixed culture are presented. Due to coronavirus medication, drugs are more important than ever. The monitoring of monoclonal antibody breakthrough curves in chromatographic downstream operations is presented in chapter seven. As a measurement system, a Raman spectrometer is used and complemented with an extended Kalman filter. It is demonstrated that this approach allows the estimation of the antibody concentrations with reduced noise and increased robustness. In the eighth chapter, a model predictive controller is presented, which can regulate the heating power of an incubator for bird eggs. It is demonstrated that the air temperature can be kept constant, although the eggshell temperatures are different in different zones of the incubator. An adaptive occupant-based predictive controller for a heating, ventilation, and air conditioning system using a predictive classification model to provide an optimal indoor climate with respect to temperature and humidity is discussed in the ninth chapter. As mentioned already, sensors are fundamental for an optimal process performance. In chapter ten, examples of well-defined conjugated macromolecules based on oligo(arylene ethynylene) skeletons are reviewed for their use in sensor applications. In chapter eleven, the most important models for physiological, biochemical, and physical properties governed by temperature are discussed. In the review, a toolset for the future exploitation of temperature as a control variable for optimization, monitoring, and control applications in bioprocess engineering is presented. The importance of the control of the specific growth rate to enable the improvement of the quality and reproducibility of bioprocesses is reviewed in the last chapter. Requirements are given that must be met to successfully implement the specific growth rate control system. Furthermore, recommendations are presented for the selection of particular control systems for specific biotechnological processes.

I would like to thank the authors of the chapters for their excellent work, as well as the stuff of MDPI, who supported me in making this book a reality.-Summer 2020 Bernd Hitzmann

Bernd Hitzmann
Editor

Editorial

Special Issue: Bioprocess Monitoring and Control

Bernd Hitzmann

Department of Process Analytics and Cereal Science, Institute of Food Science and Biotechnology, University of Hohenheim, 70599 Stuttgart, Germany; Bernd.Hitzmann@Uni-Hohenheim.de

Received: 14 July 2020; Accepted: 14 July 2020; Published: 16 July 2020

Bioprocesses can be found in different areas such as the production of food, feed, energy, chemicals, and pharmaceuticals. From bio-catalysis to fermentation processes or mammalian cell cultures, different reaction systems are applied. Due to the bio-economy initiatives in different countries, the number of bioprocesses will grow further in the future. One characteristic feature of all these different bioprocesses is a complex reaction matrix where different substances play an important role. Frequently, one must deal with a three-phase system, i.e., a liquid, a gas, and a solid phase. For the optimal operation of these processes, monitoring and supervision systems are required for all phases. Additionally—although bioprocesses have been applied for several thousand years, such as the fermentation of dough—on-line measurement systems for important process variables are still rare. Although the measurement of key variables is a challenge, the control of them to guarantee optimal yields is an even greater challenge.

This Special Issue of *Processes* entitled "Bioprocess Monitoring and Control" presents novel examples of on-line monitoring and closed loop control techniques applied to different bioprocesses. The accepted manuscripts cover a range of important topics in different bioprocess areas, where microorganisms, bird's eggs, and humans are involved. Different techniques such as those for the construction of sensors, the production of a biocontrol agent, scaling up procedures, the application of observers, closed loop control, and the model-based monitoring of a downstream process are presented. The accepted manuscripts are nine original research papers and three reviews, which are summarized below.

Lisci et al. [1] studied a model-based soft sensor, which is based on the adjustable-structure geometric observer and can be used to estimate important variables in a bioreactor. The performance of this approach was compared to the performance of an extended Kalman filter. Using simulations, they were able to show that both estimators lead to good estimation performance. The geometric observer estimation is more sensitive to measurement noise, probably because of the presence of the Lie derivative in the correction term. Lisci et al. concluded that the systematic geometric approach led to the best solution for the estimation problem, giving a structure that did not depend on the correction algorithm.

Brignoli et al. [2] present a new control method for the fed-batch cultivation of *Kluyveromyces Marxianus*. They counter the problem of noise and oscillations in the control variable and address the exponential growth dynamics more effectively. Based on dielectric spectroscopy for the on-line biomass concentration measurements, the specific growth rate was estimated. Using a feedforward-feedback controller, the authors could demonstrate that the specific growth rate could be maintained at different set point values. Therefore, the feasibility of the closed loop control of the specific growth rate of yeast in long-duration fed-batch cultures was demonstrated successfully.

Abo-Zaid et al. [3] present results from tests of twenty fluorescent *Pseudomonas* isolates for their ability to produce siderophores. The assessment of their antagonistic activity against six plant pathogenic fungal isolates is demonstrated. For the promising strains, a scaling-up production of siderophores from fluorescent *Pseudomonads* was carried out. They could show that the exponential fed-batch fermentation of *P. aeruginosa* F2 and *P. fluorescens* JY3 gave higher concentrations of siderophores and biomass than batch fermentation. Furthermore, they demonstrated that formulations of siderophore-producing fluorescent *Pseudomonads* were effective in controlling soil-borne fungi and for the stimulation of plant growth. Therefore, they concluded that bio-friendly formulations of siderophore-producing fluorescent *Pseudomonas* isolates could be used as biocontrol agents for controlling some plant fungal diseases.

Zhang et al. [4] isolated from the commercial Yanjing Natto food a bactericide-secreting Bacillus strain, i.e., *Bacillus subtilis* natto, which is potentially useful as a biocontrol agent. Upon the optimization of the growth medium for optimal bactericide secretion, the antimicrobial activity of the strain was enhanced significantly. They could demonstrate the inhibitory properties of the obtained agent against *S. aureus*, *E. coli*, and *S. typhimurium*. Using HPLC, ^{13}C-nuclear magnetic resonance, and mass spectral analyses, the structure of the purified new bactericides could be identified.

The effect of nitrate and perchlorate on selenate reduction in a batch reactor was investigated by Kim et al. [5]. They selectively enriched selenate-reducing bacteria in bench-scale sequencing batch reactors, which were seeded with activated sludge, and operated them semi-continuously in parallel for more than one and a half months. They show that complete selenate and nitrate reduction can be accomplished simultaneously. Kim et al. concluded that selenate-reducing bacteria are capable of enduring the competition associated with the reduction of other oxyanions and electron donors without significant inhibition after appropriate acclimation.

Jafari et al. [6] studied a bioleaching process for the extraction of metals from a flotation concentrate. In their contribution, results regarding the influence of two typical flotation frothers on the sensitivity of bacteria in the mesophilic mixed culture are presented. As a traditional mixed mesophilic microorganism culture, *Acidithiobacillus ferrooxidans*, *Leptospirillum ferrooxidans*, and *Acidithiobacillus thiooxidans* were used. By increasing the dosage of the frothers, they could show a negative correlation with bacterial activities. However, the mixed culture showed a lower sensitivity to the toxicity of the frothers than the examined pure cultures.

Feidl et al. [7] investigated the monitoring of monoclonal antibody breakthrough curves in chromatographic downstream operations. As a measurement system, they used a Raman spectrometer connected to the process by a self-developed flow cell. An extended Kalman filter was developed by complementing the measurement information with information coming from a lumped kinetic model. Feidl et al. demonstrate in their contribution that this approach allows the estimation of the antibody concentrations with reduced noise and increased robustness.

For the incubation of bird eggs, Youssef et al. [8] developed a model predictive controller to regulate the heating power. They used several IR radiators divided into three zones to adjust the eggshell temperatures individually in each zone. To test and implement the developed controller, four full incubation trials were performed. The authors could demonstrate that the controllers were able to follow the reference trajectory defined for each individual zone. They could keep the air temperature constant, although the eggshell temperatures within the middle zone were different from those in the sidelong zones.

That we as humans are also part of a kind of bioprocess and have to be maintained under optimal conditions is considered by Youssef et al. [9]. They propose an adaptive occupant-based predictive controller for a heating, ventilation, and air conditioning system using a predictive classification model to provide an optimal indoor climate with respect to temperature and humidity. To estimate the individual metabolic rates of 25 participants, three input variables—aural temperature, heart rate, and average skin heat-flux—were used. The least squares support vector machine technique was applied to predict the individual's thermal sensation. Based on that, they recommend an adaptive model predictive controller to adjust the indoor climate.

Without reliable sensors, the control of processes is not possible. Krywko-Cendrowska et al. [10] emphasize the importance of monitoring in general but also for bioprocesses. They review examples of well-defined conjugated macromolecules based on an oligo(arylene-ethynylene) skeleton used for sensor applications and discuss their relevance and their perspectives, not only for biological samples. In the review, they focus exclusively on examples of uniform macromolecules.

Noll et al. [11] summarize the most important models for physiological, biochemical, and physical properties governed by temperature. A timeline of the publication of different temperature models is presented as are the pro and cons of mechanistic and empirical models. In their review, a toolset for the future exploitation of temperature as a control variable for optimization, monitoring, and control applications in bioprocess engineering is presented.

Galvanauskas et al. [12] emphasize in their review the importance of the control of the specific growth rate because this enables the improvement of the quality and reproducibility of bioprocesses. Requirements are given that must be met to successfully implement the specific growth rate control system. Furthermore, recommendations are presented for the selection of particular control systems for specific biotechnological processes.

The articles in this Special Issue highlight the diversity of bioprocesses and new applications in the development and management of these processes. Especially for closed loop control, the reliability of the measurements is central. Beside the measurements, the applicability of the models is equally important. They must provide the required accuracy in estimating bioprocess variables. The articles in this Special Issue show a major step forward towards efficient bioprocesses but in the future further research is still needed. The papers from this Special Issue can be accessed at the following link: https://www.mdpi.com/journal/processes/special_issues/bioprocess_control.

References

1. Lisci, S.; Grosso, M.; Tronci, S. A geometric observer-assisted approach to tailor state estimation in a bioreactor for ethanol production. *Processes* **2020**, *8*, 480. [CrossRef]
2. Brignoli, Y.; Freeland, B.; Cunningham, D.; Dabros, M. Control of specific growth rate in fed-batch bioprocesses: Novel controller design for improved noise management. *Processes* **2020**, *8*, 679. [CrossRef]
3. Abo-Zaid, G.A.; Soliman, N.A.M.; Abdullah, A.S.; El-Sharouny, E.E.; Matar, S.M.; Sabry, S.A. Maximization of siderophores production from biocontrol agents. *Pseudomonas aeruginosa* F2 and *Pseudomonas fluorescens* JY3 using batch and exponential fed-batch fermentation. *Processes* **2020**, *8*, 455. [CrossRef]
4. Zhang, J.; Bilal, M.; Liu, S.; Zhang, J.; Lu, H.; Luo, H.; Luo, C.; Shi, H.; Hafiz, N.; Iqbal, M.; et al. Isolation, identification and antimicrobial evaluation of bactericides secreting *Bacillus subtilis* natto as a biocontrol agent. *Processes* **2020**, *8*, 259. [CrossRef]
5. Kim, H.W.; Hong, S.H.; Choi, H. Effect of nitrate and perchlorate on selenate reduction in a sequencing batch reactor. *Processes* **2020**, *8*, 344. [CrossRef]
6. Jafari, M.; Golzadeh, M.; Shafaei, S.Z.; Abdollahi, H.; Gharabaghi, M.; Chelgani, S.C. Effects of conventional flotation frothers on the population of mesophilic microorganisms in different cultures. *Processes* **2019**, *7*, 653. [CrossRef]
7. Feidl, F.; Garbellini, S.; Luna, M.F.; Vogg, S.; Souquet, J.; Broly, H.; Morbidelli, M.; Butté, A. Combining mechanistic modeling and raman spectroscopy for monitoring antibody chromatographic purification. *Processes* **2019**, *7*, 683. [CrossRef]
8. Youssef, A.; Norton, T.; Berckmans, D. Bioenvironmental zonal controlling of incubated avian embryo using localised infrared heating. *Processes* **2019**, *7*, 651. [CrossRef]
9. Youssef, A.; Caballero, N.; Aerts, J. Model-based monitoring of occupant's thermal state for adaptive HVAC predictive controlling. *Processes* **2019**, *7*, 720. [CrossRef]
10. Krywko-Cendrowska, A.; Szweda, D.; Szweda, R. Well-defined conjugated macromolecules based on oligo(arylene ethynylene)s in sensing. *Processes* **2020**, *8*, 539. [CrossRef]
11. Noll, P.; Lilge, L.; Hausmann, R.; Henkel, M. Modeling and exploiting microbial temperature response. *Processes* **2020**, *8*, 121. [CrossRef]
12. Galvanauskas, V.; Simutis, R.; Levišauskas, D.; Urniežius, R. Practical solutions for specific growth rate control systems in industrial bioreactors. *Processes* **2019**, *7*, 693. [CrossRef]

Article

A Geometric Observer-Assisted Approach to Tailor State Estimation in a Bioreactor for Ethanol Production

Silvia Lisci, Massimiliano Grosso and Stefania Tronci *

Dipartimento di Ingegneria Meccanica Chimica e dei Materiali, Università degli Studi di Cagliari, Via Marengo 2, 09123 Cagliari, Italy; s.lisci@dimcm.unica.it (S.L.); massimiliano.grosso@dimcm.unica.it (M.G.)
* Correspondence: stefania.tronci@dimcm.unica.it; Tel.: +39-070-675-5050

Received: 28 February 2020; Accepted: 16 April 2020; Published: 20 April 2020

Abstract: In this work, a systematic approach based on the geometric observer is proposed to design a model-based soft sensor, which allows the estimation of quality indexes in a bioreactor. The study is focused on the structure design problem where the set of innovated states has to be chosen. On the basis of robust exponential estimability arguments, it is found that it is possible to distinguish all the unmeasured states if temperature and dissolved oxygen concentration measurements are combined with substrate concentrations. The proposed estimator structure is then validated through numerical simulation considering two different measurement processor algorithms: the geometric observer and the extended Kalman filter.

Keywords: nonlinear state estimation; geometric observer; bioreactor; continuous system; extended Kalman filter; model-based sensor

---

1. Introduction

Bioreactors are units where a wide variety of products are made in industrial plants and where a diversity of important processes, such as fermentation, occur. Usually, the control of bioreactors is accomplished through the regulation of variables, such as pH and temperature, for optimizing the microbial growth [1,2]. Product quality indexes such as biomass, substrate, product or by product, and dissolved oxygen concentrations are not usually controlled, because they are difficult to measure in real-time [3]. Even though many works report the availability and advantages of monitoring techniques, industrial biotechnology processes have a scarce capacity for real-time monitoring, which implies a limited implementation of efficient control of the process [4,5]. Because an unpredicted perturbation may lead to significant changes in the qualitative behavior of the system [6–8], it is crucial to accurately monitor the process.

Model-driven soft sensors can be a possible approach to estimate variables from secondary measurements. They rely on first principles process models and on algorithms that reconcile the available measurements with predictions carried out by the model. Several estimation techniques have been proposed in the literature for chemical and biochemical processes. Among them, the following have been recognized to have strong potential in the online estimation of nonlinear systems: (i) extended Kalman filter [9], (ii) high gain observer [10], (iii) sliding mode observer [11], (iv) geometric observer [12,13]. Many of the strategies to estimate unmeasurable states and disturbances for partially known systems are based on the extended Kalman filter (EKF) because its design is quite simple and it is widely accepted by relevant industries [14,15].

In this paper, the problem of estimating unmeasured states in a bioreactor is addressed. The study is based on the detailed model proposed by [16], which is considered as the virtual plant. The main objective is to compare different estimation solutions depending on the available measurements and the characteristics of the sensors. An adjustable-structure geometric estimation approach is implemented, considering the estimator structure as a degree of freedom in the design with the aim of improving performance versus robustness estimation behavior [13,17,18]. The used estimation algorithm is the geometric observer (GO) with proportional innovation [19], which offers the simplicity of tuning and implementation. In order to show that the proposed procedure for choosing the estimation structure can be applied to other estimation techniques, the extended Kalman filter (EKF) is also used as the measurement processor algorithm.

2. Process Model

The biochemical process considered in the present paper is a fermentation reactor for the production of ethanol. The model was developed by [16] and, for the sake of clarity, it is hereafter reported (Equations (1)–(6)). It is assumed there is a perfect mixing in the reactor (constant pH and constant volume). The dynamics of biomass (C_X), substrate (C_S), ethanol (C_P), along with the oxygen concentration (C_{O_2}) are considered. Energy balances are also taken into account describing the reactor temperature (T_r) and jacket temperature (T_{ag}) dynamics. A low dilution rate has been considered allowing a balance between the biomass exiting from the system and the biomass produced in the reactor.

$$\frac{dC_X}{dt} = \mu_X\, C_X\, \frac{C_S}{K_S + C_S} e^{-K_P C_P} - \frac{F_e}{V} C_X \tag{1}$$

$$\frac{dC_P}{dt} = \mu_P C_X \frac{C_S}{K_{S1} + C_S} e^{-K_{P1}\, C_P} - \frac{F_e}{V}\, C_P \tag{2}$$

$$\frac{dC_S}{dt} = -\frac{1}{R_{SX}} \mu_X C_X \frac{C_S}{K_S + C_S} e^{-K_P C_P} - \frac{1}{R_{SP}} \mu_P C_X \frac{C_S}{K_{S1} + C_S} e^{-K_{P1} C_P} + \frac{F_i}{V} C_{S,in} - \frac{F_e}{V} C_S \tag{3}$$

$$\frac{dC_{O_2}}{dt} = k_l a \left(C_{O_2}^* - C_{O_2}\right) - \mu_{O_2} \frac{1}{Y_{O_2}} C_X \frac{C_{O_2}}{K_{O_2} + C_{O_2}} \tag{4}$$

$$\frac{dT_r}{dt} = \left(\frac{F_i}{V}\right)(T_{in} + 273) - \left(\frac{F_e}{V}\right)(T_r + 273) - \mu_{O_2} \frac{1}{Y_{O_2}} C_X \frac{C_{O_2}}{K_{O_2} + C_{O_2}} \frac{\Delta H_r}{32\, \rho_r\, C_{heat,r}} - \frac{K_T A_T \left(T_r - T_{ag}\right)}{V\, \rho_r\, C_{heat,r}} \tag{5}$$

$$\frac{dT_{ag}}{dt} = \left(\frac{F_{ag}}{V_j}\right)\left(T_{in,ag} - T_{ag}\right) + \frac{K_T A_T \left(T_r - T_{ag}\right)}{V_j \rho_{ag} C_{heat,ag}} \tag{6}$$

The oxygen equilibrium concentration is affected by the inorganic salts, which are added to the solution as source of inorganic nitrogen. The dependence is reported in Equation (7)

$$C_{O_2}^* = C_{O_2,\,0}^* \cdot 10^{-\sum H_i I_i} \tag{7}$$

where the equilibrium concentration as a function of temperature has been calculated using the equation proposed by [20] for distilled water

$$C_{O_2,\,0}^* = 14.6 - 0.3943 T_r + 0.007714\, T_r^2 - 0.0000646 T_r^3 \tag{8}$$

In the present work, the distillation strength $\sum H_i I_i$ is kept constant and it has been calculated using the equations reported in [16].

The model is here used to simulate a real process and to develop the model-based soft sensor (estimator). Because the aim of the work is to mimic a real situation, the simulation using the model parameters reported in [16] is considered as the real plant (hereafter referred to as the virtual plant). On the other hand, some of the parameters used for the model in the estimator algorithm have been modified. The aim was to insert modeling errors to simulate what usually happens in a real situation where parameter uncertainty is present. This is often the case when dealing with complex systems such as biological reactors [21]. A Monte Carlo method has been used to produce empirical error estimates

on the model parameters, using a uniform noise distribution with a maximum deviation equal to
±8%. Table 1 summarizes the parameters of the model used in the estimator algorithm obtained by
performing 100 simulations and leading to the maximum error calculated on the trajectories of the six
states. The other parameters of the model are the same as reported in [16]. The nominal conditions of
the virtual plant are reported in Table 2.

Table 1. Parameters for the virtual plant and for the estimator.

	Virtual Plant	Estimator
μ_p	1.790	1.7465
K_s	1.030	1.0248
K_p	0.139	0.1281
K_{s1}	1.680	1.8090
K_{p1}	0.070	0.0692
R_{sx}	0.607	0.6274
R_{sp}	0.435	0.4549

Table 2. Nominal operating conditions of the process.

C_{O2} = 2.5 mg/L	pH = 6
C_P = 13 g/L	T_{ag} = 29 °C
C_S = 27 g/L	T_{in} = 25 °C
$C_{S,in}$ = 60 g/L	$T_{in,ag}$ = 15 °C
C_X = 1 g/L	T_r = 26 °C
F_{ag} = 18 L/h	V = 1000 L
$F_i = F_e$ = 51 L/h	-

The model simulations have also been made more realistic by adding noise to the available
measurements, and the precision of the sensors is reported in Table 3.

Table 3. Noise for the different measuring sensors.

C_X	C_S	C_{O2}	T_r	T_{ag}
±2.5%	±2.5%	±2.5%	±0.1 °C	±0.1 °C

It is important to specify that all tests performed in the following have been carried out by
imposing step changes to three inputs of the system ($C_{S,in}$, $T_{in,ag}$, T_{in}). The description of input
variations is reported in Table 4.

Table 4. Step changes of input variables.

	Input	t = 0 h	t = 100 h	t = 200 h
T1	$C_{S,in}$ (g/L)	60	45	75
T2	$T_{in,ag}$ (°C)	15	10	20
T3	T_{in} (°C)	25	20	30

3. Estimation Problem

The current real-time monitoring methods used in ethanol production consist of secondary
measurements such as pH, turbidity, gas composition and temperature [4]. Even if such variables

provide important information about the process, they do not directly relate to the state of the system, making it difficult to apply advanced control strategies. Furthermore, even the best process measurements are corrupted by some amount of signal noise and their true values are somewhat uncertain. State estimation techniques can be used to improve the output signal of measured process states in the presence of uncertainty and when it is not possible to directly measure all the variables of interest.

The estimation problem consists of jointly designing the estimation structure (i.e., estimator model, sensors, innovated states and data assimilation mechanisms), and the estimation algorithm (i.e., the dynamic data processor), to infer some or all the states of the bioreactor on the basis of the available model in conjunction with available measurements, according to a specific estimation objective. In the present fermentation reactor estimation study, the emphasis has been placed on: (i) the detection of the more adequate measured outputs leading to the best performance, (ii) the selection of the innovated states, meaning the states which are updated by using the available measurement.

For simplifying the formulation of the problem, the model in Equations (1)–(6) is written in compact form as reported in Equation (9)

$$\dot{x} = f(x, u), \quad x(t_0) = x_0 \tag{9a}$$

$$y = h(x) \tag{9b}$$

where x is the n-dimensional state vector, equal to x_0 at the initial time t_0, u is the p-dimensional input vector, f is the n-dimensional vector fields, y is the m–dimensional vector of the measured outputs and h is the map relating states and measurements. The dimension of the measured outputs is less than the number of states, that is $m < n$. Using the geometric approach [19,22], it is possible to define the nonlinear estimation map ϕ as Equation (10)

$$\phi(x, u) = [\phi_1, \dots, \phi_i, \dots, \phi_m]^T \tag{10a}$$

$$\phi_i = \left(h_i, \, L_f h_i, \, \dots, L_f^{\kappa_i - 1} h_i \right) \tag{10b}$$

where the $L_f^j h_i$ is the jth Lie derivative of the time-varying scalar field h_i along the vector f, κ_i is the observability index of the ith output and κ is the estimator order defined in Equation (11)

$$\kappa_1 + \kappa_2 + \dots + \kappa_m = \kappa = n \tag{11}$$

If the map $\phi(x, u)$ is invertible with respect to x (Equation (12)), the system is observable, and the states can be reconstructed using the available model (Equations (1)–(6)) and a proper measurement processor algorithm [14].

$$rank(\partial_x \phi(x, u)) = n \tag{12}$$

If the Jacobian matrix $\partial_x \phi(x, u)$ is rank deficient, there are unobservable states. In this case, the system is detectable only if all the unobservable modes have negative real parts [19].

3.1. Robust Estimability and Robust Detectability

If all states can be fully observable, the observability matrix should be full-rank, but practical observability can be assessed if the condition number of the observability matrix (Σ) is small [23]. Furthermore, a small singular value of the observability matrix implies the worst estimate of the states [24]

$$rank(\Upsilon) = n, \Upsilon = \partial_x \phi(x, u) \tag{13a}$$

$$\frac{\bar{\sigma}(\Upsilon)}{\underline{\sigma}(\Upsilon)} = \Sigma < \Xi, \tag{13b}$$

$$avg\underline{\sigma}(\Upsilon) \geq \varepsilon_0 \tag{13c}$$

where Ξ and ε_0 are, respectively, the selected thresholds.

On the other hand, if matrix Υ is rank deficient and the unobservable states are stable, it is necessary to distinguish between states that can be innovated (distinguishable states) and states that cannot (undistinguishable states). In this case, the dimension of the map in Equation (10) is equal to the dimension of the distinguishable states, and robust detectability can be assessed if the following conditions are satisfied (Equation (14))

$$\frac{\overline{\sigma}(\Upsilon_{\mathrm{p}})}{\underline{\sigma}(\Upsilon_{\mathrm{p}})} = \Sigma_{\mathrm{i}} < \Xi_{\mathrm{p}} \tag{14a}$$

$$avg\underline{\sigma}(\Upsilon_{\mathrm{p}}) \geq \varepsilon_{\mathrm{p}0} \tag{14b}$$

$$\Upsilon_{\mathrm{p}} = \partial_x \Phi_{\mathrm{p}}(x, u) \tag{14c}$$

$$\Phi_{\mathrm{p}} = \left(h_1, \ldots, L_f^{\kappa_1 - 1} h_1, \ldots, h_m, \ldots, L_f^{\kappa_m - 1} h_m \right) \tag{14d}$$

$$\kappa_1 + \kappa_2 + \ldots + \kappa_m = \kappa = p \tag{14e}$$

The constants Ξ_p and ε_{p0} are, again, the selected thresholds.

3.2. Selection of the Estimator Structure

The performance of an estimator is obviously strongly affected by the model of the process and the quality of the available measurements. Biological processes are complex systems, therefore the presence of model uncertainty in terms of parameters and neglected dynamics is, in general, to be expected [21]. This means that the complete reconstruction of the states requires, in general, a combination of different measurements [4]. Within this framework, it is important to underline that there is still a gap between the sensors for laboratory use and large scale monitoring in real-time [4]. The selection of the estimator structure is therefore focused on the choice of the best monitoring strategies, by considering which are the most representative measured outputs and the presence of parameter errors in the model used in the estimator. It is considered that system monitoring can be expensive, in terms of both fixed and operation costs, therefore it could be useful to optimize performance with the least number of sensors. This analysis has been carried out comparing condition number and minimum singular values of the matrix Υ or Υ_p (Equations (13) and (14)). The performances have also been evaluated by simulating different trajectories, from which the convergence rate, presence of off-set and signal noise have been evaluated.

3.3. Algorithms for the Estimation Problem

In this study, two different algorithms have been selected and compared. The first one is the geometric observer [19], which is formally connected with the observability properties reported in the previous section. The geometric observer (GO) can be applied also to detectable systems [13], and it demonstrated to be simple to implement and tune [18,22]. The geometric approach is also used to select the estimator structure, in terms of the selection of measurements and states to be innovated.

The geometric observer algorithm is reported in Equation (15), where it is assumed that some states are not innovated. This choice may depend on the rank deficiency of the observability matrix or a design choice intended to improve the robustness and efficiency of the estimator.

$$\dot{\hat{x}}_i = \hat{f}_i(\hat{x}, u) + (\partial_{xi}\Phi(\hat{x}, u))^{-1} K(y - h(\hat{x})), x_{i0} = x_i(t_0) \tag{15a}$$

$$\dot{\hat{x}}_u = \hat{f}_u(\hat{x}, u), x_{u0} = x_u(t_0) \tag{15b}$$

The inverse of the observability matrix $\partial_{x_i} \phi(x, u)$ in Equation (15a) is calculated at each time step and K is a block diagonal matrix (Equation (16)), whose coefficients are constant tuning parameters. The estimated states $\hat{x}$ in Equation (15) are the innovated states ($\hat{x}_i$), where the dynamics predicted by the model are adjusted by means of the available measurements y, and the not innovated states ($\hat{x}_u$), which are only predicted by the model (referred to in the following as open-loop states).

$$K = \begin{pmatrix} B_1 & 0 & \cdots & 0 \\ 0 & B_2 & \cdots & 0 \\ \vdots & \vdots & \cdots & \vdots \\ 0 & 0 & \cdots & B_m \end{pmatrix}, \quad B_1 = \begin{bmatrix} k_{11} \\ \vdots \\ k_{1\nu_1} \end{bmatrix}, \quad B_2 = \begin{bmatrix} k_{21} \\ \vdots \\ k_{2\nu_2} \end{bmatrix}, \quad B_m = \begin{bmatrix} k_{m1} \\ \vdots \\ k_{m\nu_m} \end{bmatrix} \tag{16}$$

$$\nu_i = \kappa_{i-1}$$

Tuning guidelines are provided by [17], proving that a set of tuning parameters k_{ij} is required for every measurement. For observability indexes equal to 1 or 2 ($\kappa_i = 1, 2$), the proportional gains can be obtained by considering Equation (17).

$$k_{i1} = 2\zeta\omega_0, \quad k_{i2} = \omega_0{}^2 \tag{17a}$$

$$\omega_0 \in [10\omega_c, 30\omega_c], \quad \zeta = [1, 3] \tag{17b}$$

The GO has then been compared with the extended Kalman filter (EKF), which is the most used estimator algorithm in the industry because of its straightforward construction [17]. Even if the EKF is usually applied to complete observability systems, in this investigation it has been used also when the choice of measurements leads to a rank deficient observability matrix. The EKF algorithm has been applied in the continuous form, reported in the following Equation (18).

$$\dot{\hat{x}}_i = \hat{f}_i(\hat{x}, u) + K_{EKF}(y - h(\hat{x})), x_{i0} = x_i(t_0) \tag{18a}$$

$$\dot{\hat{x}}_u = \hat{f}_u(\hat{x}, u), x_{u0} = x_u(t_0) \tag{18b}$$

$$K_{EKF} = P(t)H^T R^{-1} \tag{18c}$$

$$\dot{P}(t) = P(t)F(t) + F^T(t)P(t) + Q(t) - K_{EKF}HP, P(t_0) = P_0 \tag{18d}$$

$F(t)$ is the Jacobian of the vector field $\hat{f}_i$, calculated with respect to the innovated states, P is the error covariance matrix of the innovated states, H is the matrix of the derivative of the map h with respect to the states, Q and R are, respectively, the covariance matrix of the model and measurements errors [9]. The constant matrix Q, R, and P_0 are tuning parameters of the estimation model and they have been calculated minimizing the error between the states calculated with the simulated plant and the estimator along a reference trajectory.

4. Results

4.1. Estimation Structure

The choice of the estimation structure has been carried out considering: (i) condition number and the minimum singular value of the Jacobian matrix Υ (or Υ_n) for a different choice of measurements and innovated states and (ii) evaluating the responses of the reconstructed states for a given trajectory. Temperature and dissolved oxygen measurements have always been considered available, according to the laboratory and industrial practice. On the other hand, sensors suited for ethanol measurements as well as substrate and biomass are not always available for large scale real-time applications [4]. According to the analysis reported in [3], two possible scenarios have been considered: (i) biomass concentration in the reactor is measured online or (ii) substrate concentration in the reactor is measured online. Using the representation in (9), the considered cases are reported in Equation (19):

$$y = \left(C_x, C_{O_2}, T_r, T_{ag}\right) \tag{19a}$$

$$y = \left(C_s, C_{O_2}, T_r, T_{ag}\right) \tag{19b}$$

where y represents the measured output vector.

According to Equation (13), it is easy to demonstrate that no combination of indexes κ_i satisfies the observability property for the output vector in Equation (19a). This implies that a full order observer is possible if the substrate concentration is measured online and therefore when using the output configuration reported in Equation (19b). In this case, the nonlinear estimation maps satisfying Equation (13a) are reported in Equation (20)

$$\Phi_1 = \left[C_S, L_f C_S, C_{O_2}, L_f C_{O_2}, T_r, T_{ag}\right], \quad \Phi_2 = \left[C_S, L_f C_S, C_{O_2}, T_r, L_f C_{T_r}, T_{ag}\right] \tag{20}$$

A first comparison of the two structures can be carried out by considering the values of condition number and minimum singular value for the Jacobian of the maps (20), calculated by averaging along the trajectories obtained with input step changes T1 and T2 (Table 4) and reported in Table 5. The structure Φ_2 seems to be more robust (lower condition number), but it shows a lower minimum singular value, indicating that changes in the states should affect the outputs to a lesser extent.

Table 5. Condition number and minimum singular value with four measurements.

	Φ_1	Φ_2
Σ	1802.6	71.37
$\underline{\sigma}$	0.086	0.03

The reconstruction capabilities of the two structures using the geometric observer are therefore calculated, using the input variations T1 and T2, reported in Table 4. Results are shown in Figures 1–4, only for the unmeasured variables, which are ethanol and biomass concentration. It is worth noticing that also the state values calculated only with the model used in the estimation algorithm (open-loop model), but without innovation are reported in order to better highlight the correction provided by the estimation algorithm.

It is possible to observe that using the map Φ_1, allows a good reconstruction of the biomass behavior (Figure 1), while there is a large mismatch between the ethanol concentration obtained with the virtual plant and the reconstructed one (Figure 2).

When using the second configuration, results worsen, both for biomass (Figure 3) and for ethanol (Figure 4) concentration. It is worth noticing that the state's values estimated with map Φ_2 are more corrupted by the measurement noise because in this case a greater observer gain has been used to decrease the offset.

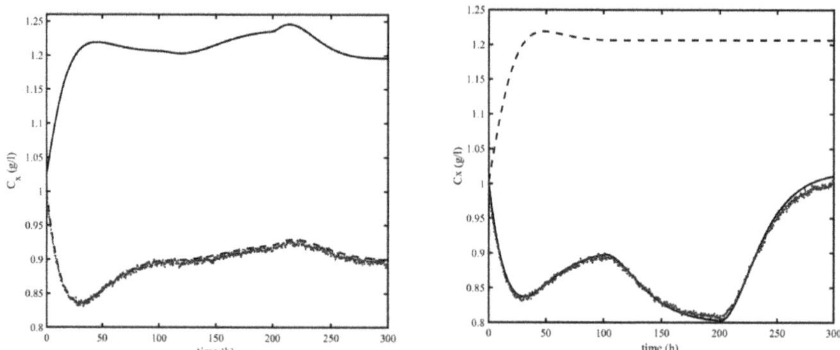

Figure 1. Dynamic response of biomass concentration calculated with the virtual plant (continuous line), open-loop model (dashed line) and geometric observer (GO) (dotted grey line) for structure ϕ_1 along trajectory T1 (**left** panel) and T2 (**right** panel).

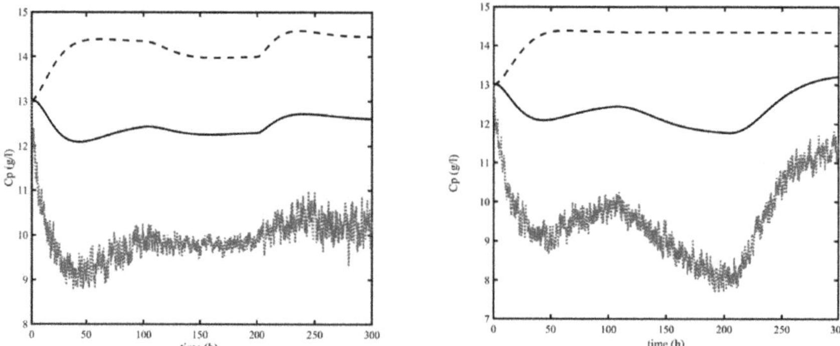

Figure 2. Dynamic response of ethanol concentration calculated with the virtual plant (continuous line), open-loop model (dashed line) and GO (dotted grey line) for structure ϕ_1 along trajectory T1 (**left** panel) and T2 (**right** panel).

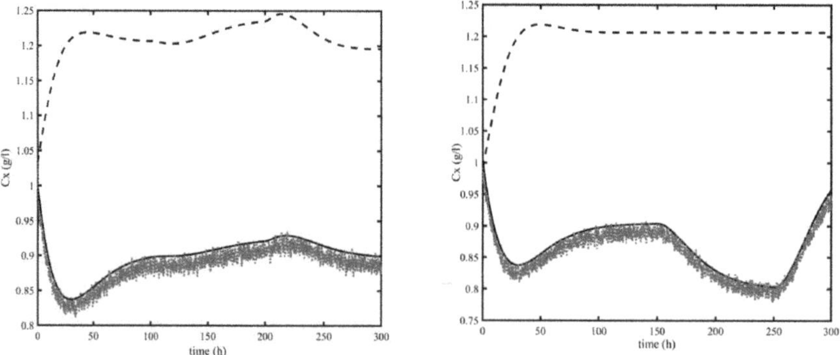

Figure 3. Dynamic response of biomass concentration calculated with the virtual plant (continuous line), open-loop model (dashed line) and GO (dotted grey line) for structure ϕ_2 along trajectory T1 (**left** panel) and T2 (**right** panel).

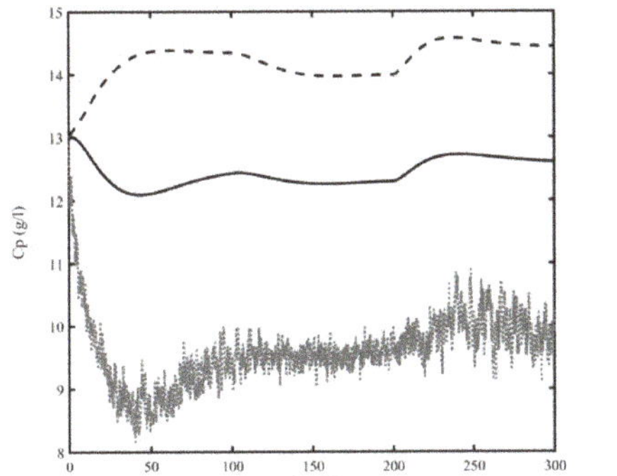 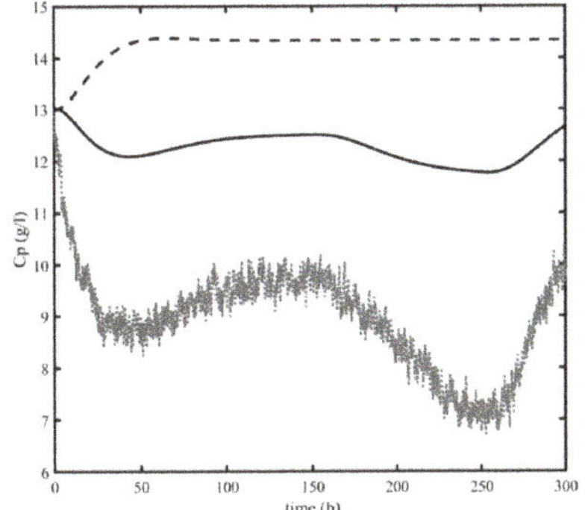

Figure 4. Dynamic response of ethanol concentration calculated with the virtual plant (continuous line), open-loop model (dashed line) and GO (dotted grey line) for structure ϕ_2 along trajectory T1 (**left** panel) and T2 (**right** panel).

The two full order structures are not able to adequately estimate the product of the reactor, therefore a different solution is required to improve ethanol concentration. Using the same measured outputs, it is possible to improve estimation performance by reducing the order of the observer using only one Lie's derivative [22]. The maps reported in Equation (21) lead to five observable states and only one detectable.

$$\phi_{p3} = \left[C_s,\, C_{O_2},\, T_r,\, L_f\, T_r,\, T_{ag} \right], \quad \phi_{p4} = \left[C_s,\, L_f C_s,\, C_{O2},\, T_r,\, T_{ag} \right] \tag{21}$$

The rank of the Jacobian of the maps ϕ_{pi} ($i = 3,\, 4$) depends on the choice of the non innovated state ($\hat{x}_u$) between the two that are not measured, which are ethanol and biomass concentration. It can be verified that the map ϕ_{p3} can be inverted only if C_x is innovated and C_p is not. On the other hand, the Jacobian of map ϕ_{p4} always has a rank equal to five, regardless of the choice of the innovated states. Recalling Equation (15), the following partitions are considered:

$$x_i = \left[C_x, C_s,\, C_{O_2}, T_r,\, T_{ag} \right], \; x_u = \left[C_p \right] \tag{22a}$$

$$x_i = \left[C_p, C_s,\, C_{O_2}, T_r,\, T_{ag} \right] \tag{22b}$$

The map ϕ_{p3} can be used with the partition in Equation (22a), while the map ϕ_{p4} can be used with both partitions in Equation (22a,b). Therefore, two different solutions are identified: $\phi_{p4,1}$ for partition (22b) and $\phi_{p4,2}$ for partition (22a). A first analysis of the possible configurations can be obtained by considering the minimum singular value and condition number reported in Table 6. The indexes' values are comparable; therefore, the evaluation of the best structure has been performed analyzing the reconstruction performance. Figures 5 and 6 represent the estimation of the unmeasured states (ethanol and biomass concentration) for the input step change T1 and T2 described in Table 4. The best reconstruction capabilities are shown by configuration ϕ_{p3} for both the states. This result may suggest that conditions calculated with Equation (14) are informative when the magnitude between the different configurations is significantly different, otherwise, it is necessary to evaluate the estimation capabilities by evaluating the estimator response for given input changes.

Table 6. Mean condition number and minimum singular value for low order structures.

	ϕ_{p3} (*C_P* open loop)	$\phi_{p4,1}$ (*C_x* open loop)	$\phi_{p4,2}$ (*C_P* open loop)
Σ	2.15	8.75	1.53
σ	0.47	0.12	0.99

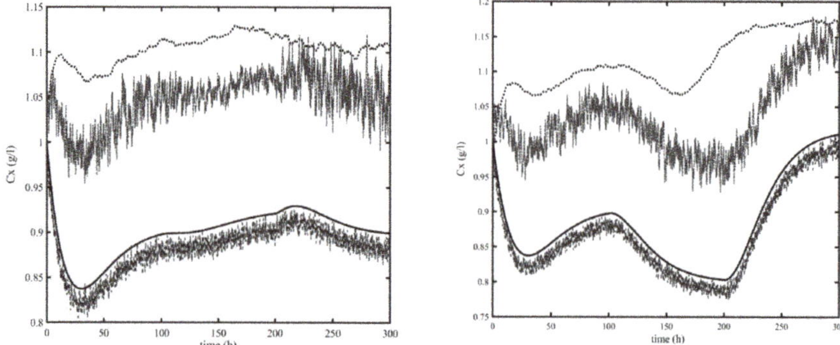

Figure 5. Dynamic response of the biomass concentration calculated with the virtual plant (continuous black line), GO with map ϕ_{p3} (dashed dark grey line), GO $\phi_{p4,1}$ (dotted black line), GO with map $\phi_{p4,2}$ (dashed-dotted grey line) along the trajectory T1 (**left** panel) and T2 (**right** panel).

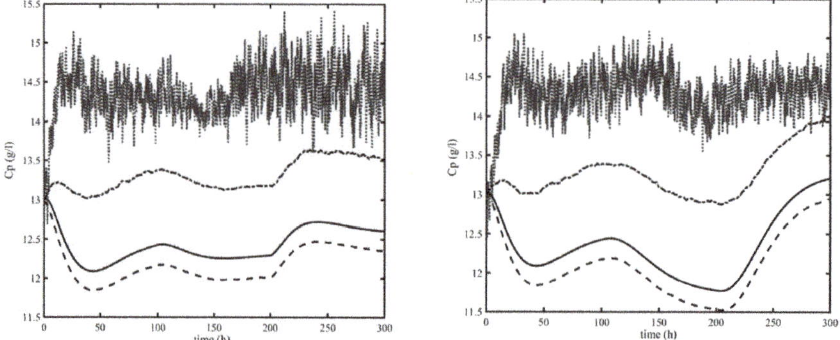

Figure 6. Dynamic response of the product concentration calculated with the virtual plant (continuous line), GO with map ϕ_{p3} (dashed black line), GO with map $\phi_{p4,1}$ (dotted black line), GO with map $\phi_{p4,2}$ (dashed-dotted grey line) along trajectory T1 (**left** panel) and T2 (**right** panel).

4.2. Validation

The analysis carried out in the previous section indicates the best estimation structure with four measured outputs. In order to validate the obtained results, a new test was carried out considering as reference trajectory the variation of the input temperature (T_{in}) as shown in Table 4 (Case T3). Figure 7 shows the dynamic behavior of biomass and product concentration and confirms that the proposed structure can effectively reconstruct the unmeasured states also with different process conditions. It is worth noticing that the ethanol concentration is not innovated, and the correction of the other states also has a positive impact on its estimation.

Using the same number and choice of measured outputs Equation (19b) and partition between innovated and not innovated states Equation (22a), the estimation task has been addressed using the extended Kalman filter (Figure 8). The main reason for using another algorithm as a measurement processor is to demonstrate that the estimator performance depends on the structure selection rather than estimation algorithm. EKF has been preferred for this validation because it is usually preferred in the industrial practice as it is easy to implement and robust if adequately calibrated [25,26].

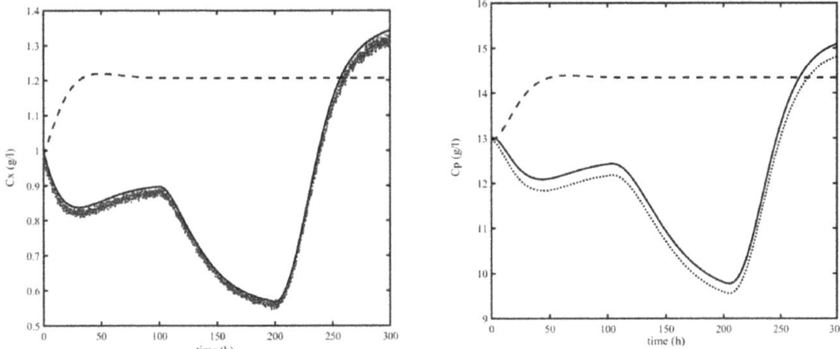

Figure 7. Dynamic response of biomass concentration (**left** panel) and ethanol concentration (**right** panel) calculated with the virtual plant (continuous line), open-loop model (dashed line) and GO (dotted grey line) for structure ϕ_{p3} along the trajectory T3.

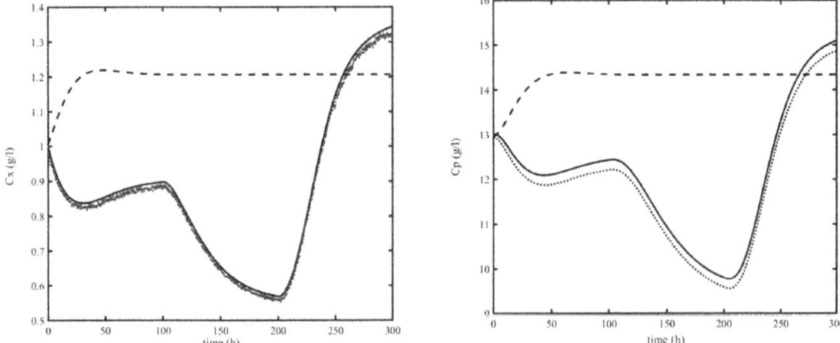

Figure 8. Dynamic response of biomass concentration and ethanol concentration calculated with the virtual plant (continuous line), open-loop model (dashed line) and extended Kalman filter (EKF) (dotted line) for structure ϕ_{p3} along the trajectory T3.

Results show that EKF can effectively reconstruct the unmeasured states, revealing that estimator structure design is the key step for a successful achievement of the estimation goals. The only difference between the two approaches is that the biomass calculated with the geometric observer is more affected by noise. This behavior can be explained by the presence of the Lie derivative in GO, which implies a higher sensitivity to measurement noise with respect to the EKF.

5. Conclusions

The problem of estimating unmeasured states in a bioreactor was addressed, and it was demonstrated that the estimation performance relies on an appropriate structure selection rather than the chosen measurement processor algorithm. An adjustable-structure geometric estimation approach was used, and the estimator structure constituted a design degree of freedom to improve its performance versus robustness behavior. The estimation structure design was based on estimability and detectability properties used together with a geometric approach. The analysis of the estimability measures showed the ill- and well-conditioned structures (condition number of the observability matrix), and the poorest estimation performance for the given structure (minimum singular value of the observability matrix). From the implementation stage with simulations, it was found that the results agreed with the ones obtained from the structural assessment when estimability measure values calculated for the different structures were significantly different. The used estimation algorithm was the geometric observer with proportional innovation, which offers simplicity of tuning and implementation. With the aim of showing that the proposed procedure for choosing the estimation structure can be applied to other estimation techniques, the extended Kalman filter was also used as measurement processor algorithm. The obtained results showed that the two estimators lead to good estimation performance, with the only difference that the geometric observer estimation is more sensitive to measurement noise, probably because of the presence of the Lie derivative in the correction term. Summarizing, the systematic geometric approach led to the best solution for the estimation problem, giving a structure that did not depend on the correction algorithm. The latter can be chosen according to the wishes of the personnel of the plant or developer experience. It is worth noticing that the systematic tuning procedure of the geometric approach was very useful for comparing the reconstruction capabilities of the different structures. The results obtained in this paper in terms of methodology could be applied to more complex biotechnological processes, such as the obtainment of ethanol from cellulosic material, where the measurement devices for real-time application in the industry are still missing. In this case, the proposed approach can be used to detect the measurements that lead to the best reconstruction capabilities and invest in them.

Author Contributions: S.L. performed the analytic calculations and performed simulations; S.T. conceived the idea, proposed the computational model and wrote the paper; M.G. contributed to the analysis of the results and provided critical feedback; S.T. and M.G. reviewed the manuscript. All authors have read and agreed to the published version of the manuscript.

Funding: This research received no external funding.

Conflicts of Interest: The authors declare no conflict of interest.

References

1. Ramaswamy, S.; Cutright, T.J.; Qammar, H.K. Control of a continuous bioreactor using model predictive control. *Process Biochem.* **2005**, *40*, 2763–2770. [CrossRef]
2. Imtiaz, U.; Assadzadeh, A.; Jamuar, S.S.; Sahu, J.N. Bioreactor temperature profile controller using inverse neural network (INN) for production of ethanol. *J. Process Control* **2013**, *23*, 731–742. [CrossRef]
3. Holzberg, T.R.; Watson, V.; Brown, S.; Andar, A.; Ge, X.; Kostov, Y.; Tolosa, L.; Rao, G. Sensors for biomanufacturing process development: Facilitating the shift from batch to continuous manufacturing. *Curr. Opin. Chem. Eng.* **2018**, *22*, 115–127. [CrossRef]
4. Cabaneros Lopez, P.; Feldman, H.; Mauricio-Iglesias, M.; Junicke, H.; Huusom, J.K.; Gernaey, K.V. Benchmarking real-time monitoring strategies for ethanol production from lignocellulosic biomass. *Biomass Bioenergy* **2019**, *127*, 105296. [CrossRef]
5. Pachauri, N.; Singh, V.; Rani, A. Two degree of freedom PID based inferential control of continuous bioreactor for ethanol production. *Isa Trans.* **2017**, *68*, 235–250. [CrossRef]
6. Tronci, S.; Grosso, M.; Alvarez, J.; Baratti, R. On the global nonlinear stochastic dynamical behavior of a class of exothermic CSTRs. *J. Process Control* **2011**, *21*, 1250–1264. [CrossRef]
7. Baratti, R.; Tronci, S.; Schaum, A.; Alvarez, J. Open and closed-loop stochastic dynamics of a class of nonlinear chemical processes with multiplicative noise. *J. Process Control* **2018**, *66*, 108–121. [CrossRef]
8. Moreno, J.A.; Alvarez, J. On the estimation problem of a class of continuous bioreactors with unknown input. *J. Process Control* **2015**, *30*, 34–49. [CrossRef]
9. Jazwinski, A.H. *Stochastic Processes and Filtering Theory*; Courier Corporation: North Chelmsford, MA, USA, 2007.
10. Ciccarella, G.; Dalla Mora, M.; Germani, A. A Luenberger-like observer for nonlinear systems. *Int. J. Control* **1993**, *57*, 537–556. [CrossRef]
11. Wang, G.B.; Peng, S.S.; Huang, H.P. A sliding observer for nonlinear process control. *Chem. Eng. Sci.* **1997**, *52*, 787–805. [CrossRef]
12. Zeitz, M. The extended Luenberger observer for nonlinear systems. *Syst. Control Lett.* **1987**, *9*, 149–156. [CrossRef]
13. Alvarez, J.; López, T. Robust dynamic state estimation of nonlinear plants. *Aiche J.* **1999**, *45*, 107–123. [CrossRef]

14. Dewasme, L.; Goffaux, G.; Hantson, A.L.; Wouwer, A.V. Experimental validation of an Extended Kalman Filter estimating acetate concentration in E. coli cultures. *J. Process Control* **2013**, *23*, 148–157. [CrossRef]

15. Longhi, L.G.S.; Marcon, S.M.; Trierweiler, J.O.; Secchi, A.R. State estimation of an experimental bioreactor using the extended Kalman filtering technology. *Ifac Proc. Vol.* **2002**, *35*, 379–382. [CrossRef]

16. Nagy, Z.K. Model based control of a yeast fermentation bioreactor using optimally designed artificial neural networks. *Chem. Eng. J.* **2007**, *127*, 95–109. [CrossRef]

17. Álvarez, J.; Fernández, C. Geometric estimation of nonlinear process systems. *J. Process Control* **2009**, *19*, 247–260. [CrossRef]

18. Porru, M.; Özkan, L. Monitoring of Batch Industrial Crystallization with Growth, Nucleation, and Agglomeration. Part 2: Structure Design for State Estimation with Secondary Measurements. *Ind. Eng. Chem. Res.* **2017**, *56*, 9578–9592. [CrossRef]

19. Alvarez, J. Nonlinear state estimation with robust convergence. *J. Process Control* **2000**, *10*, 59–71. [CrossRef]

20. Bela, S. *Bioengineering Operations*; Tankonykiado: Budapest, Hungary, 1992.

21. Spigno, G.; Tronci, S. Development of hybrid models for a vapor-phase fungi bioreactor. *Math. Probl. Eng.* **2015**, *2015*, 801213. [CrossRef]

22. Salas, S.D.; Romagnoli, J.A.; Tronci, S.; Baratti, R. A geometric observer design for a semi-batch free-radical polymerization system. *Comput. Chem. Eng.* **2019**, *126*, 391–402. [CrossRef]

23. Dochain, D.; Tali-Maamar, N.; Babary, J.P. On modelling, monitoring and control of fixed bed bioreactors. *Comput. Chem. Eng.* **1997**, *21*, 1255–1266. [CrossRef]

24. Long, R.; Qin, Y.Y.; Jia, J.C. Observable degree analysis of SINS initial alignment based on singular value decomposition. In Proceedings of the 2008 IEEE International Symposium on Knowledge Acquisition and Modeling Workshop, Wuhan, China, 21–22 December 2008; pp. 444–448.

25. Leu, G.; Baratti, R. An extended Kalman filtering approach with a criterion to set its tuning parameters; application to a catalytic reactor. *Comput. Chem. Eng.* **2000**, *23*, 1839–1849. [CrossRef]

26. Salas, S.D.; Ghadipasha, N.; Zhu, W.; Mcafee, T.; Zekoski, T.; Reed, W.F.; Romagnoli, J.A. Framework design for weight-average molecular weight control in semi-batch polymerization. *Control Eng. Pract.* **2018**, *78*, 12–23. [CrossRef]

Article

Control of Specific Growth Rate in Fed-Batch Bioprocesses: Novel Controller Design for Improved Noise Management

Yann Brignoli [1], Brian Freeland [2,3,4], David Cunningham [2] and Michal Dabros [1,*]

[1] School of Engineering and Architecture of Fribourg, HES-SO University of Applied Sciences and Arts Western Switzerland, CH-1700 Fribourg, Switzerland; yann.brignoli@gmail.com
[2] School of Biotechnology, Dublin City University, Glasnevin, Dublin 9, Ireland; brian.freeland@dcu.ie (B.F.); david.cunningham@dcu.ie (D.C.)
[3] I-Form Advanced Manufacturing Research Centre, Dublin City University, Glasnevin, Dublin 9, Ireland
[4] Advanced Processing Technology Research Centre, School of Mechanical and Manufacturing Engineering, Dublin City University, Glasnevin, Dublin 9, Ireland
* Correspondence: michal.dabros@hefr.ch; Tel.: +41-26-429-68-79

Received: 27 April 2020; Accepted: 6 June 2020; Published: 9 June 2020

Abstract: Accurate control of the specific growth rate (μ) of microorganisms is dependent on the ability to quantify the evolution of biomass reliably in real time. Biomass concentration can be monitored online using various tools and methods, but the obtained signal is often very noisy and unstable, leading to inaccuracies in the estimation of μ. Furthermore, controlling the growth rate is challenging as the process evolves nonlinearly and is subject to unpredictable disturbances originating from the culture's metabolism. In this work, a novel feedforward-feedback controller logic is presented to counter the problem of noise and oscillations in the control variable and to address the exponential growth dynamics more effectively. The controller was tested on fed-batch cultures of *Kluyveromyces marxianus*, during which μ was estimated in real time from online biomass concentration measurements obtained with dielectric spectroscopy. It is shown that the specific growth rate can be maintained at different setpoint values with an average root mean square control error of $23 \pm 6\%$.

Keywords: bioprocess monitoring and control; specific growth rate control; signal noise management; dielectric spectroscopy; PAT; microbial bioprocessing

---

1. Introduction

The field of biotechnology is experiencing continued progress, owing to advances in online process monitoring and control. The application of various process analytical technologies enables researchers and end users to have insight into the process and to monitor its substrate uptake, biomass evolution, and cell metabolism. Traditionally, cultures were mostly performed in batch mode. One of the main drawbacks of this mode is that the growth dynamics cannot be controlled during the culture [1,2]. Significant improvement in this respect can be achieved by switching to the fed-batch mode and putting in place appropriate process supervision methodologies. In addition to greater control over the product quality, the cost efficiency of the process increases along with the enhancement of the culture yield [1,3,4].

The key variable that can be controlled in a fed-batch culture is the specific growth rate, μ. This variable has a direct effect upon the cell metabolism, and controlling it can be used to supress or to induce, depending on the need, the formation of secondary metabolites [3,5–8]. The challenge to control the specific growth rate comes from the fact that the determination of its instantaneous value is dependent on the online estimation of the biomass concentration. Indeed, there does not yet exist a direct online measurement tool for cell concentration. All the currently used technologies offer an indirect estimation, where the measured signal is correlated to the biomass concentration through an external calibration [9]. Two different monitoring approaches can be identified. The first approach is noninvasive towards the process and typically involves the use of off-gas analysis combined with a metabolic model to estimate the biomass concentration and its evolution [10–12]. The second approach, invasive to the process, includes the application of an in-situ optical density probe [13–15], dielectric (capacitance) spectroscopy [16–18], or fluorescence spectroscopy [19,20] to monitor biomass. Other methods rely on a significant capital investment, including biocalorimetry [21]. The main issue with the in-situ monitoring technologies is their sensitivity to the process conditions. In particular, aeration and agitation interfere with the probe, and thus create noise in the signal [16,19]. Noise and instabilities in the biomass concentration signal are directly propagated to the estimation of μ, making control of the latter a challenging task. For these reasons, biomass measuring technologies require appropriate data treatment methods, ranging from basic signal smoothing and baseline adjustment to complex filtering algorithms such as Kalman filter [22]. Signal filtering always involves a tradeoff between reducing noise and minimizing measurement delay which are both undesirable in process control applications.

With the appropriate tools set in place to estimate the cell concentration online, several methodologies are available to control the specific growth rate. The easiest approach is the application of a predetermined exponential feed. Here, the initial conditions of the fed-batch phase are used to calculate the feed rate profile corresponding to the specific growth rate setpoint, and a feedforward controller is used to implement the profile [1]. However, no feedback action is available to intervene in case the process does not follow the desired growth rate. Feedback control typically involves the use of the following two classical types of controllers: proportional (P) and proportional-integral (PI) (and, more rarely, proportional integral derivative (PID)). They are considered to be simple gains controllers since they are not based on a model [6,23–25]. Dabros et al. [24] showed that the use of a mixed, feedforward-feedback controller reduced the mean control error by 32% as compared with applying a purely feedforward exponential feed profile. As mentioned above, the noise present in the biomass measurement and in the μ estimation, in particular, is a challenge when using feedback controllers to regulate the specific growth rate. Both the proportional and the integral parts of the controller are predisposed to amplify the noise and induce undesirable oscillations to the process. Therefore, the data must be treated and smoothed prior to use [26].

More advanced controllers are often model-based, and include the popular model predictive controller (MPC) and artificial neural networks [1,2]. These control approaches require extensive process knowledge in order to build and validate a data-driven model [1,27,28], which can be difficult to achieve as part of a new project or cell strain. Recently, successful implementation of growth rate controllers based on MPC, where noninvasive Process Analytical Technology (PAT) tools were applied towards microbial cultures, has been reported [29–32]. In addition, multisensor techniques have been applied by combining dielectric capacitance monitoring with biocalorimetry to reduce measurement noise [33]. However, these approaches add an extra layer of process complexity, resulting either from the extra modeling effort required or from the application of additional PAT tools. A critical advantage of dielectric capacitance monitoring is that it offers a direct and easily scalable method of biomass estimation [34,35] that can also be applied to single-use bioreactors [36,37]. Indeed, the signal-to-noise ratio often increases with scale, due to the reduced effect of bubbles on the probe, and therefore fast process development from bench scale to pilot scale [34]. Thus, control applications that require process scale-up would benefit from the reduced process complexity by applying dielectric capacitance as compared with utilizing multivariate approaches or biocalorimetry monitoring, where significant investment is required on scale-up [38]. Additionally, as no modeling is involved to implement direct dielectric monitoring based control, the systems developed should be easily transferable to other organisms, as previously demonstrated [24,39].

The technique used to control the specific growth rate depends on the ability to monitor the biomass and to manage the nonlinear growth dynamics of the process. A simple exponential feed cannot cope with unforeseen metabolic changes or process variability, while P or PI feedback controllers can have issues dealing with the exponential nature of cellular growth and with signal noise. Ensuring reliable biomass monitoring and a robust way of estimating μ is essential, as is the proper design of a controller capable of addressing the particularities of the process.

In this work, a revised control logic is presented to counter the above-mentioned issues. Biomass concentration measurements, provided by a dielectric spectroscopy probe, are smoothed with the Savitzky–Golay algorithm and used to estimate the current specific growth rate of the process. A modified feedforward-feedback controller is designed and optimized to maintain μ at the desired setpoint. The novel control design is adapted to the culture's exponential dynamics, resulting in improved noise management.

2. Materials and Methods

2.1. Cell Strain and Culture Conditions

In this study, the fed-batch cultures were performed with the wild type strain of *Kluyveromyces marxianus* DSMZ 5422. This strain was chosen mainly for its fast growth dynamics and easy cultivation conditions. A cell bank of 24 cryovials was prepared from four colonies of *K. marxianus* isolated on a fresh YPLA (yeast extract 10 g/L, peptone 20 g/L, lactose 40 g/L, and agar 20 g/L) petri dish. Each colony was inoculated in a sterile tube with 5 mL of YPL (yeast extract 10 g/L, peptone 20 g/L, and lactose 40 g/L), previously autoclaved at 121 °C for 15 min, and cultivated overnight at 30 °C and 150 rpm. The 5 mL of media were transferred to a 500 mL conical flask containing 100 mL of the previously described YPL medium and cultivated overnight at 30 °C and 150 rpm. Then, the cultures were transferred into four sterile tubes of 25 mL and centrifuged at 3500 rpm for 3 min. The biomass deposited in the four tubes was resuspended in a 5.5 mL solution of NaCl 0.9% and peptone 0.1%, and a 5.5 mL solution of glycerol 40%, reaching a total volume of 12 mL. Finally, sterile cryovials were filled with 2 mL each and stored in the freezer at −18 °C.

All the precultures were inoculated with a frozen vial retrieved from the cell bank. The cells were grown for 24 h at 30 °C and 120 rpm in a 1000 mL baffled conical flask containing 200 mL of YPL solution cited above. The preculture media were autoclaved for 10 min at 115 °C. After 24 h, the cells were isolated by centrifugation at 3500 rpm for 3 min and resuspended in a 4.5 mL solution of NaCl 0.9% and peptone 0.1%, then transferred to a 10 mL sterile syringe for inoculation.

The cultures were performed at 30 °C in a 3.6 L bench scale bioreactor (KLF, BioEngineering AG, Wald, Switzerland), with a working volume of 2.5 L. The agitation rate was set to 1000 rpm and the air flow rate to 250 NL/h, corresponding to 2 vvm, to maintain the dissolved oxygen (DO) levels above 20% throughout the experiments. In addition, to accommodate the strains particularly high respiration demands, a sintered sparger was used to enhance oxygen transfer and obtain a $k_L a$ value of 0.06 s^{-1}. The bioreactor was equipped with a double 6-blade Rushton type agitator, a PT-100 temperature probe, pH and DO probes (Mettler-Toledo, Columbus, OH, USA), inlet gas flow controller, exhaust gas port, acid and base ports, a feed port, an antifoam port, and a sampling port, as illustrated in Figure 1. The exhaust gas passed through a condenser cooled to 4 °C to minimize liquid loss. Temperature, pH, and pO$_2$ were controlled using the software provided by BioEngineering. Medium pH was controlled at 5 using solutions of NaOH 3 M and HCl 1 M. A solution of 10% antifoam in water was used to prevent foam formation.

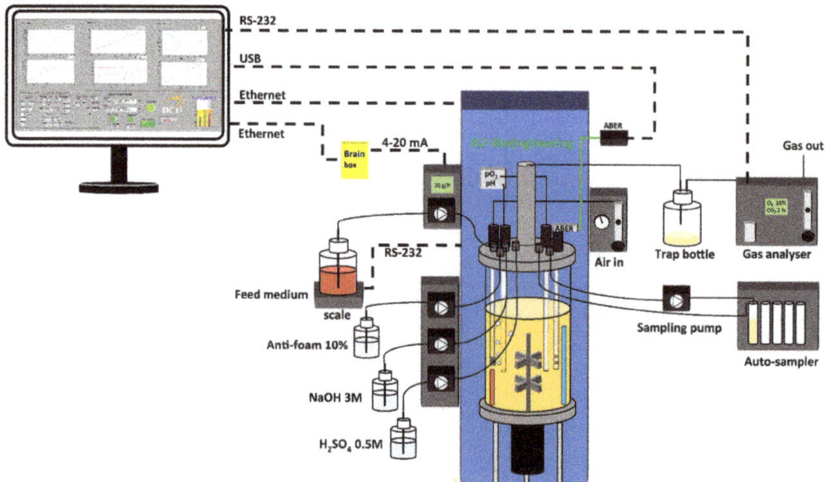

Figure 1. Fed-batch bioreactor setup.

Biomass concentration was monitored with a dielectric spectroscopy probe (ABER Instruments Ltd., Aberystwyth, UK). The probe was used in dual-frequency mode at 580 and 15,650 Hz with the polarization correction BM220 PolC tuned on and the filter value set at 140 to decrease the impact of aeration on the reading. Using the dual-frequency mode is more reliable because it minimizes the influence of baseline shifts and medium conductivity changes on the dielectric signal. Feeding was performed with a Peripex W2 pump (BioEngineering), which was controlled analogically between 4 and 20 mA by the current loop box ED-550 (Brainboxes, Liverpool, UK). Oxygen and CO_2 concentration levels in the exhaust gas were analyzed using a Tandem Pro gas analyzer (Magellan BioTech, Borehamwood, UK).

The cells were first grown in batch mode until depletion of the carbon source (which was detected by monitoring the dissolved oxygen and off-gas analysis signals), and then in fed-batch mode. Table 1 shows the semi-defined batch and feed media composition. Yeast extract was used as an iron and vitamin source [40–42], whereas peptone was used as an additional source of vitamins and protein, replacing a trace elements solution typically used [16,24,43].

Table 1. Composition of the batch and feed media.

Component	Batch Medium g/L	Feed Medium g/L
Lactose	5	300
$(NH_4)_2SO_4$	5	50
KH_2PO_4	3	25
$MgSO_4 \cdot 7H_2O$	0.5	3
Peptone	5	25
Yeast extract	5	25
Antifoam	2	3

2.2. Signal Filtering and Smoothing

A data acquisition program was developed using LabVIEW 18 (National Instruments, Austin, TX, USA) to handle all data communications between the KLF bioreactor and process analyzers (dielectric probe and off-gas analyzer). The program also treated the dielectric signal, reducing measured noise online. Basic moving average and the Savitzky–Golay smoothing algorithms were implemented and evaluated, with the latter showing better performance in terms of striking a compromise between minimizing signal delay and maximizing the accuracy of the filtered signal. A window of 121 data points and a first-order polynomial fit was used within the Savitzky–Golay algorithm to smooth the dielectric signal ahead of the specific growth rate estimation.

2.3. Dielectric Spectroscopy Adaptive Calibration

The technology of dielectric spectroscopy, when used in dual-frequency mode, measures the drop in the medium permittivity between the two electrical field frequencies. This drop is assumed to be linearly proportional to the concentration of live cells; however, the correlation is different for each cell strain and culture conditions [44]. The magnetic field used by the sensor is sensitive to the average bubble size, medium conductivity and, to some extent, fluid viscosity. In order to maintain the sensor's response accuracy to these changing parameters, the probe was recalibrated for each culture using the cell concentration data acquired at the end of batch phase. The adapted calibration model was included within the LabVIEW on-line monitoring program prior to fed-batch phase, accounting for changes in medium conductivity and fluid viscosity, and applied for the fed-batch segment of the experiment.

For the purpose of reference, cell concentration was measured off-line by dry cell weight, at intervals of 60 min during the batch phase and 3 h during the fed-batch phases. Samples of 5 mL of broth were filtered through a prescaled membrane filter. Then, the filter was dried until constant weight, the yeast mass was calculated and converted to grams per litre.

2.4. Specific Growth Rate Estimation

The filtered biomass signal was used to calculate the estimated specific growth rate. As cell growth is an exponential process, the linearization involves the use of a logarithm within the derivative. The theoretical way of calculating μ is given in Equation (1), where C_X is the current biomass concentration and V the medium volume.

$$\mu(t) = \frac{d\ln(C_X V)}{dt} \tag{1}$$

Specific growth rate values obtained with Equation (1) are, however, very sensitive to any residual noise in the biomass measurement. In order to minimize this effect, a Δt window of 15 min was used in this work to calculate μ, as shown in Equation (2). This window size was the result of a compromise between reducing noise and minimizing signal delay.

$$\mu_{est}(t) = \frac{\ln\left(\frac{C_{X,t} V_t}{C_{X,t-15min} V_{t-15min}}\right)}{\Delta t} \tag{2}$$

Throughout the batch phase, the volume of the medium remained constant; however, during the fed-batch phase, the culture volume increased as a function of the feed flow rate. The instantaneous value of the volume was updated by monitoring the mass of medium fed into the reactor and taking into account media density. Equation (3) shows the calculation used to correct the medium volume, V_t, throughout the culture. Here V_{ini} is the initial volume of the batch medium, m_{fed} is the mass of the feed solution fed into the vessel, as recorded by the scale, and ρ_{feed} is the feed density.

$$V_t = V_{ini} + \frac{m_{fed}}{\rho_{feed}} \tag{3}$$

2.5. Controller Design

The classic method of regulating the specific growth rate is open-loop (feed-forward) control; this technology implies implementing a predefined, exponential feed rate of medium to the culture. However, using this approach does not allow reacting to any potential process disturbances or deviations from the setpoint. In order to enhance the open-loop action, a feedback logic can be implemented [5,6,8]. Dabros et al. [24] proposed a new type of PI controller where the controller gains were included within the exponential term of the control equation. The advantage with this control logic is that it is well adapted to the exponential dynamics of the culture, making it more effective and robust over the course of the fed-batch culture.

In this work, we demonstrate the use of a similar feedforward-feedback controller, as the one proposed in [24], but with a slight modification. The proposed control logic alteration allows for greater noise management and makes the controller more robust, particularly for long-duration fed-batch cultures.

The feed-forward part of the controller action, for a given specific growth rate setpoint, μ_{sp}, is given by the following expression:

$$F_{FF}(t) = F_0 \exp(\mu_{sp} t) \tag{4}$$

where F_0 is the theoretical initial feed flow rate, calculated as follows:

$$F_0 = C_{X,0} V_0 \frac{\mu_{sp}}{Y_{X/S} S_F} \tag{5}$$

Here, $C_{X,0}$ is the initial biomass concentration, V_0 the initial volume of the culture, $Y_{X/S}$ is the biomass yield coefficient, and S_F is the substrate concentration in the feed medium. The initial biomass concentration is obtained with the dielectric measurement, while the biomass yield is determined at the end of the initial batch phase.

The feedback action of the controller is based on the process control error, calculated as follows:

$$\varepsilon(t) = \mu_{sp} - \mu_{est}(t) \tag{6}$$

The feedback controller contains proportional and integral gains (respectively, K_p and K_i). In order to avoid the use of adaptive gains, both K_p and K_i were included within the exponential term of the feedback control equation, F_{FB}, as shown in the following expression:

$$F_{FB}(t) = F_0 \exp\left(\left(K_p \varepsilon(t) + K_i \int_0^t \varepsilon(t)dt\right)t\right) \tag{7}$$

Both parts of the controller, feedforward and feedback, are used simultaneously and can be written in the same and final control equation given below:

$$F(t) = F_0 \exp\left(\left(\mu_{sp} + K_p \varepsilon(t) + K_i \int_0^t \varepsilon(t)dt\right)t\right) \tag{8}$$

This form of feedforward-feedback controller was used by Dabros et al. [24]. Its main drawbacks include sensitivity to noise and oscillatory behavior, observed to increase in time. Indeed, if the error term is unstable and noisy, the controller action can intensify the problem. To reduce the oscillations, the controller gains have to be chosen meticulously. In this work, two different methods were applied to tune the gains' values. The first approach involved manual tuning from the initial values reported in [24], which were 1.5 [-] for K_p and 0.5 h^{-1} for K_i. After several cultures, the gains were adapted according to a manual tuning methodology [45] as follows: K_p was reduced to 0.75 [-] and K_i increased to 1 h^{-1}. The second method used to calculate the controller gains was the Ziegler–Nichols open-loop tuning table [46]. This approach is based on the system's response following a step input applied to the control variable. In this work, a predefined feed flow rate was applied according to Equation (4), and the specific growth rate response was used to determine the proportional and integral terms. The calculated K_p was 0.66 [-] and K_i was 0.6 h^{-1}, both values relatively close to the ones obtained with manual tuning. The two sets of gains were tried in order to make the final choice: 0.75 [-] and 1 h^{-1} for K_p and K_i, respectively.

By analyzing Equation (8), it can be noted that the proportional and integral terms of the controller are time dependent. This design effectively makes the weights of the two terms increase with time. Indeed, in this work, it was noticed that the oscillatory behavior increased with time not only in the control variable, μ, but also in the manipulated variable, $F(t)$. This meant that the oscillations were time dependent, and therefore the controller logic needed to be changed to remove the gains' dependence on time. To address this issue, a novel approach is proposed as follows.

First, only the proportional term was changed, yielding the following expression:

$$F(t) = F_0 \exp\left(\left(\mu_{sp} + \frac{K_p}{t}\varepsilon(t) + K_i \int_0^t \varepsilon(t)dt\right)t\right) \tag{9}$$

By dividing K_p by the time, the effect of this term becomes constant over time. This change should increase the performance of the controller by reducing the high-frequency oscillations in the control and manipulated variables.

Similarly, the time dependence of the integral term was suspected to induce increasing oscillations in the control and the manipulated variables. In this case, the disturbance was expected to be low in frequency, since the integral term's effect is delayed in time. Thus, by making the integral term constant, the low-frequency oscillations should be reduced. To verify this hypothesis, Equation (9) was adapted as follows:

$$F(t) = F_0 \exp\left(\mu_{sp}t + K_p\varepsilon(t) + K_i \int_0^t \varepsilon(t)dt\right) \tag{10}$$

This last modification was thought to be necessary particularly for long-duration fed-batch cultures. It should be noted that with the final controller logic (Equation (10)), both parts of the feedback controller have a constant weight over time, but they are still included in the exponential term of the equation. This is important because it allows for exponential gain scheduling, making the otherwise linear PI controller robust with respect to the exponential growth dynamics of the fed-batch culture.

Each time a new specific growth rate setpoint was defined during a running culture, the initial feed flow rate, F_0, was calculated anew using Equation (5), applying the latest value of the biomass concentration, $C_{X,t}$, and the current volume, V_t. To prevent integral windup, the window of integration was limited to 3 h. This limitation allowed maintaining the controller action continuous throughout the culture without having to reset the integral term between two setpoints.

Figure 2 shows the controller block diagram used in this work. The final μ controller logic is given by Equation (10). The control loop was executed every 20 s by the LabVIEW supervision program, registering all the process measurements and sending commands to the feed pump.

The performance of the controller was assessed, at each setpoint, by calculating the root mean square error (RMSE, Equation (11)). In order to allow for adequate stabilization time, the calculation window started one hour from the moment when the setpoint was initially applied and continued until the next setpoint change. The stabilization time was necessary because the dielectric signal showed signs of increased disturbance resulting from the metabolic adjustments following each setpoint change.

$$RMSE = \sqrt{\frac{\sum_i^n \left(\mu_{est_i} - \mu_{sp}\right)^2}{n}} \tag{11}$$

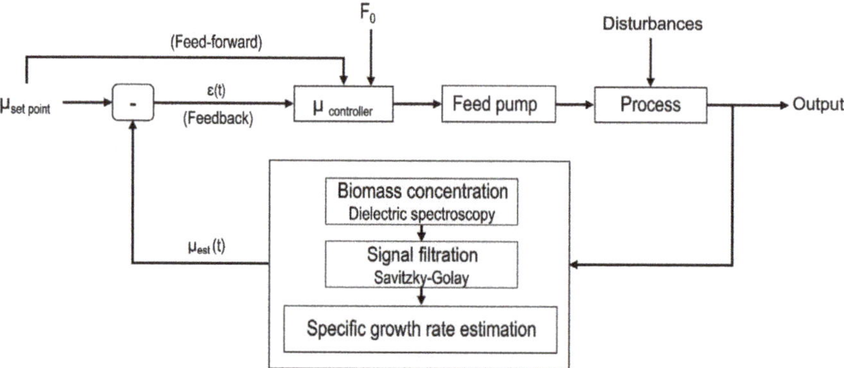

Figure 2. Controller block diagram.

3. Results

First, the performance of the original, unmodified feedforward-feedback PI controller (Equation (8), [24]) was assessed by analyzing the profiles of the specific growth rate and of the feed flow rate over the duration of a fed-batch culture. Analyzing the μ profile, shown in Figure 3, it can be observed that the process showed increasing oscillations starting after two hours of the fed-batch culture. The RMSE for this experiment was 0.063 h^{-1}, corresponding to a mean controller error of 32%.

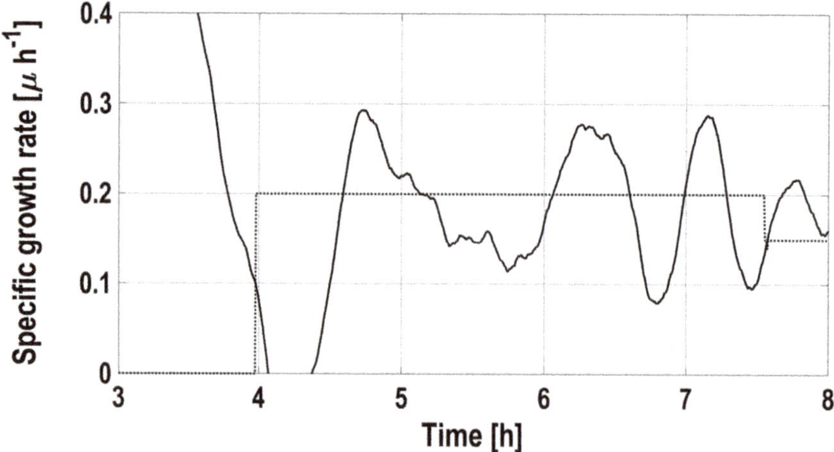

Figure 3. Setpoint (dash line) and estimated specific growth rate (solid line) during a controlled fed-batch culture, controller logic (Equation (8)).

To improve the controller and reduce the amplified oscillations, it was necessary to reduce the effect of the part of the controller action that reacts fast to the evolution of the error, $\varepsilon(t)$. Indeed, Figure 4 confirms that oscillations were equally present in the manipulated variable. This meant that the action responsible for this behavior was dictated by the proportional term.

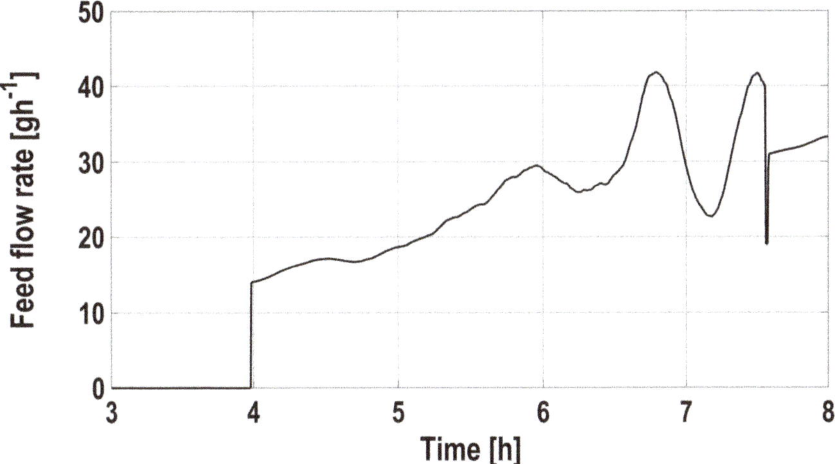

Figure 4. Feed flow rate applied during a controlled fed-batch culture, controller logic (Equation (8)).

For the subsequent experiment, the controller logic was changed to the one given by Equation (9). By maintaining the proportional gain, K_p, constant overtime, the high-frequency oscillatory behavior was reduced. In order to assess the controller's improvement, a longer step was performed. Figure 5 shows the specific growth rate profile over a nine-hour duration fed-batch culture. The process still shows oscillations, but ones that are constant and not amplified over time. In this experiment, the root mean square error (RMSE) was 0.072 h^{-1} which corresponded to a mean controller error of 36%. It was stipulated that the controller still could be improved by examining the corresponding feed flow rate profile, shown in Figure 6. The oscillatory behavior was still present in this signal, but its frequency was lower ($\sim$0.5 h^{-1}) than that observed previously (Figure 3, $\sim$1 h^{-1}). Nevertheless, the amplitude of the oscillations did increase in time.

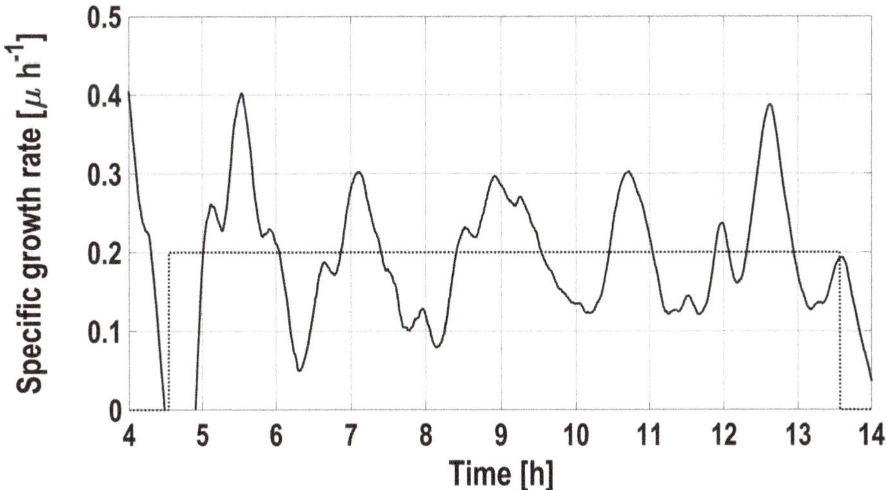

Figure 5. Setpoint (dash line) and estimated specific growth rate (solid line) during a controlled fed-batch culture, controller logic (Equation (9)).

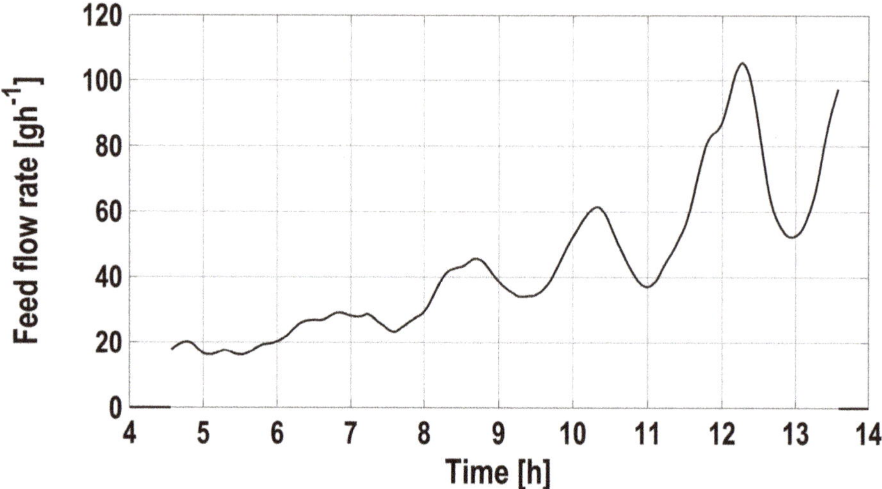

Figure 6. Feed flow rate profile obtained during a controlled fed-batch culture, controller logic (Equation (9)).

As the proportional gain remained constant, it could not be responsible for amplifying the oscillations in the feed rate. On the other hand, the integral term was still time dependent, its effect increasing over time. Moreover, since the oscillatory frequency was lower, suggesting that the term responsible for inducing them was a slow-acting term, it pointed at the integral term. Following this consideration, the controller logic was changed to that given in Equation (10).

To assess the improvement in the controller's performance, another experiment was performed where the specific growth rate setpoint was changed mid-culture. With the theoretical maximal specific growth rate of *K. marxianus* being close to 0.6 h^{-1}, it was decided to control the growth rate at a high and a low setpoint within the same culture [40,47,48].

Figure 7 shows the specific growth rate obtained during this fed-batch experiment. It was possible to control the growth rate at 0.4 h^{-1}; however, the culture could not be run at that setpoint for a long time. Indeed, the cell concentration increased too rapidly to allow for full aerobic conditions (DO > 20%). After 3 h, the limit of 20% DO was reached, and the setpoint was lowered to maintain the culture in aerobic respiration. It is unclear whether the remaining mild oscillatory behavior was caused by the residual noise in the filtered dielectric signal or by the controller itself. Figure 8 shows the feed flow rate applied throughout the culture; this time, little oscillation was observed in the manipulated variable. The resulting smooth control action allowed the controller to reach and maintain the growth very close to the setpoint throughout the experiment. The root mean square error (RMSE) for the μ setpoint of 0.4 h^{-1} was 0.061 h^{-1} and for the setpoint of 0.1 h^{-1}, the RMSE was 0.019 h^{-1}. The combined mean controller error for this culture was 17% which was the half compared to the errors observed during the cultures where the controllers given by Equations (8) and (9) were used (Figures 3 and 5, combined mean controller error of 32% and 36%, respectively).

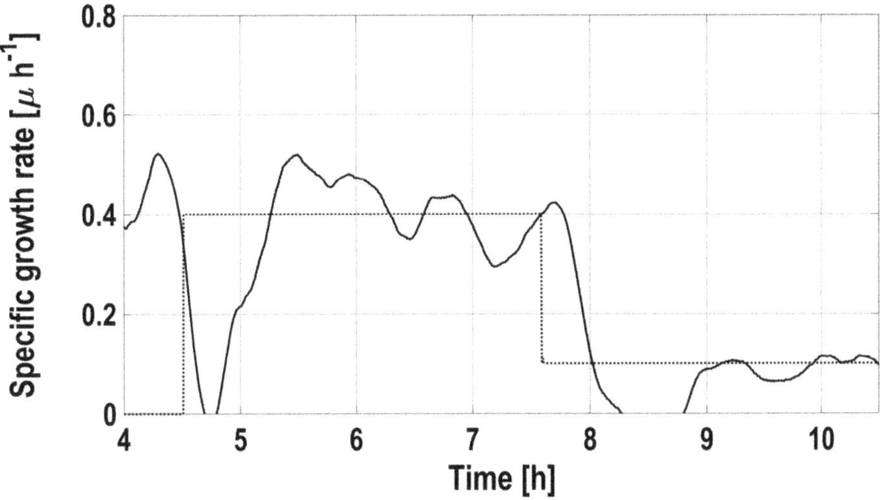

Figure 7. Setpoint (dash line) and estimated specific growth rate (solid line) during a controlled fed-batch culture, controller logic (Equation (10)).

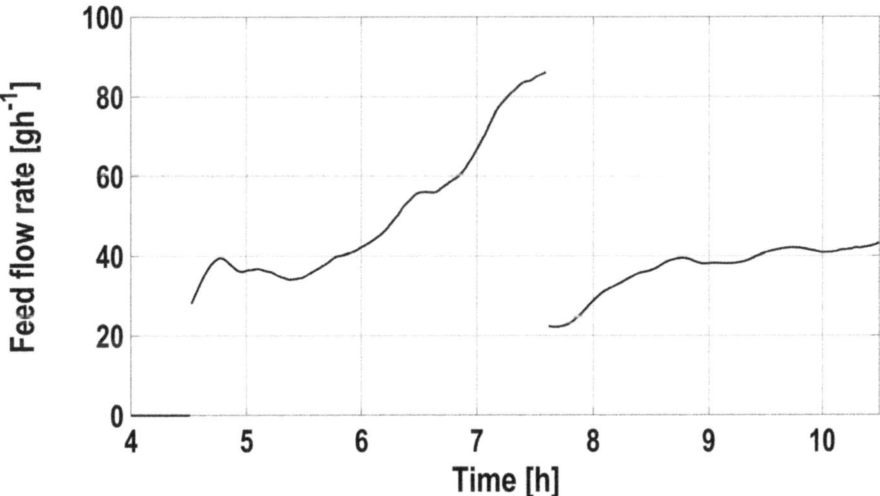

Figure 8. Feed flow rate profile during a controlled fed-batch culture, controller logic (Equation (10)).

Figure 9 shows the evolution of the biomass concentration during this fed-batch culture. The individual stages of the culture can easily be identified, i.e., the initial batch phase (until culture time of 4.4 h), followed by the two fed-batch phases at μ setpoints of 0.4 h^{-1} and 0.1 h^{-1}.

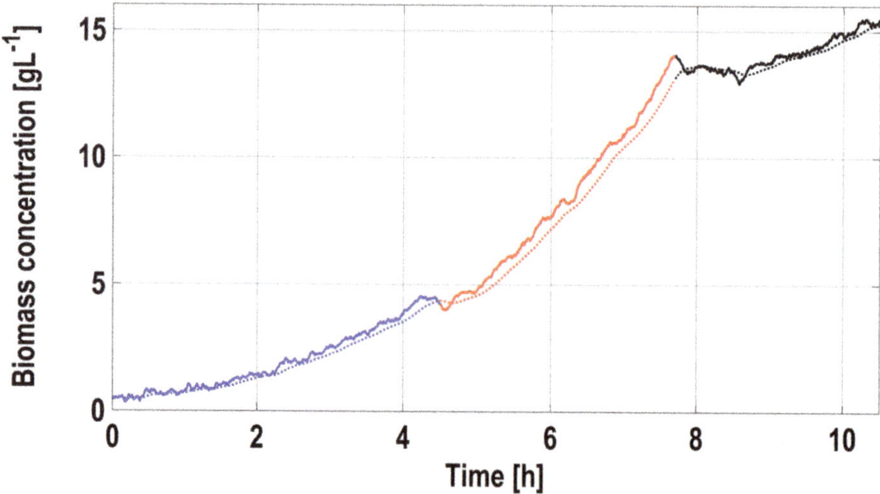

Figure 9. Evolution of biomass concentration (raw signal in solid line, filtered signal in dashed line) during the initial batch phase (blue) and two fed-batch phases at $\mu_{sp} = 0.4\,\text{h}^{-1}$ (red) and $0.1\,\text{h}^{-1}$ (black).

4. Discussion

In this work, it was demonstrated that the logic of the specific growth rate controller was equally, if not more, important than the controller gains. Indeed, the control performance changed significantly depending on the approach chosen.

As a first step in the study, we have addressed the weakest element in the control loop, namely, the biomass measurement and the specific growth rate estimation. The biomass measurement signal was inherently noisy and sensitive to process conditions, such as aeration and agitation. The filtering approach was optimized to minimize the effect of the bubbles without inducing too long a lag time. The window of 15 min chosen for the specific growth rate estimator was a trade-off with respect to the lag time of the filtering algorithm and the remaining noise on the biomass signal.

Secondly, it was illustrated that by adjusting the controller logic, the signal oscillations could be managed effectively, and the controller stability improved. Using the original controller (Equation (8)), a fed-batch culture lasting longer than 2 h was difficult to achieve because of increasing oscillations in both the control and the manipulated variables. With the new control logic, the experiment was shown to be stable in time, and the mean controller error was reduced to 17%. The remaining error seems to be limited by the residual noise amplitude on the biomass measurement.

Finally, the reproducibility of the controller's performance was assessed by comparing three fed-batch cultures run at the same specific growth setpoint ($0.1\,\text{h}^{-1}$). The result, shown in Figure 10, demonstrates that after the initial period of instability (1 h), the estimated μ tracked the assigned setpoint in all three runs. The level of accuracy was similar, showing an average relative RMSE of 23% $\pm$ 6% (1σ) despite the variable starting biomass concentration, indicating controller robustness.

All fed-batch cultures were run between 6 and 9 h in total duration, with variable μ setpoints. Throughout this work, the maximum biomass concentration allowing aerobic conditions was around 25 g/L of dry cell weight. Above this concentration, regardless of the specific growth rate setpoint, the oxygen transfer rate was insufficient to allow purely aerobic respiration of the cells. The aim of fed-batch cultures is to ensure a permanent state of substrate limitation. If the culture conditions changed from aerobic to anaerobic, the limiting factor would be a combination of substrate and oxygen. The agitation rate could be increased to enhance the oxygen transfer and allow for higher biomass concentrations. However, this change was observed to deteriorate the growth rate estimation by increasing significantly the noise in the biomass concentration measurement.

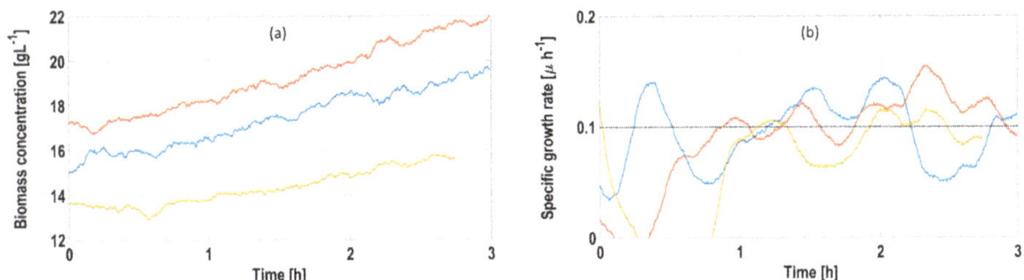

Figure 10. Biomass concentration profiles (**a**) and specific growth rate estimates (**b**) during three fed-batch phases run at $\mu_{sp} = 0.1\ \text{h}^{-1}$.

5. Conclusions

The aim of this work was to implement and optimize a novel proportional-integral (PI) feedforward-feedback controller, designed to maintain a desired specific growth rate of a microbial culture. The proposed new control logic provides robust setpoint tracking of an exponentially evolving fed-batch culture, while ensuring improved noise and oscillation management. To reduce the noise present in the biomass measurement, the filtering approach was optimized using a first-order Savitzky–Golay algorithm instead of a simple moving average. Then, to limit the oscillations in the manipulated and control variables, the time dependency of the P and I feedback terms was removed, as shown in Equation (10). This change enhanced the stability of the system for longer processes without any noticeable increase in signal oscillations, as was previously observed. Despite the remaining noise in the biomass measurement, the controller was able to cope with this issue and maintain the desired specific growth rate over the duration of the culture. In conclusion, it was shown that a strain of *K. marxianus* could be grown successfully in fed-batch mode, under substrate-limited, aerobic conditions, at different setpoints ranging from $0.1\ \text{h}^{-1}$ to $0.4\ \text{h}^{-1}$.

Some aspects could still be improved, particularly the biomass measurement signal filtration. In order to diminish the undesired effect of aeration on the dielectric signal, two possibilities become apparent. The first solution is to work under a slight pressure in the reactor, which would increase the solubility of oxygen, and thereby reach aerobic conditions with a reduced aeration rate. The second solution is to use an external bypass (flow-through cell) where the probe could be installed to measure the medium permittivity without the presence of bubbles. The latter approach would require modifications to the existing bioreactor. Finally, a more powerful signal filtration technique could be applied, such as the Kalman filter [22].

It was noticed that the biomass yield varied between the fed-batch cultures that were performed. Depending on the specific growth rate, the obtained biomass yield, $Y_{X/S}$, varied from 0.3 to 0.65 g/g. This fact complicated the use of the initial feed flow rate calculation, (F_0, Equation (5)) since both terms, μ_{sp} and $Y_{X/S}$, are present in this equation and yet, they are interdependent. It could be interesting to investigate more in depth the effect of the specific growth rate on the biomass yield, as suggested in [5]. By knowing at which growth rate a particular strain is more productive, industrial applications could be optimized and their costs reduced.

This work highlighted the feasibility of controlling the specific growth rate of yeast in long-duration fed-batch cultures and in spite of a noisy biomass measurement signal. The proposed approach should be applicable to other microbial systems, as well as to mammalian cell cultures. In the latter case, it is expected that the slower growth dynamics should allow a further reduction in signal noise and oscillations.

Author Contributions: Y.B. performed the experimental work with technical help from D.C. and under the supervision of B.F. and M.D.; Y.B. wrote the original manuscript, which was subsequently revised by M.D. and B.F. All authors have read and agreed to the published version of the manuscript.

Funding: This research received no external funding.

Acknowledgments: This publication has emanated from research supported by the School of Biotechnology, Dublin City University. The authors would like to thank Rachel Crossley, European Technical Sales Specialist at Aber Instruments Ltd. and Karin Philipp of Bioengineering AG for support.

Conflicts of Interest: The authors declare no conflict of interest.

References

1. Mears, L.; Stocks, S.M.; Sin, G.; Gernaey, K.V. A review of control strategies for manipulating the feed rate in fed-batch fermentation processes. *J. Biotechnol.* **2017**, *245*, 34–46. [CrossRef]
2. Lee, J.; Lee, S.Y.; Park, S.; Middelberg, A.P.J. Control of fed-batch fermentations. *Biotechnol. Adv.* **1999**, *17*, 29–48. [CrossRef]
3. Kim, B.S.; Lee, S.C.; Lee, S.Y.; Chang, Y.K.; Chang, H.N. High cell density fed-batch cultivation of Escherichia coli using exponential feeding combined with pH-stat. *Bioprocess Biosyst. Eng.* **2004**, *26*, 147–150. [CrossRef] [PubMed]
4. Johnston, W.; Cord-Ruwisch, R.; Cooney, M.J. Industrial control of recombinant E. coli fed-batch culture: New perspectives on traditional controlled variables. *Bioprocess Biosyst. Eng.* **2002**, *25*, 111–120.
5. Habegger, L.; Rodrigues Crespo, K.; Dabros, M. Preventing Overflow Metabolism in Crabtree-Positive Microorganisms through On-Line Monitoring and Control of Fed-Batch Fermentations. *Fermentation* **2018**, *4*, 79. [CrossRef]
6. Kesson, M.; Hagander, P.; Axelsson, J.P. Avoiding acetate accumulation in Escherichia coli cultures using feedback control of glucose feeding. *Biotechnol. Bioeng.* **2001**, *73*, 223–230. [CrossRef] [PubMed]
7. Chen, W.; Graham, C.; Ciccarelli, R.B. Automated fed-batch fermentation with feed-back controls based on dissolved oxygen (DO) and pH for production of DNA vaccines. *J. Ind. Microbiol. Biotechnol.* **1997**, *18*, 43–48. [CrossRef]
8. Soons, Z.I.T.A.; Voogt, J.A.; van Straten, G.; van Boxtel, A.J.B. Constant specific growth rate in fed-batch cultivation of Bordetella pertussis using adaptive control. *J. Biotechnol.* **2006**, *125*, 252–268. [CrossRef]
9. Junker, B.H.; Wang, H.Y. Bioprocess monitoring and computer control: Key roots of the current PAT initiative. *Biotechnol. Bioeng.* **2006**, *95*, 226–261. [CrossRef]
10. Jenzsch, M.; Simutis, R.; Eisbrenner, G.; Stückrath, I.; Lübbert, A. Estimation of biomass concentrations in fermentation processes for recombinant protein production. *Bioprocess Biosyst. Eng.* **2006**, *29*, 19–27. [CrossRef]
11. Claes, J.E.; Van Impe, J.F. Combining yield coefficients and exit-gas analysis for monitoring of the baker's yeast fed-batch fermentation. *Bioprocess Eng.* **2000**, *22*, 195–200. [CrossRef]
12. Chen, L.Z.; Nguang, S.K.; Li, X.M.; Chen, X.D. Soft sensors for on-line biomass measurements. *Bioprocess Biosyst. Eng.* **2004**, *26*, 191–195.
13. Luttmann, R.; Bracewell, D.G.; Cornelissen, G.; Gernaey, K.V.; Glassey, J.; Hass, V.C.; Kaiser, C.; Preusse, C.; Striedner, G.; Mandenius, C.F. Soft sensors in bioprocessing: A status report and recommendations. *Biotechnol. J.* **2012**, *7*, 1040–1048. [CrossRef]
14. Kiviharju, K.; Salonen, K.; Moilanen, U.; Eerikäinen, T. Biomass measurement online: The performance of in situ measurements and software sensors. *J. Ind. Microbiol. Biotechnol.* **2008**, *35*, 657–665. [CrossRef]
15. Arnoux, A.S.; Preziosi-Belloy, L.; Esteban, G.; Teissier, P.; Ghommidh, C. Lactic acid bacteria biomass monitoring in highly conductive media by permittivity measurements. *Biotechnol. Lett.* **2005**, *27*, 1551–1557. [CrossRef]
16. Schuler, M.M.; Marison, I.W. Real-time monitoring and control of microbial bioprocesses with focus on the specific growth rate: Current state and perspectives. *Appl. Microbiol. Biotechnol.* **2012**, *94*, 1469–1482. [CrossRef]
17. Kiviharju, K.; Salonen, K.; Moilanen, U.; Meskanen, E.; Leisola, M.; Eerikäinen, T. On-line biomass measurements in bioreactor cultivations: Comparison study of two on-line probes. *J. Ind. Microbiol. Biotechnol.* **2007**, *34*, 561–566. [CrossRef]
18. Austin, G.D.; Watson, R.W.J.; D'Amore, T. Studies of on-line viable yeast biomass with a capacitance biomass monitor. *Biotechnol. Bioeng.* **1994**, *43*, 337–341. [CrossRef]
19. Biechele, P.; Busse, C.; Solle, D.; Scheper, T.; Reardon, K. Sensor systems for bioprocess monitoring. *Eng. Life Sci.* **2015**, *15*, 469–488. [CrossRef]

20. Olsson, L.; Nielsen, J. On-line and in situ monitoring of biomass in submerged cultivations. *Trends Biotechnol.* **1997**, *15*, 517–522. [CrossRef]
21. Schuler, M.M.; Sivaprakasam, S.; Freeland, B.; Hama, A.; Hughes, K.M.; Marison, I.W. Investigation of the potential of biocalorimetry as a process analytical technology (PAT) tool for monitoring and control of Crabtree-negative yeast cultures. *Appl. Microbiol. Biotechnol.* **2012**, *93*, 575–584. [CrossRef] [PubMed]
22. Shimizu, H.; Araki, K.; Shioya, S.; Suga, K.I. Optimal production of glutathione by controlling the specific growth rate of yeast in fed-batch culture. *Biotechnol. Bioeng.* **1991**, *38*, 196–205. [CrossRef] [PubMed]
23. Ehgartner, D.; Hartmann, T.; Heinzl, S.; Frank, M.; Veiter, L.; Kager, J.; Herwig, C.; Fricke, J. Controlling the specific growth rate via biomass trend regulation in filamentous fungi bioprocesses. *Chem. Eng. Sci.* **2017**, *172*, 32–41. [CrossRef]
24. Dabros, M.; Schuler, M.M.; Marison, I.W. Simple control of specific growth rate in biotechnological fed-batch processes based on enhanced online measurements of biomass. *Bioprocess Biosyst. Eng.* **2010**, *33*, 1109–1118. [CrossRef]
25. Gnoth, S.; Jenzsch, M.; Simutis, R.; Lübbert, A. Control of cultivation processes for recombinant protein production: A review. *Bioprocess Biosyst. Eng.* **2008**, *31*, 21–39. [CrossRef]
26. Riesenberg, D.; Schulz, V.; Knorre, W.A.; Pohl, H.D.; Korz, D.; Sanders, E.A.; Roß, A.; Deckwer, W.D. High cell density cultivation of Escherichia coli at controlled specific growth rate. *J. Biotechnol.* **1991**, *20*, 17–27. [CrossRef]
27. Marison, I.; Hennessy, S.; Foley, R.; Schuler, M.; Sivaprakasam, S.; Freeland, B. The choice of suitable online analytical techniques and data processing for monitoring of bioprocesses. *Adv. Biochem. Eng. Biotechnol.* **2013**, *132*, 249–280.
28. Liu, C.; Gong, Z.; Shen, B.; Feng, E. Modelling and optimal control for a fed-batch fermentation process. *Appl. Math. Model.* **2013**, *37*, 695–706. [CrossRef]
29. Galvanauskas, V.; Simutis, R.; Vaitkus, V. Adaptive control of biomass specific growth rate in fed-batch biotechnological processes. A comparative study. *Processes* **2019**, *7*, 810. [CrossRef]
30. Mohan, N.; Sivaprakasam, S. Heat Compensation Calorimeter as a Process Analytical Tool to Monitor and Control Bioprocess Systems. *Ind. Eng. Chem. Res.* **2017**, *56*, 8416–8427. [CrossRef]
31. Steinwandter, V.; Zahel, T.; Sagmeister, P.; Herwig, C. Propagation of measurement accuracy to biomass soft-sensor estimation and control quality. *Anal. Bioanal. Chem.* **2017**, *409*, 693–706. [CrossRef] [PubMed]
32. Sagmeister, P.; Wechselberger, P.; Jazini, M.; Meitz, A.; Langemann, T.; Herwig, C. Soft sensor assisted dynamic bioprocess control: Efficient tools for bioprocess development. *Chem. Eng. Sci.* **2013**, *96*, 190–198. [CrossRef]
33. Paulsson, D.; Gustavsson, R.; Mandenius, C.F. A soft sensor for bioprocess control based on sequential filtering of metabolic heat signals. *Sensors (Switzerland)* **2014**, *14*, 17864–17882. [CrossRef] [PubMed]
34. Metze, S.; Ruhl, S.; Greller, G.; Grimm, C.; Scholz, J. Monitoring online biomass with a capacitance sensor during scale-up of industrially relevant CHO cell culture fed-batch processes in single-use bioreactors. *Bioprocess Biosyst. Eng.* **2020**, *43*, 193–205. [CrossRef]
35. Moore, B.; Sanford, R.; Zhang, A. Case study: The characterization and implementation of dielectric spectroscopy (biocapacitance) for process control in a commercial GMP CHO manufacturing process. *Biotechnol. Prog.* **2019**, *35*, e2782. [CrossRef]
36. Justice, C.; Brix, A.; Freimark, D.; Kraume, M.; Pfromm, P.; Eichenmueller, B.; Czermak, P. Process control in cell culture technology using dielectric spectroscopy. *Biotechnol. Adv.* **2011**, *29*, 391–401. [CrossRef]
37. Carvell, J.; Lee, M.; Bhat, A.R. Recent developments in scaling down and using single use probes for measuring the live cell concentration by dielectric spectroscopy. *BMC Proc.* **2015**, *9*, P46. [CrossRef]
38. Voisard, D.; Pugeaud, P.; Kumar, A.R.; Jenny, K.; Jayaraman, K.; Marison, I.W.; Von Stockar, U. Development of a large-scale biocalorimeter to monitor and control bioprocesses. *Biotechnol. Bioeng.* **2002**, *80*, 125–138. [CrossRef]
39. Cole, H.E.; Demont, A.; Marison, I.W. The application of dielectric spectroscopy and biocalorimetry for the monitoring of biomass in immobilized mammalian cell cultures. *Processes* **2015**, *3*, 384–405. [CrossRef]
40. Urit, T.; Li, M.; Bley, T.; Löser, C. Growth of Kluyveromyces marxianus and formation of ethyl acetate depending on temperature. *Appl. Microbiol. Biotechnol.* **2013**, *97*, 10359–10371. [CrossRef]
41. Löser, C.; Urit, T.; Keil, P.; Bley, T. Studies on the mechanism of synthesis of ethyl acetate in Kluyveromyces marxianus DSM 5422. *Appl. Microbiol. Biotechnol.* **2015**, *99*, 1131–1144. [CrossRef] [PubMed]

42. GRANT, C.L.; PRAMER, D. Minor element composition of yeast extract. *J. Bacteriol.* **1962**, *84*, 869–870. [CrossRef] [PubMed]

43. Urit, T.; Stukert, A.; Bley, T.; Löser, C. Formation of ethyl acetate by Kluyveromyces marxianus on whey during aerobic batch cultivation at specific trace element limitation. *Appl. Microbiol. Biotechnol.* **2012**, *96*, 1313–1323. [CrossRef] [PubMed]

44. Dabros, M.; Dennewald, D.; Currie, D.J.; Lee, M.H.; Todd, R.W.; Marison, I.W.; Von Stockar, U. Cole-Cole, linear and multivariate modeling of capacitance data for on-line monitoring of biomass. *Bioprocess Biosyst. Eng.* **2009**, *32*, 161–173. [CrossRef] [PubMed]

45. Howard, R.; Cooper, D. A novel pattern-based approach for diagnostic controller performance monitoring. *Control Eng. Pract.* **2010**, *18*, 279–288. [CrossRef]

46. Meshram, P.M.; Kanojiya, R.G. Tuning of PID controller using Ziegler-Nichols method for speed control of DC motor. In Proceedings of the IEEE-international conference on advances in engineering, science and management (ICAESM-2012), Nagapattinam, India, 30–31 March 2012; pp. 117–122.

47. Löser, C.; Urit, T.; Stukert, A.; Bley, T. Formation of ethyl acetate from whey by Kluyveromyces marxianus on a pilot scale. *J. Biotechnol.* **2013**, *163*, 17–23. [CrossRef]

48. Urit, T.; Löser, C.; Wunderlich, M.; Bley, T. Formation of ethyl acetate by Kluyveromyces marxianus on whey: Studies of the ester stripping. *Bioprocess Biosyst. Eng.* **2011**, *34*, 547–559. [CrossRef]

Article

Maximization of Siderophores Production from Biocontrol Agents, *Pseudomonas aeruginosa* F2 and *Pseudomonas fluorescens* JY3 Using Batch and Exponential Fed-Batch Fermentation

Gaber Attia Abo-Zaid [1],[*], Nadia Abdel-Mohsen Soliman [1], Ahmed Salah Abdullah [1], Ebaa Ebrahim El-Sharouny [2],[3], Saleh Mohamed Matar [1],[4] and Soraya Abdel-Fattah Sabry [2]

[1] Bioprocess Development Department, Genetic Engineering and Biotechnology Research Institute, City of Scientific Research and Technological Applications, New Borg El-Arab City, 21934 Alexandria, Egypt; nadiastuttgart@yahoo.com (N.A.-M.S.); abosalahelnazzer@gmail.com (A.S.A.); salehmatar@yahoo.com (S.M.M.)

[2] Botany and Microbiology Department, Faculty of Science, Alexandria University, 21526 Alexandria, Egypt; ebaaloulou@yahoo.com (E.E.E.-S.); sabrysoraya@gmail.com (S.A.-F.S.)

[3] Biology Department, Faculty of Science, Taibah University, Medina 42353, Saudi Arabia

[4] Chemical Engineering Department, Faculty of Engineering, Jazan University, Jazan 45142, Saudi Arabia

[*] Correspondence: gaberam57@yahoo.com; Tel.: +2-01226804278

Received: 15 February 2020; Accepted: 31 March 2020; Published: 12 April 2020

Abstract: Twenty fluorescent *Pseudomonas* isolates were tested for their ability to produce siderophores on chrome azurol S (CAS) agar plates and their antagonistic activity against six plant pathogenic fungal isolates was assessed. Scaling-up production of siderophores from the promising isolates, *P. aeruginosa* F2 and *P. fluorescens* JY3 was performed using batch and exponential fed-batch fermentation. Finally, culture broth of the investigated bacterial isolates was used for the preparation of two economical bioformulations for controlling *Fusarium oxysporum* and *Rhizoctonia solani*. The results showed that both isolates yielded high siderophore production and they were more effective in inhibiting the mycelial growth of the tested fungi compared to the other bacterial isolates. Exponential fed-batch fermentation gave higher siderophore concentrations (estimated in 10 µL), which reached 67.05% at 46 h and 45.59% at 48 h for isolates F2 and JY3, respectively, than batch fermentation. Formulated *P. aeruginosa* F2 and *P. fluorescens* JY3 decreased the damping-off percentage caused by *F. oxysporum* with the same percentage (80%), while, the reduction in damping-off percentage caused by *R. solani* reached 87.49% and 62.5% for F2 and JY3, respectively. Furthermore, both formulations increased the fresh and dry weight of shoots and roots of wheat plants. In conclusion, bio-friendly formulations of siderophore-producing fluorescent *Pseudomonas* isolates can be used as biocontrol agents for controlling some plant fungal diseases.

Keywords: *Pseudomonas*; siderophores; antagonism; batch fermentation; exponential fed-batch fermentation; bio-friendly formulations; biocontrol

---

1. Introduction

Siderophores are a group of low-molecular-weight molecules (400–1500 Da) with a high affinity to ferric ions. These molecules are secreted by different microorganisms in response to low-iron conditions; a siderophore molecule forms a Fe-siderophore complex, which is recognized by membrane-receptor proteins within the microorganism. After that, Fe-siderophore complex is transported into cell periplasm, wherever it facilitates iron uptake [1,2]. Siderophores are iron chelators that have many applications, such as iron chelation therapy, antibiotic carriers, inhibitors of metalloenzymes, promotion of plant growth, biocontrol of plant and fish pathogens, biocontrol of algal blooms, removal of petroleum hydrocarbons from the marine environment, soil bioremediation, recovery of unusual earth elements and modification of surfaces [3]. The role of siderophore production in the biocontrol of *Erwinia carotovora* was demonstrated by Kloepper et al., 1980, who were the first to use pseudobactin-producing *Pseudomonas fluorescens*, a type of siderophore, as a biocontrol agent. Several species of fluorescent pseudomonads are able to secrete siderophores such as *P. aeruginosa*, *P. fluorescens* and *P. syringae* [4,5]. *P. fluorescens*, which secretes a hydroxamate-type siderophore revealed a high efficiency against *Macrophomina phaseolina*, the pathogen causing peanut charcoal rot [6], while a catechol-type siderophore produced by *P. syringae* had inhibitory effects on spore germination and mycelium morphology of *Fusarium oxysporum* [7]. Maximization of siderophore production from fluorescent pseudomonad isolates is performed in the bioreactor using different fermentation strategies [8–10]. There are three common operational modes of fermentation process: batch, fed-batch and continuous fermentation [11]. Batch fermentation is a very simple process, where all the medium components are sterilized in the vessel of bioreactor; however, some sensitive components might be sterilized separately and then added after sterilization to the bioreactor. Then, the organism is inoculated into the vessel and its growth is maintained inside this closed system until the end of the run without any further modification. The main objective of the fermentation process is maximizing production of the biological product or biomass, so that the process ends when the yield or products are achieved, when consumption of nutrients and minerals has occurred, or when toxic products accumulate [11,12]. Fed-batch fermentation is started with a batch phase and accomplished with a fed-batch phase by adding a carbon source or any other nutrients to the bioreactor at the end of the exponential phase of the culture growth (after consumption of the initial carbon source) without any removal of the fermentation products until the end of the fermentation process. The medium components could be added to the vessel according to designated feed rate (according model) to avoid the inhibitory effect of the initial high concentration of any medium component that occurs in batch fermentation. This ensures that the fed-batch fermentation process achieves high concentrations of the desired products [13]. There are different types of fed-batch fermentation according to the feeding strategy: constant [12], exponential [14,15], linear [16], pulse [12], and feeding based on other models [17]. Both batch and fed-batch strategies have been used for maximizing pyocyanin production from *P. aeruginosa* JY21 [12], and to improve the production of biosurfactant from *P. aeruginosa* USM AR2 and MR01 [18,19]. In addition, these strategies have also been used for increasing the production of acetyl esterase from *Pseudomonas* sp. ECU1011 [20]. Radha et al., 2014 enhanced the production of alkaline protease using batch and fed-batch fermentation strategies from a mixed culture of *P. putida* and *Staphylococcus aureus*. They reported 1.18-fold increases in the activity of alkaline protease when the enzyme was produced in a fed-batch fermentation compared to batch fermentation process [21]. Siderophores, which have high affinity to ferric irons, are naturally produced by microorganisms under low-iron conditions. The resulting Fe-siderophore complex becomes unavailable to other organisms, but the producing strain can uptake this complex via a very specific receptor in its outercellular membrane [22,23]. By this strategy, siderophores producing fluorescent *Pseudomonas* can restrict the growth of plant pathogens in the rhizospher of plants. This mechanism is known as competition for iron nutrition, which is one of the most important mechanisms of biological control of plant pathogens. Pyoverdin, a siderophore produced by *P. putida* WCS358, was effective in controlling radish fusarium wilt disease, which is caused by *F. oxysporum* f. sp. *raphani* [24]. In addition, *P. fluorescens* strains SPs9 and SPs20 that produced siderophores were effective in the inhibition of tomato wilt pathogen, *F. oxysporum* f. sp. *lycopersici* [5]. Siderophores play an imperative role in iron nutrition of plants, and consequently, promote plant growth. Iron uptake by plants is usually enhanced by microbial siderophores, when the plant is able to recognize the bacterial ferric-siderophore complex. Plants take in iron from bacterial siderophores by means of different mechanisms, for example, chelation and release of iron, the direct uptake of siderophore-Fe complexes, or by aligand exchange reaction [25]. *Arabidopsis thaliana* took up Fe-pyoverdine complex synthesized by *P. fluorescens* C7, resulting in an increase in iron concentration inside the plant, and hence, improved plant growth [26]. The main goal of our study is maximizing the production of siderophores from two *Pseudomonas* species, *P. aeruginosa* and *P. fluorescens* using batch and exponential fed-batch fermentation strategies. We then test the ability of siderophore-producing fluorescent pseudomonads to act as biocontrol agents against soil-borne fungi such as *F. oxysporum* and *Rhizoctonia solani*, the major agents of damping off and root rot diseases for many different plant crops that cause severe loss in the productivity of food and feed crops.

2. Materials and Methods

2.1. Bacterial and Fungal Isolates

2.1.1. Isolation of Fluorescent *Pseudomonas* Isolates

Isolation was carried out by the serial dilution method from forty soil samples of wheat, corn, eggplant, cotton, pepper and clover collected from Alexandria, Monufia and Sohag in Egypt. Ten grams of soil were suspended in 90 mL of sterile double distilled water and shaken for 1 h at 200 rpm. Next, 200 µL of 10^{-5} and 10^{-6} dilutions were spread onto King's B medium plates. After inoculation, plates were incubated at 28 to 30 °C for 1 to 2 days. Fluorescent *Pseudomonas* isolates were recognized by the formation of green fluorescence pigments in and around the colonies when exposed to ultraviolet light. Fluorescent *Pseudomonas* cultures were purified using the single colony technique and examined under light microscope after staining by Gram stain to confirm their purity.

2.1.2. Bacterial and Fungal Isolates from Culture Collection

Four fluorescent pseudomonad isolates, *P. fluorescens* JY3, JY7, JY8 and JY13 (GenBank accession number, KF922490, KF922494, KF922495 and KF922500, respectively) (Table 1), and six fungal isolates, *Alternaria* spp., *F. culmorum*, *F. oxysporum* isolate A, *F. oxysporum* isolate B, *F. solani* and *R. solani* used in the current study were kindly provided by the City of Scientific Research and Technological Applications (SRTA-City).

Table 1. Fluorescent *Pseudomonas* isolates used in the current study.

Fluorescent *Pseudomonas* Isolates Isolated in the Current Study					
Isolate Code	**Plant Rhizosphere**	**Isolation Governorate**	**GPS Coordinates**		**GenBank Accession Number**
			Latitude	**Longitude**	
F1	Corn	Monufia	30.5972455	30.9876321	
F2	Eggplant	Monufia	30.5972455	30.9876321	MG210480
F3	Pepper	Alexandria	31.200092	29.918739	
F4	Pepper	Alexandria	31.200092	29.918739	
F5	Eggplant	Alexandria	31.200092	29.918739	
F6	Eggplant	Alexandria	31.200092	29.918739	
F7	Cotton	Sohag	26.549999	31.700001	MG076939
F8	Corn	Sohag	26.549999	31.700001	MG210481
F9	Cotton	Sohag	26.549999	31.700001	
F10	Corn	Sohag	26.549999	31.700001	
F11	Corn	Sohag	26.549999	31.700001	
F12	Corn	Sohag	26.549999	31.700001	
F13	Wheat	Sohag	26.549999	31.700001	
F14	Clover	Sohag	26.549999	31.700001	
F15	Wheat	Sohag	26.549999	31.700001	
F16	Clover	Sohag	26.549999	31.700001	
Pseudomonas fluorescens Isolates Provided from Culture Collection					
Isolate Code	**GenBank Accession Number**				
JY3	KF922490				
JY7	KF922494				
JY8	KF922495				
JY13	KF922500				

2.2. Screening for Siderophores Production Using CAS Assay

2.2.1. Qualitative Assay

All fluorescent *Pseudomonas* isolates were checked for their capability to secrete siderophores. Detection of siderophores was performed on chrome azurol S CAS agar plates [27] where the detection depends on the high affinity of siderophores to chelate iron, so that, in the presence of siderophores the colure of medium changes from greenish blue to orange. *Pseudomonas* isolates were grown in succinic medium contained 4 g succinic acid; 6 g KH_2PO_4; 4 g K_2HPO_4; 0.2 g $MgSO_47H_2O$; 1 g $(NH_4)_2SO_4$ and dH_2O up to 1000 mL [28] for 24 h at 200 rpm and 30 °C. Next, 10 mL of each culture was centrifuged at 10,000 rpm for 10 min and then the supernatant was filtrated throughout a 0.2 μ syringe filter. Five wells were made in each CAS agar plate and each well was filled with 80 μL of culture filtrate. Plates were then incubated at 30 °C for 24 h and the presence of siderophores was detected visually.

2.2.2. Quantitative Assay

Pseudomonas isolates were cultured in succinic medium for 24 h at 200 rpm and 30 °C. A 1.5 mL volume of each culture was centrifuged at 10,000 rpm for 10 min. The relative level of siderophores was measured in a fixed volume of supernatant (10 μL) using the CAS assay method according to Schywan and Neilands (1987) [27]. Next, a 0.5 mL CAS assay solution was added to 10 μL of culture supernatant and mixed well, then 10 μL of Shuttle solution was added, mixed, and the mixture was left at room temperature for few minutes. The disappearance of the blue color relates to the presence of siderophores. The absorbance was measured at 630 nm using the media as blank. The relative level of siderophores was calculated based on the following formula:

$$\text{Relative level of siderophores \% } = Ar - As/Ar \times 100$$

A_r refers to absorption of CAS solution plus media plus shuttle solution whereas, As refers to absorption of CAS solution plus culture supernatant plus shuttle solution.

2.3. Antagonistic Effect of Fluorescent Pseudomonad Isolates (Dual Culture Method)

Antagonistic effects of all fluorescent pseudomonad isolates were tested against *Alternaria* spp., *F. culmorum*, *F. oxysporum* isolate A, *F. oxysporum* isolate B, *F. solani* and *R. solani* using the dual culture method according to Toure et al., 2004 [29]. Antagonistic isolates were streaked as a streak line with a loopfull of 2 day-old culture on potato dextrose agar plates, and incubated for 48 h prior inoculation by any tested fungus. A mycelial disc (5 mm in diameter) of an actively growing culture of the checked fungus was placed in the center at a standard distance close the other edge of the Petri plate and incubated at 30 °C for 3–7 days. Inhibition zones (the distance among the edge of antagonistic bacterial growth and the edge of fungal growth) were measured. All experiments were carried out in three replicates for each fungus.

2.4. Molecular Identification of Bacterial Isolates

Total DNA was extracted according to Istock et al., 2001 [30] from fluorescent *Pseudomonas* isolates F2, F7 and F8, which produced a higher percentage of siderophores production than the other *Pseudomonas* isolates that were isolated in the current investigation. The complete length of the 16S rRNA gene was amplified according to Matar et al., 2009 [31] using two universal primers, Start (forward) 5′ AGAGTTTGATCMTGGCTCAG 3′ and End (reverse) 5′ TACGGYACCTTGTTACGACTT 3′. The amplified 16S rRNA gene of *Pseudomonas* isolates was purified and sequenced based on enzymatic chain terminator technique by the use of a Big Dye terminator sequencing kit. After that, the nucleotide sequences were aligned with pseudomonad 16S rRNA gene sequences obtained from GenBank database (http://www.ncbi.nlm.nih.gov). The phylogenetic tree was constructed with the UPGMA method using MEGA software version 5 and the number of bootstraps replications was 2000.

2.5. Fermentation Experiments

2.5.1. Bioreactor

Fermentation experiments were carried out in a 10-L bench-top bioreactor (Cleaver, Saratoga, CA, USA) [26]. Temperature was monitored at 25 °C while the pH was adjusted to 7 by adding HCl 2 N and NaOH 2 N using feeding pumps. Aeration was performed using sterilized air that was supplied at 0.5 VVM (air volume per broth volume per minute). Agitation speed ranged between 200 to 600 rpm to keep the percentage of dissolved oxygen more than 30%. DO percentage and pH values were recorded automatically using the online module.

2.5.2. Batch Fermentation

P. aeruginosa F2 and *P. fluorescens* JY3 (GenBank accession number, MG210480 and KF922490), which have the greatest percentage of siderophores, were cultivated in the bioreactor using a volume of 5 L of optimized medium for siderophore production. The medium contained 20 mL glycerol; 14.5 g glucose; 1 g glutamic acid; 4 g sodium succinate; 5 g asparagine; 0.1 g urea; 1 g $(NH_4)_2SO_4$; 1 g kH_2PO_4; 3.5 g K_2HPO_4; 1 g $MgSO_4$; 0.5 µM $FeCl_3$ and dH_2O up to 1000 mL, pH 7 for *P. aeruginosa* F2 [27]. However, *P. fluorescens* JY3 was cultivated on optimized medium contained 10 mL glycerol; 1 g glucose; 0.5 g glutamic acid, 3.14 g sodium succinate; 1 g asparagine; 0.1 g urea; 0.1 g $(NH_4)_2SO_4$; 6 g kH_2PO_4; 4 g K_2HPO_4; 0.1 g $MgSO_4$; 0.62 µM $FeCl_3$ and dH_2O up to 1000 mL, pH 7 [32]. Cultivation of *P. aeruginosa* F2 and *P. fluorescens* JY3 in the bioreactor started with optical density (O.D$_{600nm}$) of 0.3 as an inoculum size by inoculation from seed culture of LB broth of each strain. Several samples from each culture were taken at time intervals for the determination of relative level of siderophores, biomass, glucose and glycerol.

2.5.3. Fed-batch Fermentation

Fed-batch fermentation was initiated with batch phase using 5 L of optimized medium as previously described. At the end of exponential phase, a feeding step was created by adding the feeding medium of that every 1-L of optimized medium, including 10-fold of each medium component except kH_2PO_4 and K_2HPO_4, which were added with original weights shown in the optimized media and without 10-fold increase . The feeding strategy for each strain was accomplished using a exponential programmed feed rate that was initialized with a rate of 1.5 mL min^{-1} to finish at rate of 22.5 mL min^{-1} [15].

2.6. Analytical Procedures

2.6.1. Relative Level of Siderophores

Relative level of siderophores was determined by CAS assay method according to Schywan and Neilands (1987) as mentioned before [27].

2.6.2. Biomass Estimation

Dry cell weight was estimated according to Van Dam-Mieras et al., 1992 [33].

2.6.3. Glucose Estimation

Glucose concentration was estimated using enzymatic colorimetric kit (Diamond Diagnostics, Egypt).

2.6.4. Glycerol Estimation

Glycerol concentration was estimated by the method developed by Bok and Demain (1977) [34]. One mL of supernatant containing glycerol was added to 1 mL of 15 mM sodium metaperiodate in

0.12 M HCl in a test tube and incubated at room temperature for 10 min. After oxidation of periodate, 2 mL of 0.1% (w/v) L-rhamnose solution was added to the test tube in order to remove the excessive periodate ions. After mixing, 4 mL of Nash reagent containing ammonium acetate (150 g mL^{-1}), acetic acid (0.2%, v/v) and acetylacetone (0.2%, v/v) were added. The color of the mixture was allowed to develop for 15 min in a water bath at 53 °C. After cooling, the optical density of the mixture was measured at 412 nm using a spectrophotometer and converted into glycerol concentration (g L^{-1}) according to a calibration curve (ranging from 0 to 40 mg L^{-1}).

2.7. Application of Siderophores-Producing Pseudomonas Isolates as Biocontrol Agents

2.7.1. Formulation Experiment

Culture broth of *P. aeruginosa* F2 and *P. fluorescens* JY3 obtained from fed-batch fermentation containing $2.0{\times}10^9$ cfu mL^{-1} was formulated using talc powder (TP) as a carrier with some other additives that include glycerol or glucose as a carbon source and carboxymethylcellulose (CMC) as an adhesive. Also, calcium carbonate was added to the mixture for adjusting the pH to 7 [35].

2.7.2. Fungal Inoculum Preparation

Inoculums of *F. oxysporum* and *R. solani* were prepared using sorghum/coarse sand/water (2:1:2 v/v/m) medium. All components were combined, packed and sterilized for 2 h. Agar discs, obtained from the edge of 5-day-old culture of each investigated fungus were inoculated into the sterilized medium. After two weeks of incubation at 30 °C, fungal inoculums became available for soil infestation [36].

2.7.3. Soil Infestation

Inoculums of *F. oxysporum* and *R. solani* were added independently to the soil surface of each pot at the rate of 2% w/w, and then coated with a thin film of autoclaved soil. The infested pots were irrigated and reserved for 7 days before sowing.

2.7.4. Disease Assessment

Disease assessment was estimated as a percentage of damping-off (pre- and post-emergence) after 7 and 21 days from sowing, respectively, using the following formula:

$$\% \text{ Pre-emergence} = \text{No. of non-emerged seedlings/No. of sown seeds} \times 100$$

$$\% \text{ Post-emergence} = \text{No. of dead emerged seedlings/No. of sown seeds} \times 100$$

$$\% \text{ Damping-off} = \% \text{ Pre-emergence} + \% \text{ Post-emergence}$$

2.7.5. Greenhouse Experiment

A pots experiment was performed to study the effect of the formulated *P. aeruginosa* F2 and *P. fluorescens* JY3 for the biocontrol of *F. oxysporum* isolate A and *R. solani*. Pots (18 × 18 cm diameter) containing sterilized soil infested as previously declared. Nine treatments were performed as follows: (1) *F. oxysporum*; (2) *R. solani*; (3) untreated and uninfected control (healthy); (4) *P. aeruginosa* F2 formulation; (5) *P. fluorescens* JY3 formulation; (6) F2 formulation + *F. oxysporum*; (7) JY3 formulation + *F. oxysporum*; (8) F2 formulation + *R. solani* and (9) JY3 formulation + *R. solani*. Five wheat seeds were sown per pot, three replicate (pots) were used for each treatment. Seeds of wheat were treated with the formulations as seed drench at a dose of 10 g kg^{-1} of seeds. Formulations were applied twice at a dose of 3 kg acre^{-1}, 15 and 30 days after seed sowing as a soil drench.

2.8. Statistical Analysis

Analysis of variance (ANOVA) was used to analyze the results achieved in this study using CoStat software. The least significant difference (LSD) at $P \leq 0.05$ level of probability was utilized for detecting significant differences among treatments.

3. Results

3.1. Isolation of Fluorescent Pseudomonas Isolates

Sixteen fluorescent *Pseudomonas* isolates were isolated using King's B medium (Table 1). The green fluorescent colonies were picked up under UV light, and further purified by repeated streaking (single colony) on the same medium, and examined under light microscope.

3.2. Screening for Siderophores Production

All bacterial isolates were screened qualitatively and quantitatively for siderophore production. In a primary step, qualitative screening has been performed using CAS agar plates. All isolates that formed a yellow to orange zone around the pure colony indicated a positive siderophore production. In addition, in the agar well diffusion method, when cell-free supernatant of each culture was applied to the wells of CAS agar plates, a yellow to orange halo was observed around the wells that indicates the production of siderophores (Figure 1A). In order to confirm previously obtained results, siderophores produced by all bacterial isolates were quantitatively measured in liquid cultures. *P. fluorescens* JY3 and *P. aeruginosa* F2 showed the highest production of siderophores (1.758% and 1.749%, respectively). Also, *P. aeruginosa* F7 and F8 that were isolated in the current work gave a high percentage of siderophore production but *P. fluorescens* JY8 showed the lowest production of siderophores (0.011%) in a fixed 10 µL using succinic medium (Figure 1B).

Table 2. Antagonistic effects of fluorescent *Pseudomonas* isolates against growth of some phytopathogenic fungi using the dual culture method.

Pseudomonas Isolates	Vertical Distance of Inhibition Diameter (cm)					
	Alternaria Sp	*F. culmorum*	*F. oxysporum* Isolate A	*F. oxysporum* Isolate B	*F. solani*	*R. solani*
F1	*2.00 ± 0.17 [bcd]**	2.13 ± 0.45 [abcd]	2.20 ± 0.10 [a]	1.70 ± 0.10 [a]	2.07 ± 0.21 [abc]	1.63 ± 0.15 [e]
F2	2.23 ± 0.21 [ab]	2.47 ± 0.06 [a]	2.10 ± 0.29 [abc]	1.70 ± 0.20 [a]	2.23 ± 0.30 [a]	2.23 ± 0.21 [a]
F3	1.37 ± 0.32 [e]	1.87 ± 021 [de]	1.60 ± 0.10 [fg]	1.43 ± 0.06 [abc]	1.37 ± 0.32 [gh]	1.77 ± 0.21 [de]
F4	1.80 ± 0.26 [d]	1.90 ± 0.26 [de]	1.80 ± 0.20 [def]	1.60 ± 0.10 [ab]	1.80 ± 0.26 [cde]	2.10 ± 0.36 [abc]
F5	2.27 ± 0.25 [a]	2.10 ± 0.17 [bcd]	1.80 ± 0.10 [def]	1.60 ± 0.00 [ab]	2.13 ± 0.25 [ab]	2.10 ± 0.15 [abc]
F6	1.43 ± 0.12 [e]	1.80 ± 0.10 [de]	1.67 ± 0.15 [efg]	1.53 ± 0.15 [abc]	1.47 ± 0.15 [fg]	2.10 ± 0.10 [abc]
F7	2.13 ± 0.06 [abc]	2.27 ± 0.15 [abc]	2.17 ± 0.15 [ab]	1.57 ± 0.11 [abc]	2.13 ± 0.21 [ab]	2.10 ± 0.35 [abc]
F8	2.00 ± 0.10 [bcd]	2.13 ± 0.15 [abcd]	1.97 ± 0.12 [bcd]	1.57 ± 0.49 [abc]	2.00 ± 0.10 [abcd]	2.00 ± 0.36 [abcd]
F9	2.17 ± 0.25 [ab]	2.33 ± 0.25 [ab]	1.63 ± 0.12 [efg]	0.87 ± 0.32 [d]	2.07 ± 0.25 [abc]	2.07 ± 0.15 [abc]
F10	2.23 ± 0.12 [ab]	2.00 ± 0.10 [bcde]	1.93 ± 0.15 [cd]	1.43 ± 0.15 [abc]	2.23 ± 0.12 [a]	2.17 ± 0.20 [ab]
F11	2.10 ± 0.26 [abc]	2.07 ± 0.21 [bcd]	1.53 ± 0.15 [gh]	1.23 ± 0.06 [c]	2.10 ± 0.10 [ab]	2.23 ± 0.21 [a]
F12	1.90 ± 0.10 [cd]	1.97 ± 0.06 [cde]	1.27 ± 0.21 [ij]	1.27 ± 0.59 [bc]	1.90 ± 0.10 [bcde]	1.90 ± 0.26 [bcde]
F13	2.10 ± 0.26 [abc]	1.70 ± 0.36 [e]	1.83 ± 0.21 [de]	1.50 ± 0.10 [abc]	2.10 ± 0.26 [ab]	1.90 ± 0.20 [bcde]
F14	0.00 ± 0.00 [g]	2.10 ± 0.10 [bcd]	1.60 ± 0.10 [fg]	0.00 ± 0.00 [e]	0.00 ± 0.00 [j]	1.63 ± 0.15 [e]
F15	0.00 ± 0.00 [g]	0.00 ± 0.00 [h]	1.33 ± 0.15 [hi]	0.00 ± 0.00 [e]	0.00 ± 0.00 [j]	0.00 ± 0.00 [h]
F16	0.00 ± 0.00 [g]	1.33 ± 0.42 [f]	0.00 ± 0.00 [k]	0.00 ± 0.00 [e]	0.67 ± 0.31 [i]	1.83 ± 0.25 [cde]
P. fluorescens JY3	0.63 ± 0.15 [f]	1.07 ± 0.15 [fg]	1.07 ± 0.15 [j]	1.23 ± 0.15 [c]	1.77 ± 0.38 [de]	0.63 ± 0.10 [g]
P. fluorescens JY7	0.10 ± 0.00 [g]	0.77 ± 0.21 [g]	0.00 ± 0.00 [k]	0.30 ± 0.10 [e]	1.17 ± 0.10 [h]	0.00 ± 0.00 [h]
P. fluorescens JY8	0.13 ± 0.00 [g]	0.00 ± 0.00 [h]	0.00 ± 0.00 [k]	0.00 ± 0.00 [e]	1.40 ± 0.10 [fgh]	1.07 ± 0.06 [f]
P. fluorescens JY13	0.00 ± 0.00 [g]	0.93 ± 0.25 [g]	0.00 ± 0.00 [k]	0.00 ± 0.00 [e]	1.67 ± 0.15 [ef]	0.00 ± 0.00 [h]
L.S.D	0.266	0.360	0.222	0.342	0.276	0.296

* Means in each column followed by the identical letter do not differ significantly at $P \leq 0.05$ level; ** Significant letters.

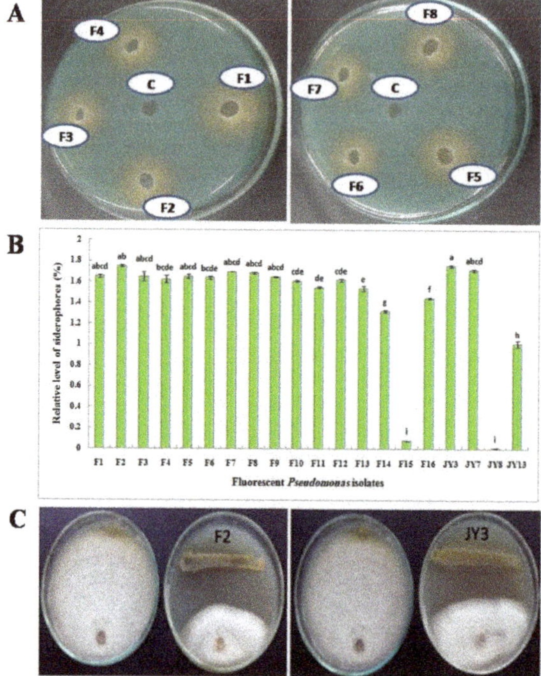

Figure 1. (**A**) Qualitative assay of siderophore production on chrome azurol S (CAS) agar plates using supernatant of some fluorescent *Pseudomonas* isolates; (**B**) Relative level of siderophores produced by fluorescent *Pseudomonas* isolates grown on succinic medium using CAS assay and (**C**) antagonistic effect of *Pseudomonas aeruginosa* F2 and *Pseudomonas fluorescens* JY3 against *Fusarium solani*, plate on the left in each photo is the corresponding control without the antagonist (data are presented in Table 2).

3.3. Antagonistic Effect of Fluorescent Pseudomonad Isolates (Dual Culture Method)

This experiment aimed to investigate the antagonistic effect of the experimental *Pseudomonas* isolates on the growth of some tested plant pathogenic fungi using dual culture technique. Isolate F2 was the most efficient isolate in inhibiting the mycelial growth of all tested pathogens. On the other hand, isolates F14, F15, F16, JY7, JY8 and JY13 showed a weak antagonistic activity (Table 2). All tested *Pseudomonas* isolates showed antagonistic activity against *Alternaria* sp. except *Pseudomonas* isolates F14, F15, F16 and JY13. *Pseudomonas* isolate F5 followed by isolates F2, F9 and F10, which showed no significant differences among each other, were the most effective isolates in inhibiting the mycelial growth of *Alternaria* sp. Meanwhile, F2 had more antagonistic activity than the other fluorescent *Pseudomonas* isolates on *F. calmorum*. All *Pseudomonas* isolates showed antagonistic activity against *F. oxysporum* isolate A except F16, JY7, JY8 and JY13, where *Pseudomonas* isolate F1 had a more-antagonistic effect on *F. oxysporum* isolate A compared to the other isolates followed by F7. *Pseudomonas* isolates F1 and F2 followed by F4 and F5 followed by F7 and F8 were the most effective isolates in inhibiting the mycelial growth of *F. oxysporum* isolate B, but *Pseudomonas* isolate F2 had more antagonistic activity than the other fluorescent *Pseudomonas* isolates on *F. solani* and *R. solani* (Figure 1C).

3.4. Molecular Identification of Bacterial Isolates

Pseudomonas isolates F2, F7 and F8, which produced a higher percentage of siderophore production than the other *Pseudomonas* isolates isolated in the current study, were identified based on sequencing of 16S rRNA gene. A database search to identify the bacterial isolates was achieved in BLAST search at the National Center for Biotechnology Information site (http://www.ncbi.nlm.nih.gov). Blast results revealed that sequences of isolates F2, F7 and F8 were almost similar to several *P. aeruginosa* strains with homology percentage of 99%. *Pseudomonas* isolates F2, F7 and F8 were identified as *P. aeruginosa* with accession numbers of MG210480, MG076939 and MG210481, respectively (Table 1). A phylogenetic tree of the 16S rRNA gene that was generated using the nucleotide sequences of the fluorescent *Pseudomonas* isolates F2, F7 and F8 (obtained in the current investigation) and other pseudomonad 16S rRNA gene sequences (obtained from GenBank database) revealed that two major clusters exist. Cluster 1 included *Cellvibrio ostraviensis*, while all fluorescent *Pseudomonas* isolates obtained in this study and provided from GenBank database were clustered in cluster 2. This cluster is divided into two groups: the first group included *P. aeruginosa* F2, F7 and F8 with the other *P. aeruginosa* strains provided from GenBank; while the second group contained *P. fluorescens, P. putida, P. chlororaphis, P. syringae* and *P. meliae* (Figure 2).

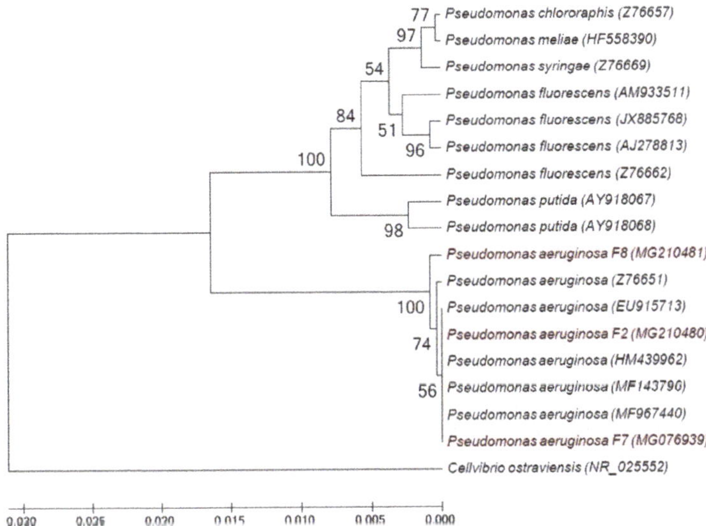

Figure 2. Phylogenetic tree of fluorescent *Pseudomonas* isolates F2, F7 and F8 obtained in the current study and validly described members of the genus *Pseudomonas* based on the nucleotide sequences of the 16S rRNA gene. Phylogenetic tree was constructed with UPGMA method using MEGA version 5 and the number of bootstraps replications is 2000.

3.5. Batch Fermentation

Batch fermentation technique in the bioreactor was used in this study to enhance siderophores production from *P. aeruginosa* F2 and *P. fluorescens* JY3 compared to cultivation in shake flasks, which produced a low relative level of siderophores in our previous study. Batch fermentation was performed using *P. aeruginosa* F2 and *P. fluorescens* JY3 in a 10-L bench-top bioreactor (Cleaver, Saratoga, CA, USA), where both isolates produced a higher percentage of siderophores production than the other *Pseudomonas* isolates that were screened in the current work. Figure 3A shows the relative level of siderophores, cell biomass, glucose and glycerol concentrations of the culture broth of *P. aeruginosa* F2 plotted against time. Cell biomass of *P. aeruginosa* F2 increased slowly during lag phase, which continued for 5 h. After that, the culture entered exponential phase (log phase) where the biomass increased rapidly with a constant specific growth rate, μ of 0.043 h^{-1}. The dissolved oxygen decreased as a result of the increasing demand for O_2 that is required for bacterial growth (Figure 3B). To assure sufficient oxygen provided, oxygen was maintained at over 30% by increasing the agitation rate. After 41 h the dissolved oxygen was increased gradually until the end of the run. The maximum biomass achieved was 10.3 g L^{-1} at 38 h. Relative level of siderophores reached 1.91% at 12 h but it accomplished its maximum value of 31.9% at 40 h. Glycerol concentration decreased slowly and it was not fully consumed as its value reached 4.5 g L^{-1} at 50 h but glucose concentration decreased rapidly and it was fully consumed at 26 h. Relative level of siderophores, biomass and glycerol concentration in the culture broth of *P. fluorescens* JY3 against time are shown in Figure 4A, where relative level of siderophores achieved was 1.97% at 12 h, reaching its maximum value of 23.9% at 48 h. Moreover, the maximum biomass achieved was 5.2 g L^{-1} at 24 h and glycerol concentration decreased gradually to reach 0.25 g L^{-1} at 52 h. *P. fluorescens* JY3 cell biomass increased exponentially with specific growth rate, μ of 0.096 h^{-1}. The dissolved oxygen decreased rapidly and reached 30% in the first hour, so that, motor speed was increased and accordingly agitation rate increased until 6 h, consequently, dissolved oxygen increased rapidly as a result of increasing agitation speed (Figure 4B).

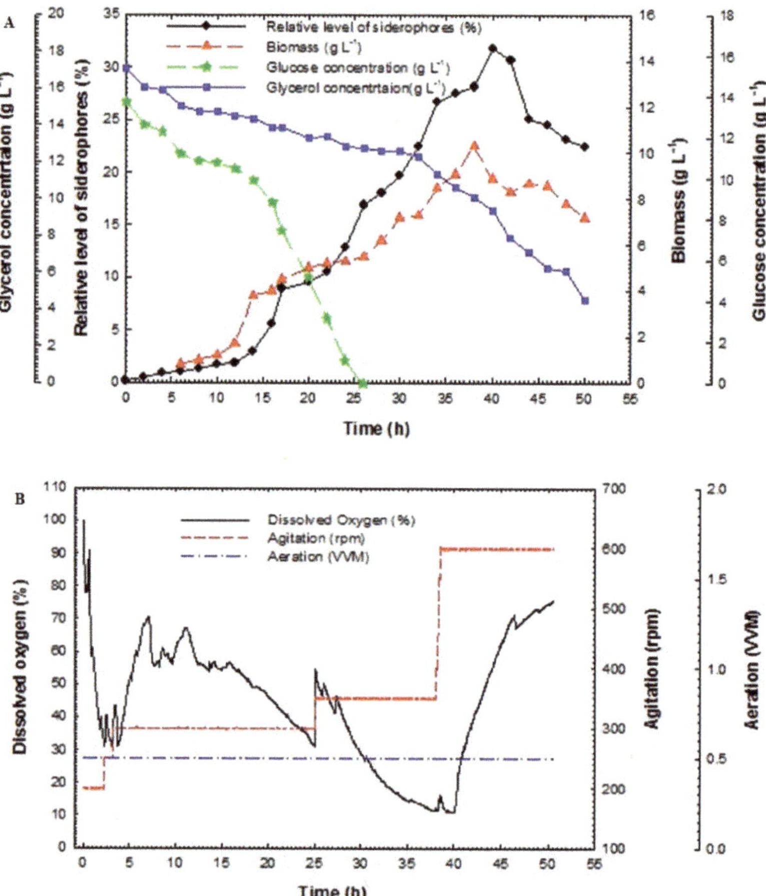

Figure 3. (**A**) Relative level of siderophores, biomass, glucose concentration and glycerol concentration as a function of time for batch fermentation culture broth of *Pseudomonas aeruginosa* F2 and (**B**) dissolved oxygen, agitation and aeration as a function of time during batch fermentation of *Pseudomonas aeruginosa* F2.

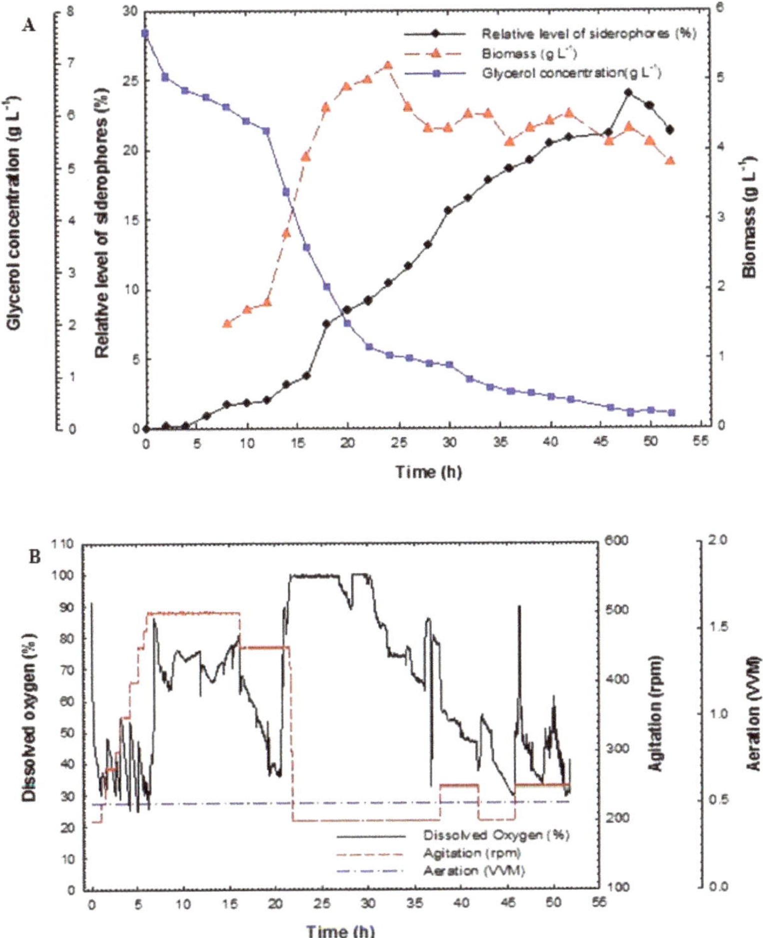

Figure 4. (**A**) Relative level of siderophores, biomass and glycerol concentration as a function of time for batch fermentation culture broth of *Pseudomonas fluorescens* JY3 and (**B**) dissolved oxygen, agitation and aeration as a function of time during batch fermentation of *Pseudomonas fluorescens* JY3.

3.6. Fed-Batch Fermentation

Maximization of siderophores production from *P. aeruginosa* F2 and *P. fluorescens* JY3 was completed using fed-batch technique. Fed-batch fermentation was carried out in a 10-L bench-top bioreactor (Cleaver, Saratoga, USA) using *P. aeruginosa* F2 and *P. fluorescens* JY3, which produced the highest percentage of siderophores production compared to other *Pseudomonas* isolates screened in this study. The fed-batch process was started with a batch phase and completed using a fed-batch phase using exponential feed rate. The fed-batch stage initiated with the addition of the feeding medium using an exponential feeding type of prearranged feed rate. The feeding started with an initial rate of 1.5 mL min⁻¹ until it reached the final rate of 22.5 mL min⁻¹. Figure 5A illustrates relative level of siderophores, biomass and glucose/glycerol concentration of the culture broth of *P. aeruginosa* F2 as a function of time. During the batch step, the culture grew exponentially with a constant specific growth rate. Dissolved oxygen decreased quickly during the exponential phase due to the increasing demand for O_2 that is required for the growing cell mass. The oxygen was held in reserve at above 30% by controlling motor speed and consequently agitation rate (Figure 5B).

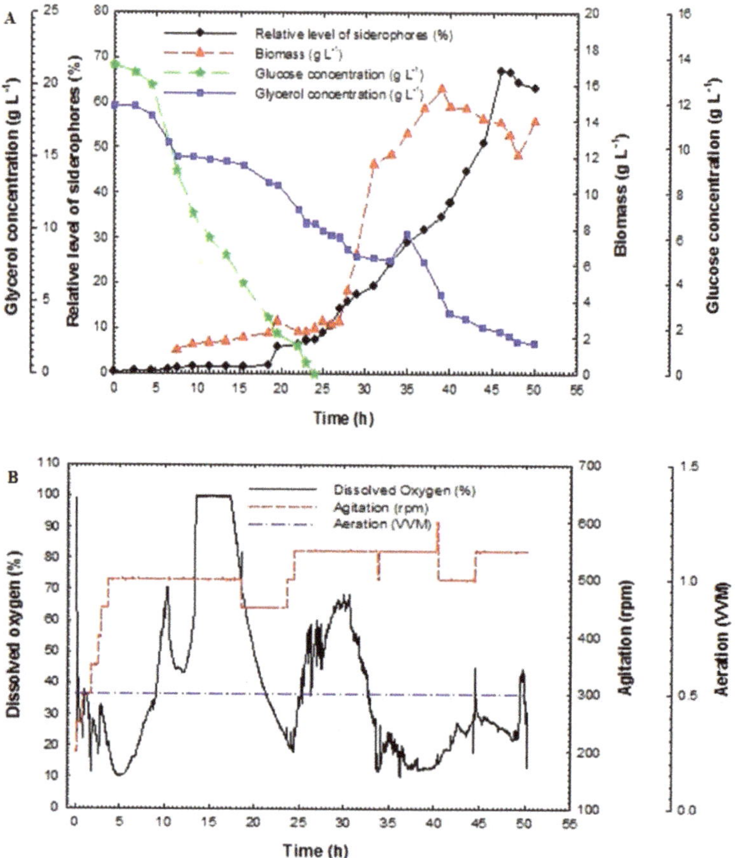

Figure 5. (**A**) Relative level of siderophores, biomass, glucose concentration and glycerol concentration as a function of time for fed-batch fermentation culture broth of *Pseudomonas aeruginosa* F2 and (**B**) dissolved oxygen, agitation and aeration as a function of time during fed-batch fermentation of *Pseudomonas aeruginosa* F2.

The relative level of siderophores reached 1.8% at 18 h and then increased rapidly to achieve 37.8% at 40 h but it reached its maximum value of 67.05% at 46 h. The maximum biomass was 15.8 g L^{-1} at 39 h. Glucose concentration decreased rapidly and it was fully consumed at 24 h. In addition, glycerol concentration decreased slowly and was not fully consumed at 24 h; its value reached 10.36 g L^{-1}. After feeding, the glycerol concentration increased for a short period and then decreased rapidly to near zero at the end of the run.

In thecase of *P. fluorescens* JY3, cell biomass increased exponentially with a specific growth rate. The relative level of siderophores, biomass and glycerol concentration against time are shown in Figure 6A. The relative level of siderophores reached 1.92% at 16 h and increased rapidly to reach 19.24% at 36 h, achieving its maximum value of 45.59% at 48 h. The maximum biomass achieved was 13.7 g L^{-1} at 44 h. Glycerol concentration decreased gradually and reached 2.89 g L^{-1} at 24 h. After feeding, glycerol concentration increased gradually for some time to reach 4.75 g L^{-1} at 42 h and decreased again to reach 2.85 g L^{-1} at the end of the process. The dissolved oxygen decreased rapidly, therefore the concentration of oxygen was restricted to a value of 30% by increasing the speed of motor to the desired agitation rate. The rise of the agitation rate resulted in an increase in the dissolved oxygen concentration in the medium until 8 h; after this point dissolved oxygen increased gradually. Once feeding starts, the dissolved oxygen decreased rapidly, and consequently the agitation speed was increased gradually (Figure 6B). The fed-batch fermentation technique produced the highest percentage of relative level of siderophores of all methods, so cultures of *P. aeruginosa* F2 and *P. fluorescens* JY3 obtained by fed-batch fermentation were used for the preparation of talc formulations to be utilized as biocontrol agents.

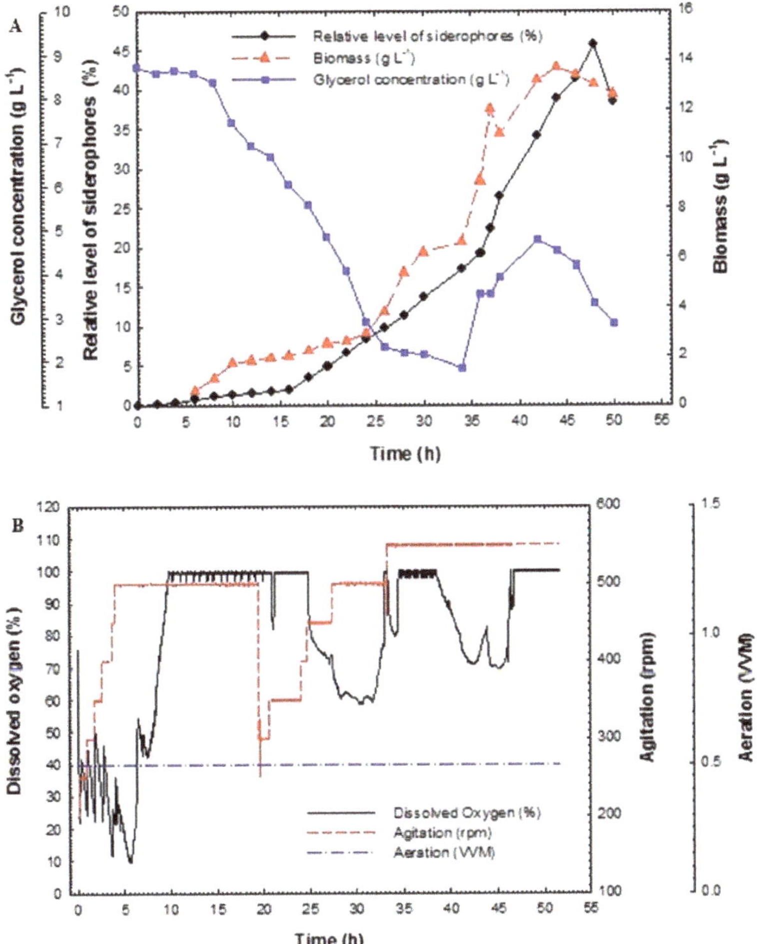

Figure 6. (**A**) Relative level of siderophores, biomass and glycerol concentration as a function of time for fed-batch fermentation culture broth of *Pseudomonas fluorescens* JY3 and (**B**) dissolved oxygen, agitation and aeration as a function of time during fed-batch fermentation of *Pseudomonas fluorescens* JY3.

3.7. In-Vivo Antagonistic Effect of P. aeruginosa F2 and P. fluorescens JY3 against Some Phytopathogenic Fungi

Formulations of siderophores-producing *Pseudomonas* isolates using talc powder as a carrier were applied as biocontrol agents in a soil infested with *F. oxysporum* and *R. solani*. Wheat seeds were treated with the formula and sown after seven days of soil infestation. Damping-off percentage and reduction percentage were evaluated. In addition, fresh and dry weights of shoots and roots were weighed. Formulations of *P. aeruginosa* F2 and *P. fluorescens* JY3, which produced a higher percentage of siderophores production than other *Pseudomonas* isolates used in this work, were effective in reducing damping-off caused by *F. oxysporum* and *R. solani* compared to the control infested with plant pathogens. Damping-off percentage caused by *F. oxysporum* and *R. solani* reached 33.3% and 53.3%, respectively. Meanwhile, treatments using formulations of F2 and JY3 in a soil infected with *F. oxysporum* were effective in reducing damping-off percentage, which was decreased to reach 6.67% with a reduction percentage of 80%. Formulation of *P. aeruginosa* F2 in a soil infected with *R. solani* was more effective in reducing damping-off percentage compared to the formulation of *P. fluorescens* JY3, as the damping-off percentage recorded 6.67% and 20% with a reduction percentage of 87.49% and 62.5%, respectively (Table 3). Both formulations stimulated the growth of wheat plants in a soil infected with *F. oxysporum* and *R. solani* compared to the other control treatments inoculated with each fungus alone. Fresh and dry weights of shoots and roots increased significantly in treatments with formulations of *P. aeruginosa* F2 and *P. fluorescens* JY3 in a soil infected with plant pathogenic fungi. Treatment using formulation of F2 in a soil infected with *F. oxysporum* increased the fresh weights of shoots and roots of wheat plants, which reached 3.77 g and 3.3 g with increase percentages of 45.09% and 49.39%, respectively. Whereas, there was a 3.33 g and 2.93 g increase in fresh weights of shoots and roots, with increased percentage to 37.84% and 43%, respectively, in the case of treatment using JY3 formulation. In comparison with *F. oxysporum* treatment alone, fresh weights of shoots and roots reached 2.07 g and 1.67 g, respectively. Fresh weights of shoots were 3.53 g and 3.27 g with increase percentage of 72.52% and 70.34% in treatments using F2 and JY3 formulation in a soil infected with *R. solani*, respectively, compared to treatment using the fungus alone that recorded 0.97 g. The fresh weights of roots reached 3.1 g and 2.83 g with an increased percentage of 80.65% and 78.8% using the same formulations in comparison with treatment using the fungus alone that reached 0.6 g (Table 4). The results of dry weights of shoots and roots obtained in this study elucidated the efficiency of treatments using formulations of *P. aeruginosa* F2 and *P. fluorescens* JY3 in increasing the dry weights of shoots in a soil infected with *F. oxysporum* and *R. solani*. The increase in shoot dry weight reached 2.83 g, 2.47 g, 2.63 g and 2.37 g, with increased percentage of 63.6%, 58.3%, 87.45% and 86.08%, respectively while the dry weights of shoots in the case of treatments using *F. oxysporum* and *R. solani* reached 1.03 g and 0.33 g, respectively. The dry weights of roots recorded 2.3 g, 2.03 g, 2.2 g and 1.83 g, with increase percentage of 55.22%, 49.26%, 91.82% and 90.16%, respectively. Whereas treatments using *F. oxysporum* and *R. solani* recorded 1.03 g and 0.18 g, respectively (Table 4).

Table 3. Effect of culture formulations of *Pseudomonas aeruginosa* F2 and *Pseudomonas fluorescens* JY3 on damping-off of wheat plants caused by *Fusarium oxysporum* and *Rhizoctonia solani*.

Treatment	Damping Off (%)	Reduction [W] (%)	Reduction [X] (%)
1. *F. oxysporum* (check)	*33.33 ± 11.55 [ab]**	0	n.d.
2. *R. solani* (check)	53.33 ± 23.09 [a]	n.d.	0
3. Control	20.00 ± 00.00 [b]	40	62.50
4. *P. aeruginosa* F2	13.33 ± 11.55 [b]	60	75
5. *P. fluorescens* JY3	13.33 ± 11.55 [b]	60	75
6. F2 + *F. oxysporum*	06.67 ± 11.55 [b]	80	n.d.
7. JY3 + *F. oxysporum*	06.67 ± 11.55 [b]	80	n.d.
8. F2 + *R. solani*	06.67 ± 11.55 [b]	n.d.	87.49
9. JY3 + *R. solani*	20.00 ± 00.00 [b]	n.d.	62.50
L.S.D	28.78		

[W] Reduction % (in case of *F. oxysporum*) = [check-treatment]/check × 100; [X] Reduction % (in case of *R. solani*) = [check-treatment]/check × 100; * Means in each column followed by the identical letter do not differ significantly at $P \leq 0.05$ level; ** Significant letters; "n.d." not determined.

Table 4. Effect of culture formulations of *Pseudomonas aeruginosa* F2 and *Pseudomonas fluorescens* JY3 on fresh and dry weight of shoots and roots of wheat plants.

Treatment	Fresh Weight of Shoots (g)	Increase [Y] (%)	Increase [Z] (%)	Fresh Weight of Roots (g)	Increase [Y] (%)	Increase [Z] (%)
1. *F. oxysporum* (check)	*2.07 ± 0.06 [c**]	0	n.d.	1.67 ± 0.21 [c]	0	n.d.
2. *R. solani* (check)	0.97 ± 0.25 [d]	n.d.	0	0.6 ± 0.10 [d]	n.d.	0
3. Control	3.63 ± 0.21 [ab]	42.98	73.28	3.23 ± 0.38 [ab]	48.30	81.42
4. *P. aeruginosa* F2	4.17 ± 0.15 [a]	50.36	76.74	3.73 ± 0.12 [a]	55.23	83.91
5. *P. fluorescens* JY3	3.60 ± 0.25 [ab]	42.50	73.06	3.10 ± 0.17 [ab]	46.13	80.65
6. F2 + *F. oxysporum*	3.77 ± 0.15 [ab]	45.09	n.d.	3.30 ± 0.20 [ab]	49.39	n.d.
7. JY3 + *F. oxysporum*	3.33 ± 0.12 [b]	37.84	n.d.	2.93 ± 0.15 [b]	43	n.d.
8. F2 + *R. solani*	3.53 ± 0.29 [ab]	n.d.	72.52	3.10 ± 0.10 [ab]	n.d.	80.65
9. JY3 + *R. solani*	3.27 ± 0.31 [b]	n.d.	70.34	2.83 ± 0.25 [b]	n.d.	78.80
L.S.D	0.75			0.78		

Treatment	Dry Weight of Shoots (g)	Increase [Y] (%)	Increase [Z] (%)	Dry Weight of Roots (g)	Increase [Y] (%)	Increase [Z] (%)
1. *F. oxysporum* (check)	1.03 ± 0.15 [c]	0	n.d.	1.03 ± 0.12 [c]	0	n.d.
2. *R. solani* (check)	0.33 ± 0.12 [d]	n.d.	0	0.18 ± 0.03 [d]	n.d.	0
3. Control	2.67 ± 0.23 [ab]	61.42	87.64	2.1 ± 0.06 [ab]	50.95	91.43
4. *P. aeruginosa* F2	3.30 ± 0.10 [a]	68.79	90	2.67 ± 0.15 [a]	61.42	93.26
5. *P. fluorescens* JY3	2.73 ± 0.21 [ab]	62.27	87.91	2.07 ± 0.10 [ab]	50.24	91.3
6. F2 + *F. oxysporum*	2.83 ± 0.21 [ab]	63.60	n.d.	2.30 ± 0.10 [ab]	55.22	n.d.
7. JY3 + *F. oxysporum*	2.47 ± 0.29 [b]	58.30	n.d.	2.03 ± 0.21 [ab]	49.26	n.d.
8. F2 + *R. solani*	2.63 ± 0.06 [ab]	n.d.	87.45	2.20 ± 0.17 [ab]	n.d.	91.82
9. JY3 + *R. solani*	2.37 ± 0.15 [b]	n.d.	86.08	1.83 ± 0.06 [b]	n.d.	90.16
L.S.D	0.68			0.66		

[Y] Increase % (in case of *F. oxysporum*) = [treatment-check]/treatment × 100; [Z] Increase % (in case of *R. solani*) = [treatment-check]/treatment × 100; * Means in each column followed by the identical letter do not differ significantly at P ≤ 0.05 level; ** Significant letters; "n.d." not determined.

4. Discussion

Siderophores are produced by several microorganisms such as *Pseudomonas*, *Enterobacter* and *Escherichia coli* [5,8,37,38]. In our study, twenty siderophores-producing *Pseudomonas* isolates were evaluated as biocontrol agents for their in-vitro antagonistic effect against the growth of some plant pathogenic fungi (*Alternaria* spp., *F. culmorum*, *F. oxysporum* isolate A, *F. oxysporum* isolate B, *F. solani* and *R. solani*). *P. aeruginosa* isolate F2 was more effective in significantly inhibiting the mycelial growth of all the tested fungi than other *Pseudomonas* isolates. These results are in harmony with the results obtained by Pàez et al., 2005 who reported that *P. aeruginosa* was more efficient than *P. fluorescens* in inhibiting some plant pathogenic fungi [39]. Also, Sasirekha and Srividya (2016) reported that siderophores-producing *P. aeruginosa* FP6 had antagonistic activity against *R. solani* [40]. Also, Islam et al., 2018 showed that *P. aeruginosa* RKA5 inhibited mycelia growth of *F. oxysporum* f. sp. *cucumerinum* [41]. On the other hand, Ahmedzedah et al., 2006 reported that 13 out of 47 fluorescent *Pseudomonas* isolates showed antagonistic effects against *Fusarium* sp. and *R. solani* [42].

The optimal performance of the upstream processing using on-line and off-line sensors in the two basic types of fermentation (batch, and fed-batch) was studied. Batch and fed-batch fermentation processes were accomplished to enhance the cell biomass and to achieve high concentrations of siderophores. In batch fermentation of *P. aeruginosa* F2 and *P. fluorescens* JY3, cells grew rapidly after the lag phase, which continued for 5 h. Biomass was subsequently increased exponentially with a constant specific growth rate of 0.043 h^{-1} and 0.096 h^{-1}, respectively, within the exponential phase. Batch fermentation of isolates F2 and JY3 in the bioreactor produced the highest concentration of biomass and siderophores. In isolate F2, maximum values of 10.3 g L^{-1} at 38 h and 31.9% at 40 h, respectively, were reached, while values of 5.2 g L^{-1} at 42 h and 23.9% at 48 h were recorded for isolate JY3, respectively. On the other hand, maximum biomass of *P. aeruginosa* PAO1 and *P. fluorescens* NCIM5096 was 1.6 and 1.96 g L^{-1}, respectively [8,43]. Maximization of biomass and siderophores concentration in the bioreactor may be related to optimal conditions of pH, agitation, and aeration, which are provided by the bioreactor for cell growth and production of siderophores. Batch fermentation could not obtain high cell density and high concentration of product because cells suffer from substrate inhibition and catabolite repression. Accumulation of by-products such as acetic acid and propionic acid during batch fermentation was documented [9,44,45]. High partial pressure of CO_2, high concentration of carbon source and high specific growth rate might be responsible for the accumulation of by-products that cause suppression of bacterial growth and production of products [46]. So, to obtain high cell density and high concentrations of siderophores, fed-batch fermentation is favored to reduce by-products production and to get rid of substrate suppression. Exponential fed-batch fermentation of *P. aeruginosa* F2 and *P. fluorescens* JY3 gave higher cell mass and siderophores concentration than batch fermentation. Biomass and siderophore estimation of *P. aeruginosa* F2 reached its maximum value of 15.8 g L^{-1} at 39 h and 67.05% at 46 h, respectively while they reached 13.7 g L^{-1} at 44 h and 45.59% at 48 h by *P. fluorescens* JY3, respectively. Sarma et al., 2010, obtained high cell density from fluorescent pseudomonad R81 using fed-batch fermentation [9].

The fact that *P. aeruginosa* is classified as a risk group 2 biological agent that rarely represents a serious health threat to human is evident [47–49]. In addition, the fact that this organism was isolated from the rhizosphere of edible plants from agricultural fields, as a member of the natural microflora found in the native soil, gave us confidence to continue working with this microorganism. Therefore, the current study used *P. aeruginosa*, as an effective biocontrol agent. Formulations of siderophores-producing *Pseudomonas* isolates, which were used as biocontrol agents in a soil infected with *F. oxysporum* and *R. solani*, were efficient in decreasing damping-off. The biocontrol mechanism of siderophores-producing fluorescent pseudomonads could be explained by their ability to chelate and reduce the amount of ferric ions available in rhizosphere (competition for iron nutrition). By this means, siderophores-producing fluorescent pseudomonads may restrict plant pathogens in the rhizosphere and reduce their ability to colonize plant roots. In addition, siderophores-producing fluorescent pseudomonads can induce plant systemic resistance, which reduces the pathogen infection. Other mechanisms, in addition to siderophores production by fluorescent pseudomonads, were reported that may help in suppressing and controlling fungal pathogens, such as the production of antifungal compounds and lytic enzymes. Solans et al., 2016 studied the role of siderophores as biocontrol agents [50] whereas Arya et al., 2018 revealed that tomato seedlings inoculated with *P. fluorescens* strains SPs9 and SPs20, which are able to produce siderophores in the soil infected with *F. oxysporum*, succeeded in controlling wilt disease with high efficiency [5]. Leeman et al., 1996 reported that siderophore-producing *P. fluorescens* induced systemic resistance against Fusarium wilt of radish [51]. Formulation of siderophores-producing *P. aeruginosa* F2 and *P. fluorescens* JY3 were sufficient in increasing fresh and dry weights of shoots and roots of wheat plants. These results may due to the ability of both isolates to induce the plant to produce some phytohormones such as auxin, cytokinin, and gibberellins. Also, may be related to the capability of plant to recognize the microbial Fe-siderophores complex and consequently iron uptake from this complex (iron nutrition). In an agreement with the current findings, Sharma et al., 2003 showed increasing iron, chlorophyll a, and chlorophyll b concentrations of *Vigna radiate* plants when they were inoculated with siderophore-producing *Pseudomonas* strain GRP3 [52]. Additionally, our results were in agreement with Sayyed et al., 2005 and Manwar et al., 2000, who documented that inoculation with *P. fluorescens* enhanced seed germination, shoot and root length of wheat [8,53]. Arya et al., 2018, reported that siderophores-producing *P. fluorescens* strain SPs9 and SPs20 were effective in increasing fresh and dry weight of tomato plants in a soil infected with *F. oxysporum* [5]. The same results were obtained by inoculating *Triticum aestivum* with fluorescent pseudomonad R62 [54].

5. Conclusions

In this study, scaling-up production of siderophores from fluorescent pseudomonads was accomplished using fermentation technology. Exponential fed-batch fermentation of *P. aeruginosa* F2 and *P. fluorescens* JY3 gave higher concentrations of siderophores and biomass than batch fermentation. Formulations of siderophores-producing fluorescent pseudomonads were effective in controlling soil-borne fungi and for stimulation of plant growth. These formulations can therefore be utilized as plant growth promoters and biocontrol agents.

Author Contributions: G.A.A.-Z. planned and designed the research; A.S.A. performed the experiments; G.A.A.-Z. and A.S.A. wrote the manuscript; N.A.-M.S., E.E.E.-S., S.M.M. and S.A.-F.S. revised the manuscript. All authors have read and agreed to the published version of the manuscript.

Funding: This research received no external funding.

Acknowledgments: The authors would like to thank the City of Scientific Research and Technological Applications for housing and facilitating this work. The authors also thank Kenawy, A.M. of the City of Scientific Research and Technological Applications, Alexandria, Egypt for critical reading and revision.

Conflicts of Interest: Authors declare there is no conflict of interest.

References

1. Neilands, J.B. Microbial iron compounds. *Annu. Rev. Biochem.* **1981**, *50*, 715–731. [CrossRef] [PubMed]
2. Hider, R.C.; Kong, X. Chemistry and biology of siderophores. *Nat. Prod. Rep.* **2010**, *27*, 637–657. [CrossRef] [PubMed]
3. Kurth, C.; Kage, H.; Nett, M. Siderophores as molecular tools in medical and environmental applications. *Org. Biomol. Chem.* **2016**, *14*, 8212–8227. [CrossRef] [PubMed]
4. Kloepper, J.W.; Leong, J.; Teintze, M.; Schroth, M.N. Enhancing plant growth by siderophores produced by plant growth-promoting rhizobacteria. *Nature* **1980**, *286*, 885–886. [CrossRef]
5. Arya, N.; Rana, A.; Rajwar, A.; Sahgal, M.; Sharma, A.K. Biocontrol efficacy of siderophore producing indigenous pseudomonas strains against fusarium wilt in tomato. *Natl. Acad. Sci. Lett.* **2018**, *41*, 133–136. [CrossRef]
6. Gupta, C.P.; Dubey, R.C.; Maheshwari, D.K. Plant growth enhancement and suppression of *Macrophomina phaseolina* causing charcoal rot of peanut by fluorescent Pseudomonas. *Biol. Fertil. Soils* **2002**, *35*, 399–405.

7. Yu, S.; Teng, C.; Liang, J.; Song, T.; Dong, L.; Bai, X.; Jin, Y.; Qu, J. Characterization of siderophore produced by *Pseudomonas syringae* BAF.1 and its inhibitory effects on spore germination and mycelium morphology of *Fusarium oxysporum*. *J. Microbiol.* **2017**, *55*, 877–884. [CrossRef]

8. Sayyed, R.Z.; Badgujar, M.D.; Sonawane, H.M.; Mhaske, M.M.; Chincholkar, S.B. Production of microbial iron chelators (Siderophores) by fluorescent Pseudomonads. *Ind. J. Biotechnol.* **2005**, *4*, 484–490.

9. Sarma, M.; Saharan, K.; Kumar, L.; Gautam, A.; Kapoor, A.; Srivastava, N.; Sahai, V.; Bisaria, V. Process optimization for enhanced production of cell biomass and metabolites of fluorescent pseudomonad r81. *Int. J. Biomed. Biol. Eng.* **2010**, *4*, 388–392.

10. Shaikh, S.; Wani, S.; Sayyed, R. Statistical-based optimization and scale-up of siderophore production process on laboratory bioreactor. *3 Biotech* **2016**, *6*, 69. [CrossRef]

11. Paulová, L.; Patáková, P.; Brányik, T. Advanced fermentation processes. In *Engineering Aspects of Food Biotechnology*; Teixeira, J., Vincente, A.A., Eds.; Taylor & Francis Group: Milton Park, UK, 2016; pp. 90–99.

12. Abo-Zaid, G.A.; Wagih, E.E.; Matar, S.M.; Ashmawy, N.A.; Hafez, E.E. Scaling-up production of pyocyanin from *Pseudomonas aeruginosa* JY21 as biocontrol agent against certain plant pathogenic fungi. *Int. J. ChemTech Res.* **2015**, *8*, 213–224.

13. Shuler, M.L.; Kargi, F. *Bioprocess Engineering Basic Concepts*, 2nd ed.; Prentice-Hall PTR: Upper Saddle River, NJ, USA, 2002.

14. Matar, S.M.; El-Kazzaz, S.A.; Wagih, E.E.; El-Diwany, A.I.; Moustafa, H.E.; El-Saadani, M.A.; Abo-Zaid, G.A.; Hafez, E.E. Bioprocessing and scaling-up cultivation of *Bacillus subtilis* as a potential antagonist to certain plant pathogenic fungi, III. *Biotechnology* **2009**, *8*, 138–143. [CrossRef]

15. Huang, H.; Ridgway, D.; Moo-Young, T.G.M. Enhanced amylase production by *Bacillus subtilis* using a dual exponential feeding strategy. *Bioprocess Biosyst. Eng.* **2004**, *27*, 63–69. [CrossRef] [PubMed]

16. Luli, G.W. The Influence of Glucose Metabolism on High Cell Density Fermentation. Ph.D. Thesis, Ohio State University, Columbus, OH, USA, 1988.

17. Lee, J.H.; Hong, J.; Lim, H.C. Experimental optimization of fed-batch culture for poly-β-hydroxybutyric acid production. *Biotechnol. Bioeng.* **1997**, *56*, 697–705. [CrossRef]

18. Salwa, M.S.; Asshifa, M.N.N.; Amirul, A.A.; Yahya, A.R.M. Different feeding strategy for the production of biosurfactant from *Pseudomonas aeruginosa* USM AR2 in modified bioreactor. *Biotechnol. Bioprocess Eng.* **2009**, *14*, 763–768. [CrossRef]

19. Lotfabad, T.B.; Tayyebi, S.; Roostaazad, R. Kinetic Measurements for *Pseudomonas aeruginosa* MR01 during biosurfactant production in two-phase system and developing a double-exponential model for viable cell profile. *World Appl. Sci. J.* **2013**, *22*, 809–816.

20. Ju, X.; Yu, H.L.; Pan, J.; Xu, J.H. Improved production of *Pseudomonas* sp. ECU1011 acetyl esterase by medium design and fed-batch fermentation. *Bioprocess Biosyst. Eng.* **2012**, *35*, 323–331. [CrossRef]

21. Radha, S.; Prasad, T.H.; Devi, N.S.; Mamata, S. Fed batch and batch submerged fermentation for alkaline protease production from mixed consortium of *Pseudomonas putida* and *Staphylococcus aureus*. *Adv. Appl. Sci. Res.* **2014**, *5*, 315–324.

22. Buyer, J.S.; Leong, J. Iron transport-mediated antagonism between plant growth-promoting and plant-deleterious *Pseudomonas* strains. *J. Biol. Chem.* **1986**, *261*, 791–794.

23. Kenawy, A.; Dailin, D.J.; Abo-Zaid, G.A.; Abd Malek, R.; Ambehabati, K.K.; Zakaria, K.H.; Sayyed, R.Z.; Enshasy, H.A. Biosynthesis of antibiotics by PGPR and their roles in biocontrol of plant diseases. In *Plant Growth Promoting Rhizobacteria for Sustainable Stress Management*; Sayyed, R.Z., Ed.; Springer: Singapore, 2019; pp. 1–36.

24. De Boer, M.; Bom, P.; Kindt, F.; Keurentjes, J.J.B.; van der Sluis, I.; Van Loon, L.C.; Bakker, P.A. Control of Fusarium wilt of radish by combining *Pseudomonas putida* strains that have different disease-suppressive mechanisms. *Phytopathology* **2003**, *93*, 626–632. [CrossRef]

25. Schmidt, W. Mechanisms and regulation of reduction-based iron uptake in plants. *New Phytol.* **1999**, *141*, 1–26. [CrossRef]

26. Vansuyt, G.; Robin, A.; Briat, J.F.; Curie, C.; Lemanceau, P. Iron acquisition from Fe-pyoverdine by *Arabidopsis thaliana*. *Mol. Plant Microbe Interact.* **2007**, *20*, 441–447. [CrossRef] [PubMed]

27. Schywn, B.; Neilands, J.B. Universal chemical assay for the detection and determination of siderophores. *Anal. Biochem.* **1987**, *160*, 47–56. [CrossRef]

28. Meyer, J.M.; Abdallah, M.A. The fluorescent pigment of *Pseudomonas fluorescens*: Biosynthesis, purification and physicochemical properties. *J. Gen. Microbiol.* **1978**, *107*, 319–328. [CrossRef]
29. Toure, Y.; Ongena, M.; Jacques, P.; Guiro, A.; Thonart, P. Role of lipopeptides produced by *Bacillus subtilis* GA1 in the reduction of grey mould disease caused by *Botrytis cinerea* on apple. *J. Appl. Microbiol.* **2004**, *96*, 1151–1160. [CrossRef] [PubMed]
30. Istock, C.A.; Ferguson, N.; Istock, N.L.; Duncan, K.E. Geographical diversity of genomic lineages in *Bacillus subtilis* (Ehrenberg) Cohn sensulato. *Org. Divers. Evol.* **2001**, *1*, 179–191. [CrossRef]
31. Matar, S.M.; El-Kazzaz, S.A.; Wagih, E.E.; El-Diwany, A.I.; Hafez, E.E.; Moustafa, H.E.; Abo-Zaid, G.A.; Serour, E.A. Molecular characterization and batch fermentation of *Bacillus subtilis* as biocontrol agent II. *Biotechnology* **2009**, *8*, 35–43. [CrossRef]
32. Abo-Zaid, G.A.; Abdullah, A.S.; Soliman, N.A.; El-Sharouny, E.E.; Sabry, S.A. Optimization of siderophores production from fluorescent pseudomonads using statistical experimental designs. *Biosci. Res.* **2018**, *15*, 3040–3051.
33. Van Dam-Mieras, M.C.E.; Jeu, W.H.; Vries, J.; Currell, B.R.; James, J.W.; Leach, C.K.; Patmore, R.A. *Techniques Used in Bioproduct Analysis*; Butterworth-Heinemann Ltd.: Oxford, UK, 1992.
34. Bok, S.H.; Demain, A.L. An Improved Calorimetric Assay for Polyols. *Anal. Biochem.* **1977**, *81*, 18–20. [CrossRef]
35. Vidhyasekaran, P.; Muthamilan, M. Development of formulation of *Pseudomonas fluorescens* for control of chickpea wilt. *Plant Dis.* **1995**, *79*, 782–786. [CrossRef]
36. Hussien, Z.N.; Mahmoud, E.Y.; Metwaly, A.H.; Sobhy, H.M. Effect of some antagonistic bacteria in reducing of peanut damping-off, root and pod rot incidence caused by *Rhizoctonia solani*. *J. Plant Prot. Pathol.* **2012**, *3*, 1173–1187.
37. Omidvari, M.; Sharifi, R.A.; Ahmadzadeh, M.; Dahaji, P.A. Role of fluorescent pseudomonads siderophore to increase bean growth factors. *J. Agric. Sci.* **2010**, *2*, 242–247. [CrossRef]
38. Sulochana, M.B.; Jayachandra, S.Y.; Kumar, S.K.A.; Dayanand, A. Antifungal attributes of siderophore produced by the *Pseudomonas aeruginosa* JAS-25. *J. Basic Microbiol.* **2014**, *54*, 418–424. [CrossRef] [PubMed]
39. Páez, M.; Martínez-Nieto, P.; Bernal-Castillo, J. Siderophores producing *Pseudomonas* as pathogenic *Rhizoctonia solani* and *Botrytis cinerea* antagonistis. *Univ. Sci.* **2005**, *10*, 65–74.
40. Sasirekha, B.; Srividya, S. Siderophore production by *Pseudomonas aeruginosa* FP6, a biocontrol strain for *Rhizoctonia solani* and *Colletotrichum gloeosporioides* causing diseases in chilli. *Agric. Nat. Res.* **2016**, *50*, 250–256. [CrossRef]
41. Islam, M.A.; Nain, Z.; Alam, M.K.; Banu, N.A.; Islam, M.R. In vitro study of biocontrol potential of rhizospheric *Pseudomonas aeruginosa* against *Fusarium oxysporum* f. sp. *cucumerinum*. *Egypt. J. Biol. Pest Control* **2018**, *28*, 90. [CrossRef]
42. Ahmadzadeh, M.; Afsharmanesh, H.; Nikkhah, M.J.; Tehrani, A.S. Identification of some molecular traits in fluorescent pseudomonads with antifungal activity. *Iran. J. Biotechnol.* **2006**, *4*, 245–253.
43. Sabra, W.; Kim, E.J.; Zeng, A.P. Physiological responses of *Pseudomonas aeruginosa* PAO1 to oxidative stress in controlled microaerobic and aerobic cultures. *Microbiology* **2002**, *148*, 3195–3202. [CrossRef]
44. Park, Y.S.; Kai, K.; Lijima, S.; Kobayashi, T. Enhanced β-galactosidase production by high cell-density culture of recombinant *Bacillus subtilis* with glucose concentration control. *Biotechnol. Bioeng.* **1992**, *40*, 686–696. [CrossRef]
45. Yee, L.; Blanch, H.W. Recombinant trypsin production in high cell density fed-batch cultures in *Escherichia coli*. *Biotechnol. Bioeng.* **1993**, *41*, 781–790. [CrossRef]
46. Lee, S.Y. High cell-density culture of *Escherichia coli*. *Trends Biotechnol.* **1996**, *14*, 98–105. [CrossRef]
47. Chosewood, L.C.; Wilson, D.E. *Biosafety in Microbiological and Biomedical Laboratories*, 5th ed.; HHS Publication: Washington, DC, USA, 2009; pp. 9–19.
48. OECD. Bacteria: Pathogenicity factors. In *Safety Assessment of Transgenic Organisms in the Environment*; OECD Consensus Documents; OECD Publishing: Paris, France, 2016; Volume 5, pp. 27–79.
49. Blacksell, S.D.; Robinson, M.T.; Newton, P.N.; Ruanchaimun, S.; Salje, J.; Wangrangsimakul, T.; Wegner, M.D.; Abdad, M.Y.; Bennett, A.M.; Richards, A.L.; et al. Biosafety and biosecurity requirements for *Orientia* spp. diagnosis and research: Recommendations for risk-based biocontainment, work practices and the case for reclassification to risk group 2. *BMC Infect. Dis.* **2019**, *19*, 1044. [CrossRef] [PubMed]

50. Solans, M.; Scervino, J.M.; Messuti, M.I.; Vobis, G.; Wall, L.G. Potential biocontrol actinobacteria: Rhizospheric isolates from the argentine pampas lowlands legumes. *J. Basic Microbiol.* **2016**, *56*, 1289–1298. [CrossRef] [PubMed]

51. Leeman, M.; Den Ouden, F.M.; Van Pelt, J.A.; Dirkx, F.P.M.; Steijl, H.; Bakker, P.A. Iron availability affects induction of systemic resistance to Fusarium wilt of radish by *Pseudomonas fluorescens*. *Phytopathology* **1996**, *86*, 149–155. [CrossRef]

52. Sharma, A.; Johri, B.N.; Sharma, A.K.; Glick, B.R. Plant growth-promoting bacterium *Pseudomonas*sp. strain GRP3 influences iron acquisition in mung bean (*Vigna radiata*L. Wilzeck). *Soil Biol. Biochem.* **2003**, *35*, 887–894. [CrossRef]

53. Manwar, A.V.; Vaigankar, P.D.; Bhonge, L.S.; Chincholker, S.B. In vitro suppression of plant pathogens by siderophores of fluorescent Pseudomonads. *Ind. J. Microbiol.* **2000**, *40*, 109–112.

54. Saharan, K.; Sarma, M.V.; Srivastava, R.; Sharma, A.K.; Johri, B.N.; Prakash, A.; Sahai, V.; Bisaria, V.S. Development of non-sterile inorganic carrier-based formulations of fluorescent pseudomonad R62 and R81 and evaluation of their efficacy on agricultural crops. *Appl. Soil Ecol.* **2010**, *46*, 251–258. [CrossRef]

Article

Isolation, Identification and Antimicrobial Evaluation of Bactericides Secreting *Bacillus subtilis* Natto as a Biocontrol Agent

Jing Zhang [1], Muhammad Bilal [1], Shuai Liu [1], Jiaheng Zhang [1], Hedong Lu [1], Hongzhen Luo [1], Chuping Luo [1], Hao Shi [1], Hafiz M. N. Iqbal [2] and Yuping Zhao [1,*]

[1] School of Life Science and Food Engineering, Huaiyin Institute of Technology, Huai'an 223003, China; zhangjing12273932@163.com (J.Z.); bilaluaf@hotmail.com (M.B.); liushuai-007@163.com (S.L.); zjh2018herb@163.com (J.Z.); luhd100@163.com (H.L.); hzluo@hyit.edu.cn (H.L.); luochuping@163.com (C.L.); ilyshihao@163.com (H.S.)

[2] Tecnologico de Monterrey, School of Engineering and Sciences, Campus Monterrey, Ave. Eugenio Garza Sada 2501, Monterrey, N.L., CP 64849, Mexico; hafiz.iqbal@tec.mx

* Correspondence: zhaoyuping@hyit.edu.cn

Received: 3 January 2020; Accepted: 20 February 2020; Published: 25 February 2020

Abstract: Herein, a bactericide-secreting *Bacillus* strain, potentially useful as a biocontrol agent, was isolated from the commercial Yanjing Natto food. Following the biochemical and physiological evaluation, the molecular identification was performed using 16S rDNA sequencing of polymerase chain reaction-amplified DNA that confirmed the natto isolate as *Bacillus subtilis* natto (*B. subtilis* natto). The biocontrol (microbial inhibitory) capability of *B. subtilis* natto was investigated against *Staphylococcus aureus*, *Escherichia coli*, *Salmonella typhimurium*, and yeast (*Yarrowia lipolytica*) and recorded. The antimicrobial activity of *B. subtilis* natto was further enhanced by optimizing the growth medium for optimal bactericides secretion. Under optimized conditions, *B. subtilis* natto exhibited much higher inhibitory activity against *S. aureus* with a zone of inhibition diameter up to 27 mm. After 48 h incubation, the optimally yielded *B. subtilis* natto broth was used to extract and purify the responsible bactericides by silica gel column chromatography, gel column chromatography, and semi-preparative high-performance liquid chromatography. Structural identification of purified bactericides (designated as NT-5, NT-6, and NT-7) from *B. subtilis* natto was performed by ^{13}C-nuclear magnetic resonance (NMR) and mass spectral analyses. The NMR comparison also revealed that NT-5, NT-6, and NT-7 had identical structures, except for the fatty chain. In summary, the present study suggests the improved biocontrol and/or microbial inhibitory potential of newly isolated bactericides secreting *B. subtilis* natto.

Keywords: biocontrol agent; *Bacillus subtilis* natto; isolation; molecular identification; medium optimization; antimicrobial activity; bactericides; spectral analyses

1. Introduction

Natto as a traditional food with a history of two thousand years has various healthcare functions such as prevention of osteoporosis, procoagulant, antimicrobial, anti-aging, anti-cancer, and gastrointestinal activities [1–3]. *Bacillus subtilis* natto (*B. subtilis* natto) is an aerobic Gram-positive probiotic that possesses a strong heat and acid-base stability [4]. This edible strain is claimed to be the organism primarily responsible for natto fermentation. It has a variety of functions including promoting the absorption of iron, calcium, and vitamin D by generating intestinal acidification [5]. *B. subtilis* natto metabolites also have antihypertensive, anti-tumor, anti-oxidation, thrombolytic and other functions. Among all these properties, potent antibacterial activity is one of the most important functions of

B. subtilis natto and might be due to the production of bacitracin, polymyxin, 2,6-pyridine dicarboxylic acid, and other antibiotics. It has the role of inhibiting pathogens such as *Salmonella typhimurium* and dysentery bacteria. Along with a broad antimicrobial spectrum, *B. subtilis* natto-derived antimicrobial substances are safe for humans as compared to other drugs [3,6,7].

A pronounced antibacterial activity against *Helicobacter pylori* has been found in *B. subtilis* natto cells. Through the minimum inhibitory concentration (MIC) tests, it was confirmed that *B. subtilis* natto possessed anti-platelet aggregation and anti-*H. pylori* activity due to the presence of dipicolinic acid [8]. Based on morphological studies, a bacterial isolate BH072 from a honey sample exhibited a broad-spectrum inhibitory activity against a range of molds including *Aspergillus niger*, *Botrytis cinerea*, and *Pythium* [9]. A lipopeptide biosurfactant was purified from *B. natto* TK-1 by acidic precipitation, methanol extraction, and thin-layer chromatography (TLC) system, and its in-vitro anti-adhesion activity was investigated. Results indicated that lipopeptide presented noteworthy antimicrobial properties and significantly inhibited the adhesion of *Escherichia coli*, *Staphylococcus aureus*, and *S. typhimurium* [3]. *Vibrio parahaemolyticus* is an emerging foodborne pathogen in seafood products. Effective measures to prevent *V. parahaemolyticus* in seafood would help minimize the probability of foodborne illness and large outbreaks of *Vibrio*. *B. subtilis* NT-6 isolated from natto secretes a novel antibacterial peptide AMPNT-6 that is a potent inhibitor of *V. parahaemolyticus* (up to 15.5 mm diameter). These results manifested that AMPNT-6 might be potentially employed as a natural inhibitor of *V. parahaemolyticus* on shrimp substrates [10].

Due to increasing people's health requirements, food safety standards have become more stringent, and considerable attention has been focused on food preservatives. Low-cost natural and broad-spectrum food preservatives are an emerging trend around the world. Compared with chemical preservatives, food preservatives produced using traditional fermented foods are newer, safer and more effective. Antibacterial substances in natto not only have the advantages of high safety, broad antibacterial spectrum but also rich in nutrition and functionality [11,12]. Nevertheless, there are few reports on the antibacterial activity of *B. subtilis* natto. Moreover, the main components and structure of the antibacterial substances are still unclear, and the antibacterial ability is weak. This study aims to isolate and purify the bactericides secreting *B. subtilis* natto. The biocontrol capability of *B. subtilis* natto was also evaluated against *E. coli*, *S. aureus*, *S. typhimurium*, and yeast (*Y. lipolytica*). Moreover, advanced instrumental techniques were used to elucidate the structural features of bactericides extracted from *B. subtilis* natto. The research of antibacterial baring compounds extraction and purification promoted its application in medicines and preservatives and expected to develop into new antibacterial drugs.

2. Materials and Methods

2.1. Materials, Chemicals, and Reagents

The natto food was obtained from Beijing Yanjing Zhongwu Material Technology Co., Ltd. (Beijing, China). Ethylenediaminetetraacetic acid (EDTA), glucose, cornflour, sucrose, soluble starch, soy flour, NaCl, tryptone, beef cream, ammonium nitrate, $ZnSO_4$, $FeSO_4$, and K_2SO_4 were purchased from Sinopharm Chemical Reagent Co., Ltd. (Shanghai, China). Shanghai Aladdin Biochemical Technology Co., Ltd. (Shanghai, China) provided *n*-hexane, ethyl acetate, methylene dichloride, and methanol. All aqueous solutions were prepared with deionized water. All other chemicals/reagents used in this study were the highest quality and used as received without any further processing unless otherwise stated.

2.2. Microbial Strains

Escherichia coli, *S. aureus*, *S. typhimurium*, and *Y. lipolytica* were procured from Shanghai Luwei Microbial Science and Technology Co. Ltd. (Shanghai, China). *B. subtilis* natto strain was isolated from natto food. All collected and isolated microbial cultures were maintained on nutrient agar slopes at 37 °C for 24 h. After the stipulated period of 24 h incubation, the viable cultures were preserved at 4 °C and sub-cultured periodically every two weeks to maintain the viability.

2.3. Isolation and Screening of B. subtilis Natto

As received natto was washed twice with warm water to eliminate the dust particles. One-gram natto was soaked in 10 mL sterile saline solution and heated in a water bath at 80 °C for 1.0 h. The natto mixture with the elimination of solid matters was allowed to cool at room temperature (25 ± 3 °C) and used as a bacterial suspension. The bacterial suspension was then serially diluted from 10^{-1} to 10^{-9} to reduce the concentration of microorganisms and spread on a sterile agar plate containing 20 mL of solidified screening medium. The freshly inoculated plates were placed at 37 °C for a time of 24 h. The resulting bacterial colonies were further purified by streaking on a screening medium, Gram-stained and observed under an optical microscope using the oil-immersion lens. A single bacterial colony was further subjected to a series of physiological and biochemical identification tests. Following Bergey's Manual of Determinative Bacteriology, catalase test, starch hydrolysis, salt tolerance (i.e., 2%, 5%, 7%, and 10%), gelatin liquefaction, acetyl methyl, Voges-Proskauer (VP), D-glucose and D-mannitol fermentation, and nitrate reduction were performed.

2.4. Molecular Identification: DNA Extraction and PCR Amplification

A bacterial genomic DNA extraction kit (*EasyTaq*® DNA Polymerase, Transgen Biotech, Beijing, China) was used for molecular identification purposes. Briefly, an overnight grown *B. subtilis* natto was centrifuged at 4000× *g* for 10 min. The extracted DNA was used as a template for polymerase chain reaction (PCR) to amplify about 500 or 1500 bp segment of the 16S rRNA gene sequence using the universal primers 27F (5′-AGAGTTTGTCCTGGCTCAG-3′), 1492R (5′-TACGGCTACCTT GTTACGACTT-3′). PCR amplification system (50 µL) contains; 1 µL genomic DNA, 1 µL each of upstream and downstream primers, 1 µL EasyTaq DNA Polymerase, 4 µL High pure dNTPs (EasyPure Genomic DNA Kit, Transgen Biotech), 5 µL 10× EasyTaq Buffer and ddH$_2$O. The amplification was performed in a thermal cycler (Thermo Fisher Scientific, Waltham, MA, USA) by DNA denaturation at 94 °C for 3 min followed by 30 cycles of 30 s at 94 °C, 30 s at 55 °C (Tm), and 30 s at 72 °C and an ultimate extension of 5–10 min at 72 °C.

2.5. S rDNA Sequencing and Data Analysis

Samples were prepared and sequenced by the Shanghai Bioengineering Co., Ltd. China accredited laboratory. The target gel band was purified using the TaKaRa MiniBEST Purification Kit. The sequences obtained with the 1492R primer were aligned and compared with other partial 16S rDNA sequences in the GenBank database (http://www.ncbi.nlm.nih.gov/BLAST). The obtained 16S rDNA sequence was deposited to GenBank for assigning accession numbers by NCBI.

2.6. Microbial Cultural Conditions

B. subtilis natto strain was cultivated in seed culture medium (pH 7.2–7.4) containing peptone 10 g/L; beef extract 5 g/L, NaCl 6 g/L, and glucose 5 g/L at 37 °C for 24 h. The screening medium includes skim milk powder 50 g/L, NaCl 5 g/L, glucose 10 g/L, agar 20 g/L. The medium was sterilized in a laboratory-scale autoclave at 115 °C and 15 psi for 30 min. The agar medium contained 0.1 MPa and water addition and sterilized at 121 °C for 15 min. After cooling slightly, both were uniformly mixed. Constant temperature and humidity culture at 37 °C for 24 h. The fermentation medium comprises peptone 5 g/L, sodium chloride 5 g/L, glucose 20 g/L, KH$_2$PO$_4$ 2 g/L, and K$_2$HPO$_2$ 4 g/L. The medium was sterilized at 115 °C for 30 min. The culture was incubated at 37 °C at 150 rpm. Potato dextrose agar (PDA) containing potato 200 g/L, glucose 20 g/L and agar 15~20 g/L was used for the cultivation of yeast. Whereas medium comprised of tryptone 10 g/L, yeast extract 5 g/L, and NaCl 10 g/L was used for *S. typhimurium* cultivation.

2.7. Antimicrobial Activity of B. subtilis Natto Extract

B. subtilis natto has a unique potential to secrete bactericidal compounds of high interest. For such reason, a cell-free extract of freshly grown *B. subtilis* natto was used to test the inhibitory activity towards *E. coli*, *S. aureus*, *S. typhimurium*, and yeast. Briefly, around 100 mL liquid seed medium was inoculated with a 5% of freshly grown *B. subtilis* natto inoculum suspension. The inoculated seed medium was incubated at 37 °C for up to 48 h under uninterrupted shaking at 160 rpm. After 48 h incubation, the fermented broth was centrifuged at $4000 \times g$ for 15 min, and the cell-free supernatant was further filtered through a 0.45 μm polysulphonate membrane filter. The resulting filtered cell-free supernatant was considered as a bactericidal containing crude extract and tested using the agar well diffusion assay against selected strains. Briefly, sterilized discs were gently placed onto the center of the agar plate with sterile forceps following the addition of 20 μL of bactericidal containing cell-free supernatant to the discs. The inhibition zone diameter (mm) formed around the disks was recorded after the plate's incubation at 37 °C for 24 h [13]. Kanamycin at the concentration of 30 μg/mL, and water ware used as a positive and negative control, respectively.

2.8. Minimal Inhibitory Concentration of B. subtilis Natto against Pathogenic Bacteria

The broth dilution method, in test tubes, was adopted to record the minimum inhibitory concentration (MIC) of pathogenic bacteria, according to the procedure described earlier [14–16] with some modifications. Briefly, different extract preparations were serially diluted using sterile nutrient broth medium as a diluent to prepare a final crude extract concentration between 0.578 and 2400 μg/mL. All the tubes were inoculated with the bacterial suspension (adjusted to 10^7 CFU/mL broth), mixed and incubated for 24 h at 37 °C. The lowest concentration of *B. subtilis* natto (highest dilution), displaying no noticeable growth of pathogenic bacteria was defined as MIC. A tube with an equal volume of water was used as a control in parallel to evaluate the influence of the sterile medium on the growth of the test bacteria.

2.9. Optimization of Growth Medium to Induce the Antibacterial Activity of B. subtilis Natto

The types of carbon, nitrogen and inorganic salts in the fermentation medium also have a certain influence on the metabolism of the bacteria. Therefore, in this experiment, different carbon sources (2.0% corn flour, glucose, sucrose, and soluble starch), nitrogen source (0.5% soy flour, tryptone, beef extract, ammonium nitrate, and inorganic salts (0.5% potassium dihydrogen phosphate, zinc sulfate, ferrous sulfate, potassium sulfate) were screened and optimized by a single factor optimization. The size of the inhibition zone was measured after 48 h of culture.

2.10. Extraction and Purification of Bactericides from B. subtilis Natto

According to the results of orthogonal experiments, *B. subtilis* natto was aerobically grown in an Erlenmeyer flask (250-mL capacity) containing sucrose 1%, soy flour 0.8%, potassium sulfate 0.2%, sodium chloride 0.5%, and water 1 L under optimized culture conditions at 37 °C for 36 h on a rotary shaker (120 rpm). After cultivation, the fermentation broth was harvested by centrifugation at $4000 \times g$ and 4 °C for 20 min. The fermentation broth was added with ethyl acetate in a ratio of 1:1, thoroughly mixed, and allowed to stand in a separating funnel for two h. The extract was collected and concentrated using a rotary evaporator. The resulting brown extract (10.1 g) was purified by silica gel vacuum liquid chromatography using a 20 × 8 cm column eluted with gradient solvents of increasing polarity (n-hexane–EtOAc, 9:1, 7:3, 1:1, 3:7; CH_2Cl_2–MeOH, 15:1, 9:1, 7:3, 0:10; 600 mL each gradient) yielding a total of three fractions. Fractions A-4 and A-5 exhibiting antibacterial activity against *E. coli* and *S. aureus* were chosen for additional separation. Fraction A-4 (280.7 mg) and A-5 were passed through a Sephadex LH-20 column (60 × 3 cm) using methanol as a mobile phase to remove pigments, and then purified by a semi-preparative high-performance liquid chromatography (Eurospher C18 column) giving NT-5 (15.3 mg), NT-6 (45.5 mg) and NT-7 (51 mg) (MeOH−H$_2$O: 0 min, 20%; 2 min: 30%; 7 min: 50%; 10 min-15 min: 100%).

2.11. Structural Identification of Bactericides from B. subtilis Natto

The samples were dissolved in 2% C5D5N or CDCl3. 13C-NMR spectra were obtained at 150 MHz on a Bruker Avance III NMR spectrometer (Bremen, Germany). HRESIMS (High-resolution electrospray ionization mass spectrometry) data were obtained from an ultra-high-resolution quadrupole time-of-flight (UHR-qTOF) Maxis 4G mass spectrometer (Bruker Daltonics, Bremen, Germany).

3. Results and Discussion

3.1. Isolation and Purification of B. subtilis Natto

For initial isolate screening, Gram staining was performed on a 24 h *B. subtilis* natto culture incubated at 37 °C under uninterrupted shaking (150-rpm) using LB liquid medium. According to the characteristics of protease secreted by *Bacillus natto*, protease-producing *Bacillus* was isolated using a milk culture medium (Figure 1A). The Gram-stained isolate was then subjected to a microscopic observation that revealed following characteristics, Gram+, rod-like in shape, flagella to form buds (full or sub-elastic), and elliptical with no obvious swelling (Figure 1B). After that, the second screening was performed covering various physiological and biochemical parameters and the results obtained are listed in Table 1. Based on the recorded characteristics during the first and second screening, the isolate was assumed to be *Bacillus subtilis* natto, which was further subjected to a molecular identification via agarose gel electrophoresis and 16S rDNA sequencing data analysis.

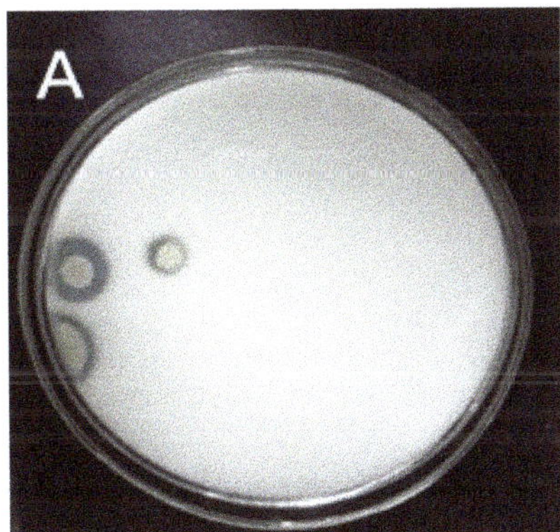
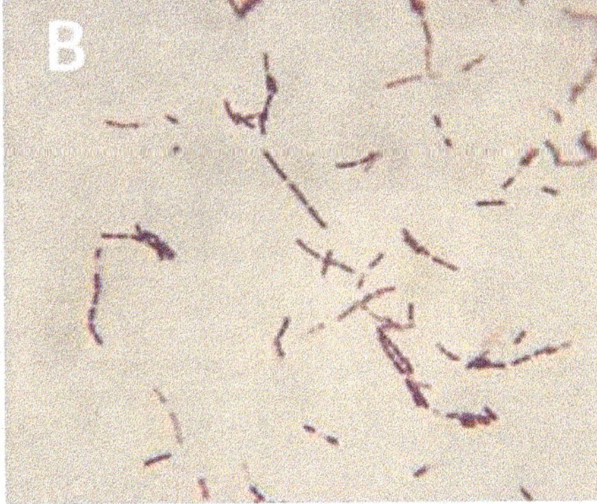

Figure 1. (**A**) Milk screening medium at 37 °C for 48 h; (**B**) Microscopic morphology of bacterial isolate cultured in LB liquid medium at 37 °C, 150 rpm for 24 h.

Table 1. Physiological and biochemical identification tests to identify *B. subtilis* natto.

Parameter	Reaction Response	Parameter	Reaction Response
Sportiness	+	Peroxidase	+
Anaerobic culture	-	Hydrogen sulfide	
Salt tolerance 1%	+	Lecithinase	+
3%	+	Using citrate	+
5%	+	Use of malonate	+
7%	+	Nitrate reduction	+
9%	-	Indigo matrix production	-
V-P experiment	+	Sticky brushed	+
Starch hydrolysis	+	Catalase	+
Gelatin liquefaction	+		

3.2. Molecular Identification: 16S Ribosomal DNA Sequence Analysis

The molecular identification of bacterial isolate from natto food was carried out based on 16S rDNA sequencing data analysis. The 16S rDNA sequence of the amplified PCR product was found to be about 1500 bp long. Comparative analysis of the 16S rDNA nucleotide sequence from the newly isolated strain with the GenBank database sequences showed 99% similarity with the equivalent sequences from the species *B. subtilis*. According to the Berger's Bacterial Identification Manual and 16s rRNA results, the bacterium is *B. subtilis* natto.

3.3. Antibacterial Activity of B. subtilis Natto

The antibacterial activity spectrum of *B. subtilis* natto was evaluated based on the extent of growth-inhibiting potential of four microbes, i.e., *S. aureus*, *E. coli*, *S. typhimurium*, and yeast by direct antagonism on agar plates using a two-layer plate method. Results revealed that *B. subtilis* natto has significant inhibitory activity against *S. aureus* and *E. coli* (Figure 2). Whereas, the antibacterial activity against *S. typhimurium* was observed to be very weak, and there was almost no inhibitory activity on yeast (Figure 2). In addition, MIC test results showed that the MIC value of *B. subtilis* natto against *E. coli* was 6.1 µg/mL, and the MIC value against *S. aureus* was 8.9 µg/mL.

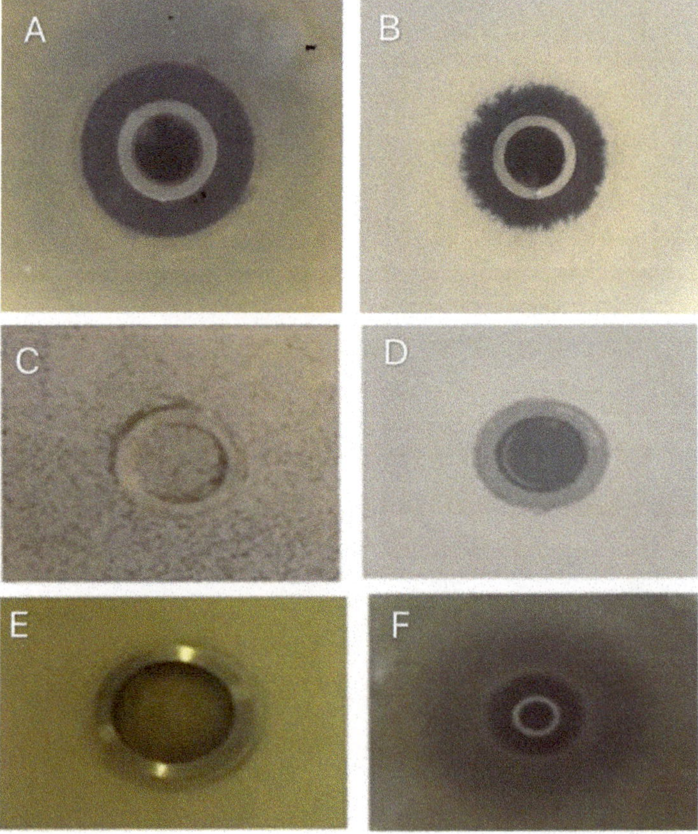

Figure 2. Inhibition zone of pathogenic bacteria, (**A**) *Staphylococcus aureus*, (**B**) *Escherichia coli*, (**C**) Yeast, (**D**) *Salmonella typhimurium* (**E**) Water, and (**F**) Kanamycin.

3.4. Optimization of Growth Medium to Induce the Antibacterial Activity of B. subtilis Natto

A classical optimization method (one factor at a time) was performed to induce the growth of *B. subtilis* natto for optimum performance. For such reason, the influence of various carbon sources, nitrogen sources and inorganic salt on the growth of bactericide secreting *B. subtilis* natto was evaluated, and the results obtained are shown in Figure 3A–C. Among different carbon sources, sucrose had the greatest effect on the antibacterial activity of *B. subtilis* natto (inhibition zone, 21.1 mm) followed by soluble starch, D-glucose and corn flour (Figure 3A). Similarly, the addition of ammonium nitrate and potassium dihydrogen phosphate showed a profound inhibitory effect with a diameter zone of 23.0 mm and 20.1 mm among the nitrogen sources (Figure 3B) and inorganic salts (Figure 3C), respectively. Based on the single factor screening results, ammonium nitrate (nitrogen source) had the greatest influence on the antimicrobial activity of different pathogenic strains, followed by sucrose (carbon source) and KH_2PO_4 (inorganic salt) (Figure 3C). Therefore, these selected factors were further optimized using three levels for each factor in nine experimental runs designed according to the Taguchi's orthogonal method (orthogonal layout L9). The three variables and their respective levels (coded as 1, 2, 3) are listed in Table 2. Response results in terms of the inhibition zone diameter of all these different combinations of the three factors were appraised using the analysis of variance (ANOVA). Scores (response results) are summed up for each factor and each level (k 1, k 2 and k 3 for levels 1, 2 and 3, respectively) as well as averaged (K 1, K 2 and K3 for levels 1, 2 and 3, respectively). It can be observed from the orthogonal array experiment that the inhibition zone reached 27 mm under the optimized medium combination of experiment 6 (Table 3). The optimization scheme was AB_3C_1 indicating that sucrose 1%, 2%, 3% can be used, whereas soy flour 0.8%, and potassium sulfate 0.2% showed the best antibacterial effect. Notably, the bacteriostatic effect was significantly improved after the optimization of medium components. Most of the previous reports have shown that *B. natto* has the ability to inhibit *E. coli* and *S. aureus*, but its antibacterial activity was found to be very weak. Through the optimization of the medium, *B. subtilis* natto exhibited a potent inhibitory activity towards *S. aureus* with an antibacterial diameter of up to 27 mm (Figure 4, Table 3). Earlier studies have mainly focused on the antifungal or antitumor activity of *B. natto*, and there are few reports on inhibition of *S. aureus* and *E. coli*.

Table 2. Factors and their respective assigned levels.

Level	Factor		
	Sucrose	Soy Flour	Potassium Sulfate
1	1%	0.20%	0.20%
2	2%	0.50%	0.50%
3	3%	0.80%	0.80%

Table 3. Results of the L9 (3 × 3) orthogonal array experiment.

Factor	Sucrose (A)	Ammonium Nitrate (B)	KH_2PO_4 (C)	Inhibition Zone (mm)
Experiment 1	1.0	1.0	1.0	19
Experiment 2	1.0	2.0	2.0	23
Experiment 3	1.0	3.0	3.0	22
Experiment 4	2.0	1.0	2.0	16
Experiment 5	2.0	2.0	3.0	20
Experiment 6	2.0	3.0	1.0	27
Experiment 7	3.0	1.0	3.0	21
Experiment 8	3.0	2.0	1.0	22
Experiment 9	3.0	3.0	2.0	20
K1	6.0	5.5	6.3	-
K2	6.0	6.1	5.7	-
K3	6.0	6.4	6.0	-

Table 3. *Cont.*

Factor	Sucrose (A)	Ammonium Nitrate (B)	KH$_2$PO$_4$ (C)	Inhibition Zone (mm)
k1	2.0	1.83	2.1	-
k2	2.0	2.03	1.9	-
k3	2.0	2.13	2.0	-
R	0.0	0.90	0.6	-

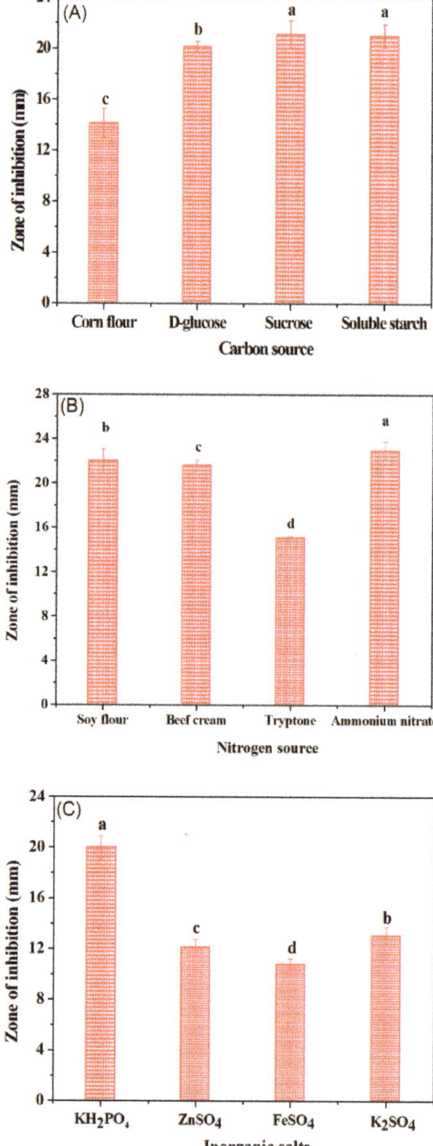

Figure 3. Effects of different (**A**) carbon sources (**B**) nitrogen sources and (**C**) inorganic salts on the antibacterial potential of *B. subtilis* natto.

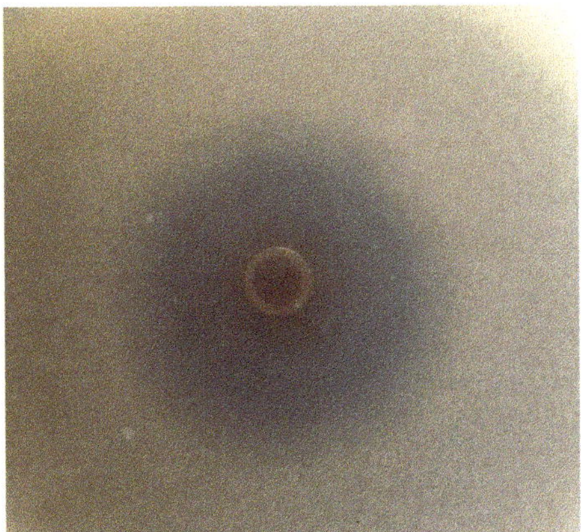

Figure 4. The optimal medium of fermentation broth against the inhibition zone diameter of *Staphylococcus aureus*.

3.5. Purification of Bactericides from B. subtilis Natto

The fermentation broth obtained from *B. subtilis* natto under optimized cultivation conditions was harvested by centrifugation at 4000× *g* and 4 °C for 15 min. The resultant cell-free supernatant was mixed with ethyl acetate and subjected to the rotary evaporator to obtain a concentrated extract. The resulting extract was subjected to silica gel vacuum liquid chromatography, which yielded three fractions. Among these, fractions A-4 and A-5 presenting inhibitory activity towards *E. coli* and *S. aureus* were further separated using a Sephadex LH-20 column to remove pigments and then purified by a semi-preparative HPLC into three fractions designating as NT-5, NT-6, and NT-7 (Figure 5). As evident from the HPLC pattern, the fractions NT-5, NT-6 and NT-7 were 100% purified with the retention time of 30.613, 35.583 and 36.067 min, respectively.

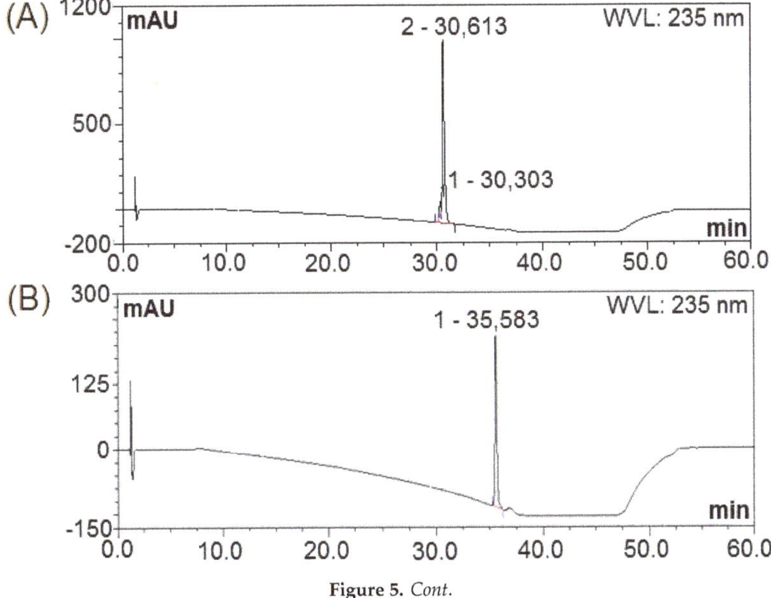

Figure 5. *Cont.*

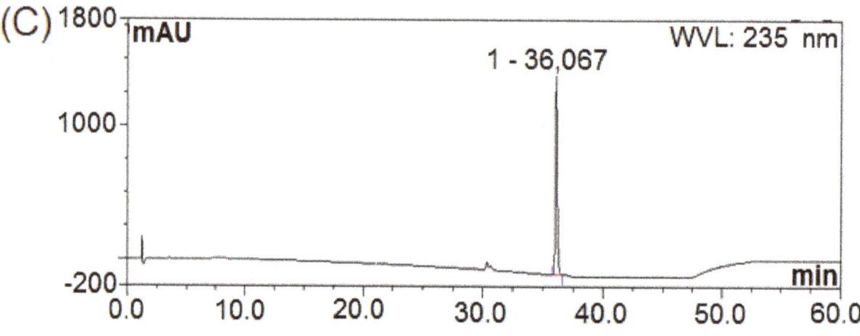

Figure 5. HPLC patterns; (**A**) NT-5, (**B**) NT-6 and (**C**) NT-7.

3.6. Structural Elucidation by ^{13}C-NMR and Mass Spectral Analyses

The structural elucidation of purified bactericides designated as NT-5, NT-6 and NT-7 were further confirmed by ^{13}C NMR analysis at 150 MHz. ^{13}C-NMR displayed different peaks for each tested fraction, which are summarized in Table 4. Compound NT-5 eluting at 30.613 min in the HPLC analysis displayed a protonated molecular ion [M+H]$^+$ at *m/z* 1008.6 (Figure 6A). Compound NT-6 eluting at 35.583 min in the HPLC analysis showed a protonated molecular ion [M + H]$^+$ at *m/z* 1022.5 (Figure 6B). Compound NT-7 eluting at 36.067 min in the HPLC analysis and revealed a protonated molecular ion [M + H]$^+$ at *m/z* 1036.8 (Figure 6C). The NMR data of NT-5, NT-6, and NT-7 summarized in Table 4 were consistent with the literature [17] and verified that NT-5, NT-6, and NT-7 has an inhibitory effect on *E. coli* and *Staphylococcus aureus* (Figure 7). The proposed structures of three active compounds are portrayed in Figure 8.

Table 4. ^{13}C-NMR data for NT-5, NT-6, and NT-7 fractions.

Position	NT6	NT-7	NT-5
1 Glu C = O	171.0	171.0	171.0
α-C	52.4	52.4	52.4
β-C	27.1	27.1	27.1
γ-C	29.7	29.7	29.7
δ-C = O	173.9	173.9	173.9
-OCH$_3$	-	-	-
2 Leu C = O	172.6	172.6	172.6
α-C	51.9	51.9	51.9
β-C	40.0	40.0	40.0
γ-C	24.2	24.2	24.2
δ-C	23.0/22.9	23.0/22.9	23.0/22.9
3 Leu C = O	172.9	172.9	172.9
α-C	51.8	51.8	51.8
β-C	38.9	38.9	38.9
γ-C	24.2	24.2	24.2
δ-C	22.9/22.6	22.9/22.6	22.9/22.6

Table 4. *Cont.*

Position	NT6	NT-7	NT-5
4 Val C = O	170.7	170.7	170.7
α-C	58.6	58.6	58.6
β-C	30.1	30.1	30.1
γ-C	19.1/17.9	19.1/17.9	19.1/17.9
5 Asp C = O	169.9	169.9	169.9
α-C	49.6	49.6	49.6
β-C	35.7	35.7	35.7
γ-C = O	171.7	171.7	171.7
-OCH$_3$	-	-	-
6 Leu C = O	171.6	171.6	171.6
α-C	50.8	50.8	50.8
β-C	41.6	41.6	41.6
γ-C	24.2	24.2	24.2
δ-C	21.7/21.7	21.7/21.7	21.7/21.7
7 Leu/Val C = O	171.9	171.9	171.9
α-C	50.8	50.8	50.8
β-C	38.8	38.8	38.8
γ-C	24.0	24.0	24.0
δ-C	21.2/21.1	21.2/21.1	21.2/21.1
Fatty acid-1	169.9	169.9	169.9
2	41.1	41.1	36.0
3	71.4	71.4	71.4
4	33.6	33.6	33.6
5	24.3	24.3	24.3
6	28.6-29.3	28.6-29.4	28.6-29.4
7	28.6-29.3	28.6-29.4	28.6-29.4
8	28.6-29.3	28.6-29.4	28.6-29.4
9	28.6-29.3	28.6-29.4	28.6-29.4
10	26.8	28.6-29.4	33.7
11	38.5	36.0	26.4
12	27.4	33.7	11.2
13	22.5	26.4	19.1
14	22.5	11.2	-
15	-	19.1	-

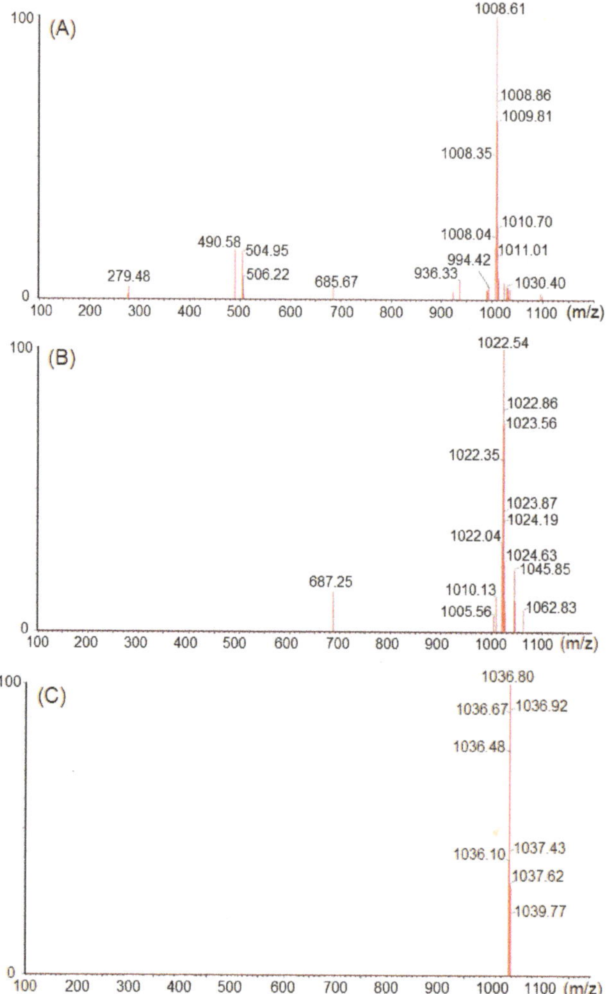

Figure 6. Mass spectral analyses (**A**) NT-5, (**B**) NT-6 and (**C**) NT-7.

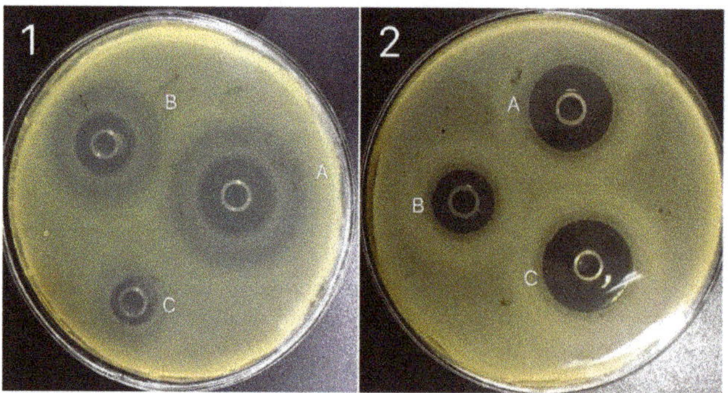

Figure 7. Inhibition zone of pathogenic bacteria, (**1**) *Staphylococcus aureus*, (**2**) *Escherichia coli.* (**A**) NT-5, (**B**) NT-6 and (**C**) NT-7.

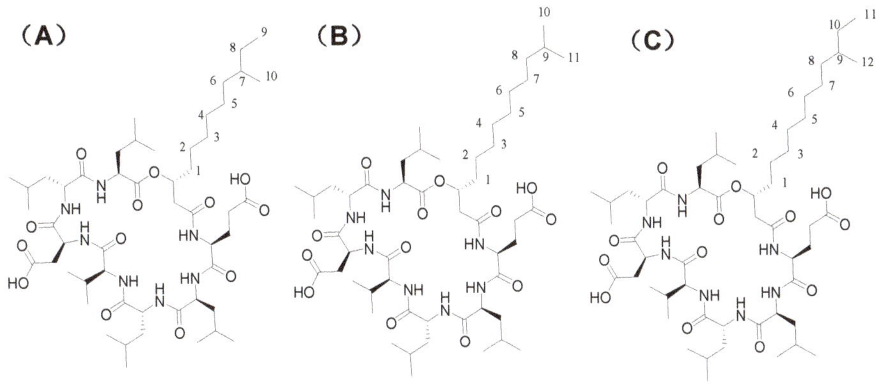

Figure 8. Proposed structures of three active compounds.

Antagonism is omnipresent in Nature among diverse species. People have remained interested for a long time in realistically using it in the areas of agricultural defense, disease therapy, and food preservation. Bacterial reproduction is challenging to control and can cause massive food losses. Extensive use of food additives is another serious problem that may cause harm to people's bodies. Using a biological antibacterial agent to control bacterial diseases is a widespread theme that has been widely investigated [18–24]. In contrast to chemical biocides, numerous antibiotics produced by antagonistic strains present the advantage of being decomposable leaving no detrimental residues. The rapid emergence of resistant pathogenic strains and the identification of adverse chemical moieties in the food chain has rekindled the researcher's attention towards green approaches to effectively tackling pathogenic bacteria. Among the alternatives, biological control by means of natural antagonistic microorganisms has been widely deliberated, and some *Bacillus* strains are effective against various microbial pathogens under the optimal medium composition [3,9,20]. For example, Tabbene et al. [25] found that nutrients such as carbon, nitrogen sources, and inorganic salts enhanced the antimicrobial activity by *B. subtilis* B38 against pathogenic *Salmonella enteridis*, *Listeria monocytogenes*, and methicillin-resistant *Staphylococcus* species. In comparison to basal medium, the antibacterial activity was 2- to 4-fold improved in the modified culture medium consisting of 0.15% (w/v) ammonium succinate, 1.5% (w/v) lactose, and 0.3 mg/L manganese. The results indicate that the nutrients act as environmental factors, qualitatively and quantitatively influencing the synthesis of antimicrobial compounds by *B. subtilis* B38.

Antimicrobial peptides secreted by *Bacillus* spp. have been demonstrated as potent biocontrol candidates against a plethora of phytopathogens [26,27]. This study isolated and identified a strain that has shown significant inhibitory activity against various bacterial strains, and also has shown potential for the synthesis of antimicrobial agents. A novel antibacterial peptide, AMPNT-6, secreted by *B. subtilis* NT-6 showed an evident inhibitory activity against *Vibrio parahaemolyticus* with a diameter and MIC of 15.5 mm and 1.25 mg/mL, respectively. In comparison to the control, the significant bactericidal activity through the changes of *V. parahaemolyticus* growth curve indicate that AMPNT-6 can potentially be used as a natural inhibitor to reduce the likelihood of foodborne outbreaks of *Vibrio* [10]. Wang et al. [28] isolated a *B. subtilis natto* CSUF5 strain and its antifungal metabolites were extracted from the medium of the inhibition zone on the dual culture plate. The identified V7-surfactins and I/L7-surfactins exhibited slight antifungal activities against Aspergillus Niger, and their MIC50 reduced in the order V7 > I/L7. Microscopic analysis revealed that surfactin variants delayed the fungal spore germination and thus suppressed the hyphae growth, mainly displayed in hyphal shriveling and distortion [28]. Surfactin is a biosurfactant produced by *B. subtilis* was first observed in 1968 in a culture broth of *B. subtilis* and was initially purified as a fibrin-clotting inhibitor. It is one of the most effective and potent lipopeptide-type biosurfactants produced by different Gram-positive and endospore-producing *B. subtilis* strains [28–30]. It is a cyclic lipopeptide characterized by a 9-hydroxycarbonic acid moiety with profound surface activities as well as antibiotic activity [31]. Cyclic lipopeptides including fengycin, iturin, lichenysin, and surfactin are among the major categories of biosurfactants secreted by *Bacillus* species [32]. Biosurfactants from *Bacillus* strains have been reported to exhibit profound antibacterial and antifungal activities, and the antimicrobial activity was increased with increasing concentration of biosurfactants. A lipopeptide biosurfactant produced by *B. natto* TK-1 showed strong antimicrobial activity against *Botrytis cinerea*, *Fusarium moniliforme*, *Micrococcus luteus* and *S. typhimurium*. The biosurfactant also significantly reduced the tumor cell viability in a dose-responsive manner [3]. Various lipopeptide biosurfactants secreted by *B. licheniformis* [33,34] and *B. subtilis* [35,36] have also revealed antimicrobial activities. For instance, antimicrobial activities of surfactin and iturin produced by *Bacillus* strains to inhibit phytopathogenic fungi have been described [34,37,38]. Standard surfactin, which was originally purified from *B. subtilis*, contained macrolide with the heptapeptide sequenced Glu-Leu-Leu-Val-Asp-Leu-Leu and a lipid fraction that consists of a mixture of several hydroxy-fatty acids (chain length of 13-15 carbon atoms) [39]. Taken together, the present results revealed that the lipopeptide surfactants produced by the *Bacillus* genus display a high perspective for biopharmaceutical and biotechnological applications owing to their biocontrol activities.

4. Conclusions

In conclusion, a Gram+ bacterial strain isolated from the commercially available natto and identified as *B. subtilis* natto has considerable inhibitory properties against *S. aureus*, *E. coli*, and *S. typhimurium*. The bactericide's secretion potential of *B. subtilis* natto was further optimized via the classical approach under different growth conditions. Through the optimization of the growth medium, its bacteriostatic performance was reaffirmed and found the highest against *S. aureus*. The previous reports have shown that *B. subtilis* natto can inhibit *E. coli* and *S. aureus*, but its antibacterial potential is weak. However, herein, we have eliminated this limited/weak activity drawback, and under-optimized environment, *B. subtilis* natto exhibited much higher inhibitory activity towards *S. aureus* with a zone of inhibition diameter up to 27 mm. The fermentation broth of *B. subtilis* natto was purified via silica gel and Sephadex column chromatography to extract bactericide baring active compounds and designated as NT-5, NT-6, and NT-7. HPLC, MS, and ^{13}C-NMR further analyzed the active compound fractions to elucidate their structural features. Comparison of the NMR data of NT-5, NT-6, and NT-7 indicated that they share the same structure except for the fatty chain. Future studies are ongoing in our lab to scrutinize the exact chemical structure and cellular toxicity of these compounds as potential candidates for biomedical purposes.

Author Contributions: Conceptualization, J.Z. (Jing Zhang) and Y.Z.; Data curation, J.Z. (Jing Zhang), H.L. (Hedong Lu), and H.S.; Formal analysis, J.Z. (Jiaheng Zhang) and S.L.; Investigation, H.L. (Hongzhen Luo); Methodology, M.B.; Project administration, Y.Z.; Supervision, Y.Z.; Validation, C.L.; Writing—original draft, M.B., H.M.N.I., and Y.Z.; Writing—review & editing, M.B., H.M.N.I., and Y.Z. All authors have read and agreed to the published version of the manuscript.

Funding: This work was financially supported by Young academic leaders in Jiangsu Province, Six talent peaks project in Jiangsu Province (2015-SWYY-026). The authors also acknowledge the support from the Postgraduate Research & Practice Innovation Program of Jiangsu Province (SJCX17_0700), a study on highly efficient biotransformation of oleic acid and linoleic acid to γ-decalactone in *Yarrowia lipolytica* based on synthetic biology (21606097), Huai'an Agricultural and Social Development Project (HAN201812), and Huai'an Special Fund Project for the Transformation of Scientific and technological achievements (HA201803). We would also like to express sincere appreciation to Professor Peter Proksch (Institute of Pharmaceutical Biology and Biotechnology, Heinrich-Heine-Universität Düsseldorf, 40225 Düsseldorf, Germany) for his assistance during the experiment.

Conflicts of Interest: The authors declare no conflict of interest.

References

1. Tsukamoto, Y.; Ichise, H.; Kakuda, H.; Yamaguchi, M. Intake of fermented soybean (natto) increases circulating vitamin K 2 (menaquinone-7) and γ-carboxylated osteocalcin concentration in normal individuals. *J. Bone Miner. Metab.* **2000**, *18*, 216–222. [CrossRef]

2. Ju, H.K.; Jun, P.Y.; Lei, Z.; Yi, P.Q. Identification of two novel fibrinolytic enzymes from Bacillus subtilis QK02. Comparative. *Biochem. Physio. Part C* **2004**, *137*, 65–74.

3. Cao, X.-H.; Liao, Z.-Y.; Wang, C.; Yang, W.-Y.; Lu, M.-F. Evaluation of a lipopeptide biosurfactant from Bacillus natto TK-1 as a potential source of anti-adhesive, antimicrobial and antitumor activities. *Braz. J. Microbiol.* **2009**, *40*, 373–379. [CrossRef]

4. Sumi, H.; Hamada, H.; Nakanishi, K.; Hiratani, H. Enhancement of the Fibrinolytic Activity in Plasma by Oral Administration of Nattokinases. *Acta Haematol.* **1990**, *84*, 139–143. [CrossRef]

5. Tam, N.K.M.; Uyen, N.Q.; Hong, H.A.; Duc, L.H.; Hoa, T.T.; Serra, C.; Henriques, A.; Cutting, S. The Intestinal Life Cycle of Bacillus subtilis and Close Relatives. *J. Bacteriol.* **2006**, *188*, 2692–2700. [CrossRef]

6. Tomohiro, H.; Rieko, H.; Shizue, S.; Akio, A.; Kan, K.; Shuichi, K. Cytokine responses of human intestinal epithelial-like Caco- 2 cells to the nonpathogenic bacterium Bacillus subtilis (natto). *Inter. J. Food. Microbiol.* **2003**, *82*, 255–264.

7. Itaya, M.; Nagasaku, M.; Shimada, T.; Ohtani, N.; Shiwa, Y.; Yoshikawa, H.; Kaneko, S.; Tomita, M.; Sato, M. Stable and efficient delivery of DNA toBacillus subtilis(natto) using pLS20 conjugational transfer plasmids. *FEMS Microbiol. Lett.* **2019**, *366*, 032. [CrossRef] [PubMed]

8. Sumi, H.; Ohsugi, T.; Ikeda, S.; Kawahara, T.; Naito, S.; Yatagai, C. ID: 101 Anti-platelet aggregation and anti-Helicobacter pylori activity of dipicolinic acid contained in Bacillus subtilis natto. *J. Thromb. Haemost.* **2006**, *4*, 217. [CrossRef]

9. Zhao, X.; Zhou, Z.-J.; Han, Y.; Wang, Z.-Z.; Fan, J.; Xiao, H.-Z. Isolation and identification of antifungal peptides from Bacillus BH072, a novel bacterium isolated from honey. *Microbiol. Res.* **2013**, *168*, 598–606. [CrossRef]

10. Xu, D.; Wang, Y.; Sun, L.; Liu, H.; Li, J. Inhibitory activity of a novel antibacterial peptide AMPNT-6 from Bacillus subtilis against Vibrio parahaemolyticus in shrimp. *Food Control.* **2013**, *30*, 58–61. [CrossRef]

11. Mo, H.; Zhu, Y.; Chen, Z. Microbial fermented tea—A potential source of natural food preservatives. *Trends Food Sci. Technol.* **2008**, *19*, 124–130. [CrossRef]

12. Schmidt, S.E.; Taylor, T.M.; Davidson, P.M. Chemical Preservatives and Natural Antimicrobial Compounds. *Food Microbiol.* **2013**, 765–801.

13. Goñi, P.; López, P.; Sanchez, C.; Gómez-Lus, R.; Becerril, R.; Nerín, C. Antimicrobial activity in the vapour phase of a combination of cinnamon and clove essential oils. *Food Chem.* **2009**, *116*, 982–989. [CrossRef]

14. Andrews, J.M. Determination of minimum inhibitory concentrations. *J. Antimicrob. Chemother.* **2001**, *48*, 5–16. [CrossRef]

15. Fazeli, M.R.; Amin, G.R.; Ahmadian-Attari, M.M.; Ashtiani, H.; Jamalifar, H.; Samadi, N. Antimicrobial activities of Iranian sumac and avishan-e shirazi (Zataria multiflora) against some food-borne bacteria. *Food Control.* **2007**, *18*, 646–649. [CrossRef]

16. Vaquero, M.R.; Alberto, M.; De Nadra, M.C.M. Antibacterial effect of phenolic compounds from different wines. *Food Control.* **2007**, *18*, 93–101. [CrossRef]

17. Pang, X.; Zhao, J.; Fang, X.; Liu, H.; Zhang, Y.; Cen, S.; Yu, L. Surfactin derivatives from Micromonospora sp. CPCC 202787 and their anti-HIV activities. *J. Antibiot.* **2016**, *70*, 105–108. [CrossRef]

18. Raaijmakers, J.M.; Vlami, M.; De Souza, J.T. Antibiotic production by bacterial biocontrol agents. *Antonie Leeuwenhoek* **2002**, *81*, 537–547. [CrossRef]

19. Asthana, R.K.; Deepali; Tripathi, M.K.; Srivastava, A.; Singh, A.P.; Singh, S.P.; Nath, G.; Srivastava, R.; Srivastava, B.S. Isolation and identification of a new antibacterial entity from the Antarctic cyanobacterium Nostoc CCC 537. *Environ. Boil. Fishes* **2008**, *21*, 81–88.

20. Yang, L.; Quan, X.; Xue, B.; Goodwin, P.H.; Lu, S.; Wang, J.; Du, W.; Wu, C. Isolation and identification of Bacillus subtilis strain YB-05 and its antifungal substances showing antagonism against Gaeumannomyces graminis var. tritici. *Boil. Control.* **2015**, *85*, 52–58. [CrossRef]

21. Bilal, M.; Rasheed, T.; Iqbal, H.M.; Hu, H.; Zhang, X. Silver Nanoparticles: Biosynthesis and Antimicrobial Potentialities. *Int. J. Pharmacol.* **2017**, *13*, 832–845. [CrossRef]

22. Bilal, M.; Rasheed, T.; Iqbal, H.M.; Hu, H.; Wang, W.; Zhang, X. Macromolecular agents with antimicrobial potentialities: A drive to combat antimicrobial resistance. *Int. J. Boil. Macromol.* **2017**, *103*, 554–574. [CrossRef]

23. Rasheed, T.; Bilal, M.; Iqbal, H.M.; Li, C. Green biosynthesis of silver nanoparticles using leaves extract of Artemisia vulgaris and their potential biomedical applications. *Colloids Surfaces B Biointerfaces* **2017**, *158*, 408–415. [CrossRef] [PubMed]

24. Rasheed, T.; Bilal, M.; Li, C.; Iqbal, H.M. Biomedical Potentialities of Taraxacum officinale-based Nanoparticles Biosynthesized Using Methanolic Leaf Extract. *Curr. Pharm. Biotechnol.* **2017**, *18*, 1116–1123. [CrossRef] [PubMed]

25. Tabbene, O.; Benslimen, I.; Djebali, K.; Mangoni, M.-L.; Urdaci, M.C.; Limam, F. Optimization of medium composition for the production of antimicrobial activity by Bacillus subtilis B38. *Biotechnol. Prog.* **2009**, *25*, 1267–1274. [CrossRef]

26. Bais, H.P.; Fall, R.; Vivanco, J.M. Biocontrol of Bacillus subtilis against Infection of Arabidopsis Roots by Pseudomonas syringae Is Facilitated by Biofilm Formation and Surfactin Production1. *Plant Physiol.* **2004**, *134*, 307–319. [CrossRef]

27. Caldeira, A.T.; Feio, S.; Arteiro, J.; Coelho, A.V.; Roseiro, J. Environmental dynamics of Bacillus amyloliquefaciens CCMI 1051 antifungal activity under different nitrogen patterns. *J. Appl. Microbiol.* **2008**, *104*, 808–816. [CrossRef] [PubMed]

28. Wang, Q.-Y.; Lin, Q.; Peng, K.; Cao, J.-Z.; Yang, C.; Xu, N. Surfactin Variants from Bacillus subtilis natto CSUF5 and Their Antifungal Properities Against Aspergillus niger. *J. Biobased Mater. Bioenergy* **2017**, *11*, 210–215. [CrossRef]

29. Wei, Y.; Wang, L.; Chang, J.-S. Optimizing Iron Supplement Strategies for Enhanced Surfactin Production with Bacillus subtilis. *Biotechnol. Prog.* **2004**, *20*, 979–983. [CrossRef]

30. Yeh, M.-S.; Wei, Y.-H.; Chang, J.-S. Enhanced Production of Surfactin from Bacillussubtilis by Addition of Solid Carriers. *Biotechnol. Prog.* **2008**, *21*, 1329–1334. [CrossRef]

31. Cooper, D.G.; Macdonald, C.R.; Duff, S.J.B.; Kosaric, N. Enhanced Production of Surfactin from Bacillus subtilis by Continuous Product Removal and Metal Cation Additions. *Appl. Environ. Microbiol.* **1981**, *42*, 408–412. [CrossRef] [PubMed]

32. Das, P.; Mukherjee, S.; Sen, R. Genetic Regulations of the Biosynthesis of Microbial Surfactants: An Overview. *Biotechnol. Genet. Eng. Rev.* **2008**, *25*, 165–186. [CrossRef] [PubMed]

33. Yakimov, M.M.; Timmis, K.N.; Wray, V.; Fredrickson, H.L. Characterization of a new lipopeptide surfactant produced by thermotolerant and halotolerant subsurface Bacillus licheniformis BAS50. *Appl. Environ. Microbiol.* **1995**, *61*, 1706–1713. [CrossRef] [PubMed]

34. Hiradate, S.; Yoshida, S.; Sugie, H.; Yada, H.; Fujii, Y. Mulberry anthracnose antagonists (iturins) produced by Bacillus amyloliquefaciens RC-2. *Phytochemistry* **2002**, *61*, 693–698. [CrossRef]

35. Sandrin, C.; Peypoux, F.; Michel, G. Coproduction of surfactin and iturin A, lipopeptides with surfactant and antifungal properties, by Bacillus subtilis. *Biotechnol. Appl. Biochem.* **1990**, *12*, 370–375.

36. Vollenbroich, D.; Pauli, G.; Ozel, M.; Vater, J. Antimycoplasma properties and application on cell cultures of surfactin, a lipopeptide antibiotic from Bacillus subtilis. *Appl. Environ. Microbiol.* **1997**, *63*, 44–49. [CrossRef]

37. Yu, G.; Sinclair, J.; Hartman, G.; Bertagnolli, B. Production of iturin A by Bacillus amyloliquefaciens suppressing Rhizoctonia solani. *Soil Boil. Biochem.* **2002**, *34*, 955–963. [CrossRef]

38. Cho, S.-J.; Lee, S.K.; Cha, B.J.; Kim, Y.H.; Shin, K.-S. Detection and characterization of the Gloeosporium gloeosporioides growth inhibitory compound iturin A from Bacillus subtilis strain KS03. *FEMS Microbiol. Lett.* **2003**, *223*, 47–51. [CrossRef]

39. Lim, J.-H.; Park, B.-K.; Kim, M.-S.; Hwang, M.-H.; Rhee, M.H.; Park, S.-C.; Yun, H.-I. The anti-thrombotic activity of surfactins. *J. Veter Sci.* **2005**, *6*, 353–355. [CrossRef]

Article

Effect of Nitrate and Perchlorate on Selenate Reduction in a Sequencing Batch Reactor

Hyun-Woo Kim [1], Seong Hwan Hong [2] and Hyeoksun Choi [3,*]

[1] Department of Environmental Engineering, Soil Environment Research Center, Jeonbuk National University, Jeonju 54896, Korea; hyunwoo@jbnu.ac.kr
[2] Gangto Engineering Co. Ltd., Anyang 14058, Korea; hmh5hot@naver.com
[3] Department of Civil and Environmental Engineering, Wonkwang University, Iksan 54538, Korea
* Correspondence: choihs@wku.ac.kr; Tel.: +82-63-850-6713

Received: 12 February 2020; Accepted: 13 March 2020; Published: 16 March 2020

Abstract: Selenate removal from a water body is being vigorously debated owing to severe health impact, but inhibitions of coexisting anions have been reported. To suggest a viable treatment option, this study investigates the effect of nitrate and perchlorate on selenate reduction in a laboratory-scale sequencing batch reactor. The experimental design tests how competing electron acceptors (NO_3^- and ClO_4^-) and electron donor (acetate) limitations affect selenate reduction in the reactor. Results show that the reactor achieves almost complete selenate reduction within the initial concentration ranges of 0.1–1 mM by enriching selenate-reducing bacteria with appropriate temperature (30 °C) and acclimation period (50 days). We monitored simultaneous selenate and nitrate reduction in the reactor without specific inhibition due to a difference in microbial growth strategy related to electron donor status. Lack of perchlorate-reducing bacteria makes perchlorate addition (0.2 mM) not to be closely associated with dissimilative perchlorate reduction. These results provide information that can help us to understand the effect of competing electron acceptors on selenate reduction and the kinetics of potential parallel reactions in the reactor.

Keywords: biological selenate reduction; electron donor competition; nitrate; perchlorate; sequencing batch

---

1. Introduction

Selenium (Se) is an essential micronutrient but can cause adverse health effects (e.g. hair loss, fingernail loss, numbness in fingers or toes, and circulatory problems) with long-term and heavy exposure [1,2]. Since Se in water originates from not only geological sources such as weathering of seleniferous soils/rocks but also anthropogenic processes such as mining, fossil fuel combustion, and other industrial activities [3], the World Health Organization has set a provisional total Se guideline of 40 µg/L in drinking water [4]. The United States Environmental Protection Agency permits the maximum concentration limit (MCL) of total Se as 50 µg/L and the regulations of national primary drinking water as 5 µg/L [2]. Likewise, the Korean Ministry of Environment is reducing the MCL to 10 µgSe/L in drinking water [5].

Se has four oxidation states (−II, 0, IV, VI) and forms several organic complexes [6]. In surface water, most Se primarily exists either selenate (SeO_4^{2-}) or selenite (SeO_3^{2-}). Both oxyanions are toxic to living organisms thus various treatment technologies have been investigated to remove Se from water [7]. Although physicochemical technologies effectively separate Se from the water supplied for domestic and industrial use, eventual post-treatments for the byproducts are required and technical limitations are still existing [8]. Fortunately, biological treatment can reduce selenate and selenite to insoluble elemental Se (Se^0) via anaerobic microbial metabolisms [6,9]. From a wide variety of environments, selenate- or selenite-reducing bacteria have been isolated [10,11].

When selenate or selenite coexist with other anions such as nitrate (NO_3^-), sulfate (SO_4^{2-}), and perchlorate (ClO_4^-), biological Se reduction can be inhibited by the electron scavenging of denitrifying bacteria, sulfate-reducing bacteria, or perchlorate-reducing bacteria because most selenate-reducing bacteria are heterotrophic facultative anaerobes which compete for electron donors under anoxic or anaerobic conditions. Another limiting factor might be the drastic change of selenate in the water body due to irrigated agricultural drainage [11], sedimentary soil erosion [3], surface mining [12], coal-fired power plants [13], and so on.

Most biological selenate reductions are targeted for either pure culture or up-flow anaerobic sludge blanket process [14]. Relatively little reports are available about the simultaneous reduction of selenate in a mixed culture when competing anions exist [14–17]. This study, therefore, investigates the feasibility of simultaneous selenate, nitrate, and perchlorate reduction in a sequencing batch reactor (SBR) and evaluate the inhibitory effects of nitrate and perchlorate on biological selenate reduction.

2. Materials and Methods

2.1. Selective Enrichment of Selenate-Reducing Bacteria

To selectively enrich selenate-reducing bacteria, bench-scale SBRs were semi-continuously operated in parallel for more than one and a half months. Seed sludge was activated sludge taken from a local municipal wastewater treatment plant with a treatment capacity of 30,000 m^3/d in the northern part of I-city, Korea. Using selenate as a sole electron acceptor, the enrichment period was kept under anoxic conditions. To support selective pressure on selenate-reducing bacteria, the temperature was controlled to 30 °C by aquarium heaters following previous literature [18].

2.2. Operating Condition of SBRs

Figure 1 shows the schematic diagram of the triplicate SBRs. The working volume of each SBR was 5 L. To verify the proper temperature condition (25 °C and 30 °C), SBRs were continuously monitored for more than 200 h until complete selenate reduction at the first batch. And then all the reactors were operated with 24 h sequence with the optimal temperature condition using the pre-acclimated biomass for 30 °C. Each SBR was completely mixed for 23 h. And then, an hour of settling period followed by rapid draw sequence of the upper liquid (2.5 L) and fill sequence with fresh feed solution. The feed solution contains selenate, acetate (CH_3COO^-), buffer, and essential minerals: 50 mg/L of SeO_4^{2-}, 200 mg/L of CH_3COO^-, 46 mg/L of $(NH_4)_2SO_4$, 13.7 mg/L of K_2HPO_4, 84 mg/L of $NaHCO_3$, 51.3 mg/L of $MgSO_4 \cdot 7H_2O$, 43 mg/L of $CaSO_4 \cdot 2H_2O$, and 2.5 mg/L of $FeSO_4 \cdot 7H_2O$. Other micronutrients were available from inoculum and endogenous cell decay. Acetate was a sole carbon source (electron donor). To test the effects of nitrate and perchlorate on selenate reduction, we designed the experiments as shown in Table 1.

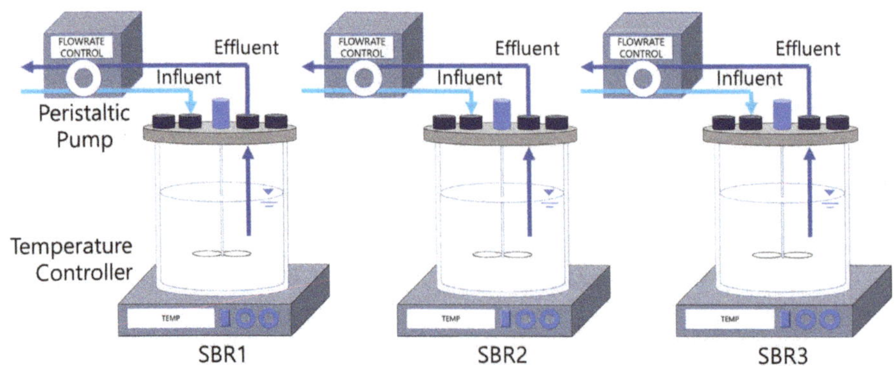

Figure 1. Schematic diagram of triplicate sequencing batch reactors (SBRs).

Table 1. Operating conditions of SBRs according to experimental design.

Division	Initial Concentration of Target Contaminants				C: N (CH₃COO⁻-C: NO₃-N)
	Selenate (mM SeO$_4^{2-}$)	Nitrate (mM NO$_3^-$)	Perchlorate (mM ClO$_4^-$)	Acetate (mM CH$_3$COO$^-$)	
Phase 0 [a]	0.35	0.00	0.0	3.4	N.A.
Phase 1	0.1	1.0	0.0	3.4	6.7: 1
Phase 2	1.0	1.0	0.0	5.1	11.1: 1
Phase 3	0.1	1.0	0.0	0.6	1.2: 1
Phase 4	1.0	1.0	0.0	0.9	2.3: 1
Phase 5	0.1	1.0	0.2	3.4	5.8: 1

[a] Initial acclimation period.

2.3. Analytical Methods

Influent and effluent liquid samples were filtered using a 0.2 µm syringe filter (Whatman, GE Healthcare Life Sciences, Marlborough, MA, USA) and kept in a refrigerator at 4 °C before analysis. Selenate was determined by using an ion chromatograph (Dionex ICX-1100, Dionex, Sunnyvale, CA, USA) equipped with an IonPac AS15 analytical column and AG15 guard column. The used eluent was a 36.5 mM NaOH solution (Daejung Chemicals, Siheung, Korea). The volume of the used sample loop for selenate determination was 100 µL. For perchlorate determination, we used the same ion chromatograph equipped with IonPac AS16 analytical column and AG16 guard column (Dionex, Thermo Fisher Scientific, Waltham, MA, USA). In this case, we used the sample loop volume of 1000 µL with the same 50 mM NaOH eluent. Nitrate and acetate concentrations were monitored by using an IonPac AS9-HC analytical and AG9-HC guard column with 9 mM Na$_2$CO$_3$ eluent and a 25 µL sample loop. The detection limits for selenate and perchlorate were 5 µg/L each. And those of acetate and nitrate were 0.5 mg/L. All the regressions for experimental data were conducted by Sigmaplot software (Systat Software Inc., San Jose, CA, USA) based on the assumption of first-order removal [19].

3. Results and Discussion

3.1. Appropriate Temperature for Selenate-Reducing Bacteria Acclimation in SBRs

To increase the activity of selenate-reducing bacteria in the seed sludge, initial acclimation (phase 0) was conducted for about 50 days using two sets of triplicate SBRs. Figure 2a shows the variations of selenate concentrations at the very first batch of the SBRs. During nine days of phase 0, only 27% of initial selenate (0.72 mM SeO$_4^{2-}$) was reduced on average at the SBRs of 25 °C. However, in the SBRs at 30 °C selenate was reduced to below detection level after nine days. This result indicates that 30 °C, higher than room temperature, is more appropriate for the growth of selenate-reducing bacteria, which is consistent with previous literature [14,20,21]. With the revealed temperature condition, all the SBRs enriched selenate-reducing bacteria at 30 °C for the rest of phase 0 for further experiments.

At the end of phase 0, monitoring results indicate that SBRs could reduce selenate (0.9 mM) to below detection level in less than four hours. This enhancement indicates that phase 0 must have made the selenate-reducing bacteria successfully acclimated to start instantaneous selenate reduction right after fill-sequence without lag-period. Figure 2b demonstrates that enriched microorganisms actively reduce selenate to Se^0 biologically at the last batch of phase 0, consistent with the literature [6,14,22,23]. Regression indicates that the observed selenate reduction rate was revealed as rapid as 0.96 h^{-1}.

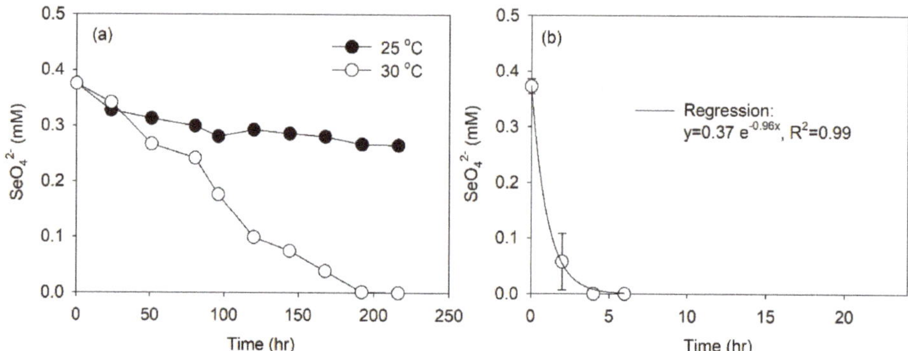

Figure 2. Dynamics of selenate concentration in SBRs of phase 0: (**a**) before acclimation, (**b**) after acclimation.

3.2. Effect of Nitrate on Selenate Reduction

At phase 1 and phase 2, this study tests the effect of most probable electron-competing anion, nitrate, on selenate reduction (Table 1). We artificially constitute low (0.1 mM, phase 1) and high (1 mM, phase 2) selenate conditions for better interpretation. Figure 3 illustrates the dynamics of average (n = 3) selenate and nitrate in a whole sequence of SBRs at a steady state. When 3.8 mM CH_3COO^- was added to 0.1 mM SeO_4^{2-} (phase 1) as an excess electron donor in the presence of 0.96 mM NO_3^--N (approximately 1:10 of influent SeO_4^{2-}: NO_3^- mole ratio), selenate and nitrate were simultaneously reduced to below detection level within six hours in SBRs (Figure 3a). In the case of phase 2, nitrate was completely reduced to below detection level, whereas a small amount of selenate was detected (0.02 mM, 98% reduction) after six hours in SBRs (Figure 3b). Close to the end of the sequence, the selenate concentration decreased to below detection level.

Within the ratio of SeO_4^{2-}: NO_3^- between 1:1 and 1:10 tested in this study, both selenate and nitrate could be simultaneously reduced without significant inhibition. The selenate reduction rate was maintained at 0.55–0.57 h^{-1} regardless of initial concentration. This result indicates that selective enrichment and long acclimation (>30 days) could make selenate-reducing bacteria endure competitive inhibition, described previously [24]. In addition, it was noticed that the denitrification rate was not interrelated with the selenate concentration and kept the rate as 0.88 h^{-1} almost constantly, which supports simultaneous selenate and nitrate reduction under excess electron donor condition.

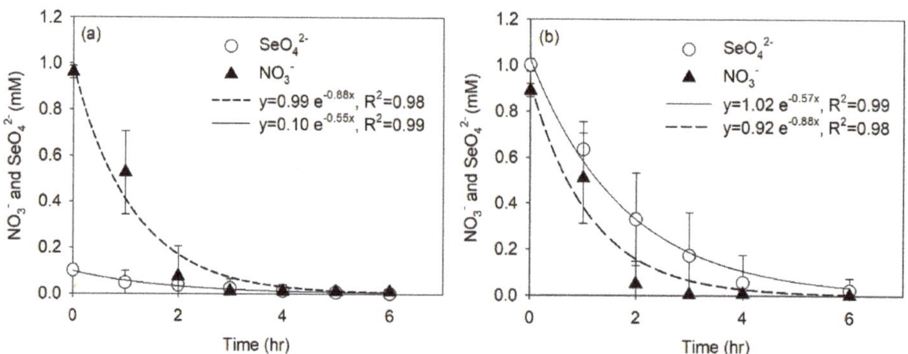

Figure 3. Dynamics of SeO_4^{2-} and NO_3^- in SBRs: (**a**) SeO_4^{2-}:NO_3^- = 1:10 (phase 1) and (**b**) SeO_4^{2-}:NO_3^- = 1:1 (phase 2).

3.3. Effect of External Carbon Limitation on Selenate Reduction

Two sets of experiments were performed to investigate the effect of carbon source limitation on simultaneous selenate and nitrate reduction in the SBRs under low (0.1 mM, phase 3) and high selenate (1 mM, phase 4) conditions. Acetate concentration was limited to 0.8 mM for phase 3 when the initial SeO_4^{2-} concentration was 0.1 mM. Keeping the nitrate concentration as 1.0 mM results in the decrease of C:N ratio from 6.7:1 to 1.2:1 compared to phase 1. Phase 4 was conducted with 1 mM of SeO_4^{2-} reducing C:N ratio from 11.1:1 (phase 2) to 2.3:1 (phase 4). Phase 3 and phase 4 were directly comparable to phase 1 and phase 2, respectively. Figure 4a,b demonstrate the variations of selenate, nitrate, and acetate concentrations in SBRs at phase 3 and 4, respectively, as described in Table 1.

Figure 4a (phase 3) shows that all the selenate was reduced instantaneously within two hours but the accompanying nitrate reduction significantly decelerates when the acetate was depleted at around 3 h. Figure 4b (phase 4) illustrates that nitrate reduction similarly stops when the acetate was depleted, but selenate reduction gradually progressed further despite the depletion of external carbon sources. This result indicates that denitrifying bacteria are more sensitive to electron donor compared to selenate-reducing bacteria. The increase of acetate concentration from 0.7 mM to 1.2 mM enhanced the nitrate reduction rate about 80% (from 0.79 hr^{-1} to 1.42 hr^{-1}) at phase 4 but the nitrate reduction rate drastically ceased as the carbon source depleted. Selenate reduction rate was also decreased by 27.4% (from 0.95 h^{-1} to 0.69 h^{-1}) possibly owing to inhibition associated with carbon source competition.

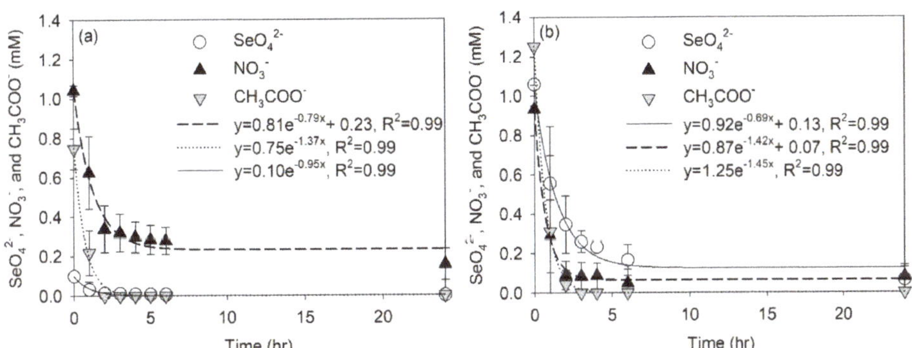

Figure 4. Dynamics of SeO_4^{2-}, NO_3^-, and CH_3COO^- under carbon limitation condition: (a) phase 3, (b) phase 4.

When the selenate and nitrates are coexisting, selenate-reducing bacteria might present the ability to compete successfully for limited carbon resources like K-strategist microorganisms [25] while nitrate-reducing bacteria exploit relative offspring trends like r-strategist microorganisms [26] in this study. This result suggests that selenate-reducing bacteria has a more competitive advantage over withstanding harsh carbon-limiting condition than nitrate-reducing bacteria.

3.4. Nitrate and Perchlorate Effect on Selenate Reduction in SBRs

To investigate the effect of another oxyanion, perchlorate, on the simultaneous selenate reduction, SBRs were operated with a feed solution containing selenate (0.1 mM), nitrate (1.0 mM), and perchlorate (0.15 mM) with an excess amount of external carbon source (3.4 mM).

Figure 5 demonstrates that selenate and nitrate reduction are not affected by perchlorate significantly. It was observed that 38% of perchlorate (reduction rate of 0.02 h^{-1}) can be reduced together with selenate and nitrate in the SBRs during 24 h of a sequence. This result indicates that dissimilatory perchlorate-reducing bacteria can grow together with selenate- and nitrate-reducing bacteria under anaerobic conditions if the carbon source (electron donor) is not limiting [27]. In this study, an insufficient population of perchlorate-reducing bacteria might have prevented the perchlorate from being a competitive inhibitor of selenate or nitrate reduction under excess electron donor conditions. Owing to perchlorate, the reduction rate of nitrate was significantly reduced from $0.9{\sim}1.4\ h^{-1}$ (excess electron donor condition) to $0.5\ h^{-1}$ at phase 5. However, that of selenate did not decline but maintained to around $0.7{-}1.3\ h^{-1}$, which evidences the ability of selenate-reducing bacteria to endure harmful perchlorate as well as electron donor competition without significant inhibition. This result also means that selenate-reducing bacteria can be dominantly enriched from activated sludge within a reasonable period of time if the carbon source is not limiting.

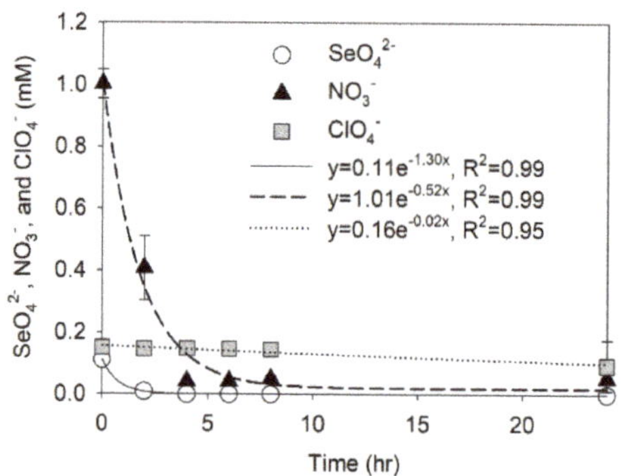

Figure 5. Dynamics of $SeO_4{}^{2-}$, $NO_3{}^-$, and $ClO_4{}^-$ with excess carbon source in SBRs.

4. Conclusions

This research provides information about how competing anions, nitrate, and perchlorate, affect selenate reduction in SBRs which are seeded with activated sludge. Based on the observed data from this research, the following conclusions are drawn as below:

(1) SBRs can rapidly enrich selenate-reducing bacteria from the activated sludge by using the selective pressure of temperature (30 °C) and sufficient acclimation period of >40 days.

(2) Complete selenate and nitrate reduction can be accomplished simultaneously in anaerobic SBRs by supplying the excess amount of electron donor. Limitation of electron donor may decrease the activity of nitrate-reducing bacteria instantaneously while selenate-reducing bacteria responds slowly using the limited resources more efficiently.

(3) Coexistence of perchlorate in the feed did not affect selenate reduction significantly owing to the shortage of dissimilatory perchlorate reducing bacteria. However, together with selenate and nitrate, 38% of perchlorate could be reduced without acclimation when electron donor is not limited.

Overall, these results evidence that selenate-reducing bacteria are capable of enduring competitions associated with other oxyanions reduction and electron donor without significant inhibition after appropriate acclimation. This study may contribute to understanding biological Se reduction better in relation to competing anions and electron donor conditions.

Author Contributions: H.C. conceived and designed the study. S.H.H. conducted the whole experiment. H.-W.K. and H.C. analyzed the data and wrote the paper. All authors have read and agreed to the published version of the manuscript.

Funding: This research was supported by Basic Science Research Program through the National Research Foundation of Korea (NRF) funded by the Ministry of Education (NRF-2016R1D1A1B03933921).

References

1. Stranges, S.; Navas-Acien, A.; Rayman, M.P.; Guallar, E. Selenium status and cardiometabolic health: state of the evidence. *Nutr. Metab. Cardiovasc. Dis.* **2010**, *20*, 754–760. [CrossRef] [PubMed]

2. USEPA. National Primary Drinking Water Regulations. Available online: https://www.epa.gov/ground-water-and-drinking-water/national-primary-drinking-water-regulations (accessed on 9 September 2019).

3. Lemly, A.D. Aquatic selenium pollution is a global environmental safety issue. *Ecotox. Environ. Safe* **2004**, *59*, 44–56. [CrossRef]

4. World Health Organization. *Guidelines for Drinking-Water Quality: Fourth Edition Incorporating the First Addendum*; World Health Organization Press: Geneva, Switzerland, 2017.

5. MOE. *Drinking Water Regulation*; Ministry of Environment: Seoul, Korea, 2019.

6. Nancharaiah, Y.V.; Lens, P.N. Selenium biomineralization for biotechnological applications. *Trends Biotechnol.* **2015**, *33*, 323–330. [CrossRef] [PubMed]

7. CH2MHill, Inc. *Review of Available Technologies for the Removal of Selenium from Water*; Selenium Working Group, North American Metals Council: Washington, WA, USA, 2010.

8. Snyder, M.M.; Um, W. Adsorption mechanisms and transport behavior between selenate and selenite on different sorbents. *Int. J. Waste Resources* **2014**, *4*, 1–8. [CrossRef]

9. Buchs, B.; Evangelou, M.W.; Winkel, L.H.; Lenz, M. Colloidal properties of nanoparticular biogenic selenium govern environmental fate and bioremediation effectiveness. *Environ. Sci. Technol.* **2013**, *47*, 2401–2407. [CrossRef]

10. Siddique, T.; Okeke, B.C.; Zhang, Y.; Arshad, M.; Han, S.K.; Frankenberger, W.T. Bacterial Diversity in Selenium Reduction of Agricultural Drainage Water Amended with Rice Straw. *J. Environ. Qual.* **2005**, *34*, 217–226. [CrossRef]

11. Zhang, Y.; Okeke, B.C.; Frankenberger, W.T. Bacterial reduction of selenate to elemental selenium utilizing molasses as a carbon source. *Bioresour. Technol.* **2008**, *99*, 1267–1273. [CrossRef]

12. Khamkhash, A.; Srivastava, V.; Ghosh, T.; Akdogan, G.; Ganguli, R.; Aggarwal, S. Mining-Related Selenium Contamination in Alaska, and the State of Current Knowledge. *Minerals* **2017**, *7*, 46. [CrossRef]

13. Tang, Q.; Liu, G.; Yan, Z.; Sun, R. Distribution and fate of environmentally sensitive elements (arsenic, mercury, stibium and selenium) in coal-fired power plants at Huainan, Anhui, China. *Fuel* **2012**, *95*, 334–339. [CrossRef]

14. Lenz, M.; Enright, A.M.; O'Flaherty, V.; van Aelst, A.C.; Lens, P.N. Bioaugmentation of UASB reactors with immobilized Sulfurospirillum barnesii for simultaneous selenate and nitrate removal. *Appl. Microbiol. Biotechnol.* **2009**, *83*, 377–388. [CrossRef]

15. Lai, C.Y.; Yang, X.; Tang, Y.; Rittmann, B.E.; Zhao, H.P. Nitrate shaped the selenate-reducing microbial community in a hydrogen-based biofilm reactor. *Environ. Sci. Technol.* **2014**, *48*, 3395–3402. [CrossRef] [PubMed]

16. Lai, C.Y.; Wen, L.L.; Shi, L.D.; Zhao, K.K.; Wang, Y.Q.; Yang, X.; Rittmann, B.E.; Zhou, C.; Tang, Y.; Zheng, P.; et al. Selenate and Nitrate Bioreductions Using Methane as the Electron Donor in a Membrane Biofilm Reactor. *Environ. Sci. Technol.* **2016**, *50*, 10179–10186. [CrossRef] [PubMed]

17. Chung, J.; Rittmann, B.E.; Wright, W.F.; Bowman, R.H. Simultaneous bio-reduction of nitrate, perchlorate, selenate, chromate, arsenate, and dibromochloropropane using a hydrogen-based membrane biofilm reactor. *Biodegradation* **2007**, *18*, 199–209. [CrossRef] [PubMed]

18. Hageman, S.P.; van der Weijden, R.D.; Weijma, J.; Buisman, C.J. Microbiological selenate to selenite conversion for selenium removal. *Water Res.* **2013**, *47*, 2118–2128. [CrossRef]

19. Fujita, M.; Ike, M.; Kashiwa, M.; Hashimoto, R.; Soda, S. Laboratory-scale continuous reactor for soluble selenium removal using selenate-reducing bacterium, Bacillus sp. SF-1. *Biotechnol. Bioeng.* **2002**, *80*, 755–761. [CrossRef]

20. Wan, H.S.; Hao, O.J.; Kim, H. Environmental factors affecting selenite reduction by a mixed culture. *J. Environ. Eng.-Asce* **2001**, *127*, 175–178. [CrossRef]

21. Fujita, M.; Ike, M.; Nishimoto, S.; Takahashi, K.; Kashiwa, M. Isolation and characterization of a novel selenate-reducing bacterium, Bacillus sp. SF-1. *J. Ferment. Bioeng.* **1997**, *83*, 517–522. [CrossRef]

22. Astratinei, V.; van Hullebusch, E.; Lens, P. Bioconversion of Selenate in Methanogenic Anaerobic Granular Sludge. *J. Environ. Qual.* **2006**, *35*, 1873–1883. [CrossRef]

23. Lenz, M.; Smit, M.; Binder, P.; van Aelst, A.C.; Lens, P.N. Biological alkylation and colloid formation of selenium in methanogenic UASB reactors. *J. Environ. Qual.* **2008**, *37*, 1691–1700. [CrossRef]

24. Kashiwa, M.; Nishimoto, S.; Takahashi, K.; Ike, M.; Fujita, M. Factors affecting soluble selenium removal by a selenate-reducing bacterium Bacillus sp. SF-1. *J. Biosci. Bioeng.* **2000**, *89*, 528–533. [CrossRef]

25. Martín-Hernández, M.; Carrera, J.; Pérez, J.; Suárez-Ojeda, M.E. Enrichment of a K-strategist microbial population able to biodegrade p-nitrophenol in a sequencing batch reactor. *Water Res.* **2009**, *43*, 3871–3883. [CrossRef] [PubMed]
26. Dytczak, M.A.; Londry, K.L.; Oleszkiewicz, J.A. Activated sludge operational regime has significant impact on the type of nitrifying community and its nitrification rates. *Water Res.* **2008**, *42*, 2320–2328. [CrossRef] [PubMed]
27. Nozawa-Inoue, M.; Scow, K.M.; Rolston, D.E. Reduction of perchlorate and nitrate by microbial communities in vadose soil. *Appl. Environ. Microbiol.* **2005**, *71*, 3928–3934. [CrossRef] [PubMed]

 *processes*

Article

Effects of Conventional Flotation Frothers on the Population of Mesophilic Microorganisms in Different Cultures

Mohammad Jafari [1], Mehdi Golzadeh [2], Sied Ziaedin Shafaei [1], Hadi Abdollahi [1,*],
Mahdi Gharabaghi [1] and Saeed Chehreh Chelgani [3,*]

[1] School of Mining Engineering, College of Engineering, University of Tehran, Tehran 16846-13114, Iran;
    mo.jafari1664@gmail.com (M.J.); zshafaie@ut.ac.ir (S.Z.S.); gharabaghi@ut.ac.ir (M.G.)
[2] Software Engineering Lab, University of Mons, 7000 Mons, Belgium; mehdi.golzadeh@umons.ac.be
[3] Minerals and Metallurgical Engineering, Department of Civil, Environmental and Natural Resources
    Engineering, Luleå University of Technology, SE-971 87 Luleå, Sweden
*   Correspondence: h_abdollahi@ut.ac.ir (H.A.); Saeed.chelgani@ltu.se (S.C.C.); Tel.:+46-073-073-7927 (S.C.C.)

Received: 1 September 2019; Accepted: 23 September 2019; Published: 25 September 2019

Abstract: Bioleaching is an environment-friendly and low-investment process for the extraction of metals from flotation concentrate. Surfactants such as collectors and frothers are widely used in the flotation process. These chemical reagents may have inhibitory effects on the activity of microorganisms through a bioleaching process; however, there is no report indicating influences of reagents on the activity of microorganisms in the mixed culture which is mostly used in the industry. In this investigation, influences of typical flotation frothers (methyl isobutyl carbinol and pine oil) in different concentrations (0.01, 0.10, and 1.00 g/L) were examined on activates of bacteria in the mesophilic mixed culture (*Acidithiobacillus ferrooxidans, Leptospirillum ferrooxidans,* and *Acidithiobacillus thiooxidans*). For comparison purposes, experiments were repeated by pure cultures of *Acidithiobacillus ferrooxidans* and *Leptospirillum ferrooxidans* in the same conditions. Results indicated that increasing the dosage of frothers has a negative correlation with bacteria activities while the mixed culture showed a lower sensitivity to the toxicity of these frothers in comparison with examined pure cultures. Outcomes showed the toxicity of Pine oil is lower than methyl isobutyl carbinol (MIBC). These results can be used for designing flotation separation procedures and to produce cleaner products for bio extraction of metals.

Keywords: flotation; bioleaching; frother; mixed culture; machine learning

1. Introduction

Pyrometallurgy and high-pressure leaching are two typical methods used for the extraction of metals from concentrates of flotation separation [1–3]. These methods have several disadvantages such as high investment and operation costs, environmental pollution (chemical reagents in the waste waters of hydrometallurgical plants and SO_2 gas generation from pyrometallurgical plants), high energy consumption in the pyrometallurgy processes, high technology requirements for pyro/hydro-metallurgy process, and finally special expertise for system operators [4,5]. Variations in the metal price have caused very intense competition among the high prestigious mining companies (Anglo American, BHP, Rio Tinto, Glencore, etc.) to revise their feasibility studies where the feasibility of mining projects has significantly depended on the project costs. Moreover, the problem of global warming and environmental pollution has led the mineral processing industry to focus on the use of low-cost, low-energy, and environmentally friendly methods [4,6–10]. Thus, several investigations have been focused on the operation and optimization of the bioleaching processes for the extraction of metals from low-grade deposits (by heap bioleaching [11,12]), waste (by columns [13,14]), and concentrates (by bioleaching tanks [13,15,16]) which considerably have lower costs and environment effects [11,16–24].

However, on one hand, few investigations studied the effects of flotation reagents on the bioleaching of sulfide flotation concentrates [25–32]. On the other hand, those studies (Table 1) are mainly focused on the cultures consisted one specific microorganism while in the industry, mixed microorganisms are mostly used for the bioleaching process [33]. Where using a mixed culture with different microorganisms can lead to the cooperative effects and bioleaching may show a higher efficiency than pure cultures [34–40].

Table 1. Investigations about the effects of flotation reagents on the bioprocess.

No.	Microorganism	Goal Metal	Process	Reagent	Description	Ref
1	*Acidithiobacillus ferrooxidans*	Copper	Bioleaching	Butyl amine	−45.0% (Cu-Recovery)	[41]
				Ethyl-xanthate	−36.7% (Cu-Recovery)	
				Isoamyl-xanthate	−20.0% (Cu-Recovery)	
				Butyl-xanthate	−11.7% (Cu-Recovery)	
2	*Acidithiobacillus ferrooxidans*	Copper	Bioleaching	Isopropyl-xanthate	−30.0% (Cu-Recovery)	[42]
		Iron (Pyrite)	Biooxidation		−50.0% (Fe-Oxidation)	
3	*Sulfolobus metallicus*	Copper	Bioleaching	Hostaflot X23	−14.0% (Cu-Recovery)	[43]
				Aero 3477	−34.0% (Cu-Recovery)	
				Flotanol C-7	−27.0% (Cu-Recovery)	
				Montanol 800	−30.0% (Cu-Recovery)	
4	*Leptospirillum ferrooxidans*	Iron (ferrous)	Biooxidation	Potassium amyl-xanthate	The inhibition effect of reagents (collector): NaEX > KAX > KIBX > KIPX > Aero3477 For frothers: MIBC > PO	[26]
				Potassium isobutyl-xanthate		
				Sodium ethyl-xanthate		
				Potassium isopropyl-xanthate		
				Dithiophosphate (Aero 3477)		
				Methyl isobutyl carbinol (MIBC)		
				Pine oil		
5	*Acidithiobacillus ferrooxidans*	Copper	Bioleaching	Isobutyl-xanthate	−53.0% (Cu-Recovery)	[29]
				Amyl-xanthate	−77.0% (Cu-Recovery)	
6	*Acidithiobacillus albertensis*	Sulfur	Biooxidation	Sodium isobutyl-xanthate	+ on the growth and S^0 oxidation	[31]
				Tween-80	+ on the growth and S^0 oxidation	
7	*Penicillium simplicissimum*	Cellulose	Decomposition	Tween-80	+11.60%	[44]
		Hemicellulose			+ 8.00%	

Table 1. *Cont.*

No.	Microorganism	Goal Metal	Process	Reagent	Description	Ref
8	*Acidithiobacillus ferrooxidans*	Iron (ferrous)	Biooxidation	Potassium amyl-xanthate	The inhibition effect of reagents (collector): KAX > KIPX > KIBX > Aero3477 > NaEX For frothers: MIBC > PO	[32]
				Potassium isobutyl-xanthate		
				Sodium ethyl-xanthate		
				Potassium isopropyl-xanthate		
				Dithiophosphate (Aero 3477)		
				MIBC		
				Pine oil		
9	*Ferroplasma*			Sodium ethyl xanthate		
10	*Acidithiobacillus ferrooxidans*			Sodium (alkyl) dithiocarbamate		
11	*Leptospirillum*	Iron (ferrous)	Biooxidation	Xanthate (Mix)	All reagents have a negative effect on the biooxidation of iron.	[45]
				Sodium ethyl-xanthate		
				Sodium n-propyl xanthate		
				Sodium isobutyl xanthate		
				Potassium amyl xanthate		
				Potassium n-butyl xanthate		
				Sodium (alkyl) dithiocarbamate		
				Sodium (alkyl) dithiocarbamate and sodium di-(alkyl) dithiophosphate		
				Isopropylthionocarbamate		
				Dithiophosphate(mixture)		
				Sodium 2-mercaptobenzthiazole		

This study investigated influences of two typical flotation frothers pine oil (PO) and methyl isobutyl carbinol (MIBC) on a population (microorganisms count) of a traditional mixed mesophilic microorganisms culture (*Acidithiobacillus ferrooxidans*, *Leptospirillum ferrooxidans*, and *Acidithiobacillus thiooxidans*). This is because there is a direct relationship between bioleaching rate (recovery of valuable metals from ores) and the population of microorganisms [46]. Three different concentrations of frothers were examined (0.01, 0.10, and 1.00 g/L). For comparison purposes, outcomes were compared with results of the same conditionings on pure cultures of *Acidithiobacillus ferrooxidans* and *Leptospirillum ferrooxidans*. Various parameters were measured in the control tests: pH, ORP (oxidation-reduction potential), total iron (Fe^T) in the solution, and DO (dissolved oxygen in the media). Mutual Information (MI) assisted by Pearson correlation was used to explore the relationship among these measured variables and select the most important parameters for further assessments. Outputs of this investigation can be used for mineral processing plants which flotation separation is their main beneficiation method to design ambient conditions. This method helps to produce cleaner products by a leaching tank. The most efficient route for processing of flotation concentrate is a leaching tank where it can process high-grade feeds and it has a high process recovery [15], for the downstream processes and environment.

2. Materials and Methods

2.1. Bacterial Strain and Growth Conditions

Pure strains of *Acidithiobacillus ferrooxidans* (T.f), *Leptospirillum ferrooxidans* (L.f), and *Acidithiobacillus thiooxidans* (T.t) which have different oxidation abilities (Table 2) were obtained from the research and development center of Sarcheshmeh mine, Kerman, Iran. Microorganism strains were cultivated in the environment presented in Table 3. 5 cc of each pure culture was selected to build the mixed culture.

Table 2. Oxidation ability of microorganisms.

Microorganism	$S^0 \rightarrow SO_4$	$Fe^{II} \rightarrow Fe^{III}$
Acidithiobacillus ferrooxidans (T.f)	√	√
Leptospirillum ferrooxidans (L.f)	-	√
Acidithiobacillus thiooxidans (T.f)	√	-

Table 3. Cultivating environment parameters.

(NH$_4$)$_2$SO$_4$	MgSO$_4$ ·7H$_2$O	9K Culture K$_2$HPO$_4$	KCL	Ca (NO$_3$)$_2$ ·H$_2$O	pH [1]	FeSO$_4$ ·7H$_2$O	S^0	Incubation Temperature	Rotation Speed
3.00 g/L	0.50 g/L	0.50 g/L	1.00 g/L	0.01 g/L	1.80	44.22 g/L	10.00 g/L	34.00 °C	140 rpm

[1] pH adjusted by 98% acid sulfuric.

2.2. Flotation Reagents

Flotation frothers (MIBC and PO) were prepared in the mineral processing laboratory at the University of Tehran, Iran. A wide range of their concentrations which are common in the various flotation plants (0.01, 0.10, and 1.00 g/L) was used and their influences were explored by different analyses.

2.3. Experimental Procedure

Twenty-one experiments, nine tests for each frother (three different dosages and three different cultures) and one control test (without frother) for each culture, were conducted. To do experiments, microorganisms were cultivated in a 9K medium containing five different mineral salts ((NH$_4$)$_2$SO$_4$: 3 g/L, MgSO$_4$·7H$_2$O: 0.5 g/L, K$_2$HPO$_4$: 0.5 g/L, KCl: 1 g/L, and Ca (NO$_3$)$_2$·H$_2$O: 0.01 g/L). The initial pH of the media was adjusted to 1.8 with H$_2$SO$_4$. As a source of energy 44.22 g/L FeSO$_4$·7H$_2$O and 10 g/L sulfur were added to the media. Incubation was performed at 34 °C in an incubator shaker having the rotation speed of 140 rpm. Effects of MIBC and PO on microorganisms count and their activities in the various cultures are assessed by measuring different parameters (pH, ORP, Fe^T, and DO) (Table 4). All the cultures were monitored and the mentioned parameters were measured for 21 days. To save time and cost, mutual information and Pearson correlation were used for the feature selection (FS). FS indicates the relationship among parameters and can be used to rank them.

Table 4. Methods for measuring of various parameters.

Parameter	Definition
pH ORP	The pH and ORP value of tests were measured by pH-ORP analyzer (Mettler Toledo).
DO	An oxygen-meter (Model JENWEY) was used to measure the amount of dissolved oxygen in the media.
Fe^T	The amounts of Fe^T were determined by atomic adsorption Spectro-photometer (AAS).
Count	The bacterial number (growth) was determined by using a Neubauer lam and 100 × magnification under a Zeiss biological microscope (Bacterial count per mL = N × 400 × 10^4), it could be indirect evidence of cell activity and cannot capture the non-culturable cells.

2.4. Feature Selection

Feature or variable selection is used to select the most effective variables on specific responses. It assists to optimize the number of variables which typically have to be measured during a process, reduce the number of parameters, and to save cost and time. In other words, collinearity may lead to measuring various parameters that show the same concept [47–49]. Therefore, FS was used through the value of measured parameters (pH, ORP, Fe^T, and DO) in the control tests to find the most effective parameters on microorganisms count (MC). The selected parameters were used as indicative factors for further assessments.

2.4.1. Pearson Correlation

Pearson correlation (r) categorizes the magnitude and value of the linear relationship between two variables. "r" statistically determines the strength of a correlation and donates negative values ($-1 \leq r < 0$) when by increasing one variable another one decreases and positive values ($0 < r \leq 1$) when they have the same orientation. "r" close to 0 means there is no relationship [50,51]. Pearson correlation was used to explore linear correlations between the measured parameters (pH, ORP, Fe^T, and DO) in the control tests through 21 days of monitoring.

2.4.2. Mutual Information

Mutual information (MI) is a unique method which can determine both the linear and nonlinear correlation between variables. MI between two variables (x;y) is non-negative and is defined as:

$$\text{MI}(x;y) = \sum_{y \in R} \sum_{x \in S} p(x,\ y) \log_2 \frac{p\ (x,\ y)}{p(x)p(y)}, \tag{1}$$

where $p(x)$ and $p(y)$ are probability density functions and $p(x,y)$ means the joint probability of a given stimulus.

3. Results

3.1. Control Test

Exploring MC in three different cultures and in the absence of frothers (Figure 1a) shows that the MC is increasing during 21 day activities, and the $\frac{MC \text{ after 21 days}}{Initial\ MC}$ ratio for the mixed culture is higher than T.f and L.f cultures ($\frac{6.4}{0.66}$ vs. $\frac{6.8}{0.78}$ and $\frac{3.2}{0.76}$, respectively). These results indicate that the MC grows faster in the mixed culture than two other cultures and/or bacteria may have higher activities in the mixed culture. Figure 1b shows that the $\frac{ORP\ after\ 21\ days}{Initial\ ORP}$ ratio is higher for the mixed culture than T.f and L.f cultures ($\frac{680}{335}$ vs. $\frac{652}{408}$ and $\frac{675}{376}$, respectively). The mixed culture has the lowest DO and Fe^T while T.f culture has the lowest pH through assessments (Figure 1c–e). These results show correlations among these measured variables (pH, ORP, Fe^T, and DO). Statistical analyses were used to do variable importance measurement (VIM) and to select the most representative parameters for further evaluations.

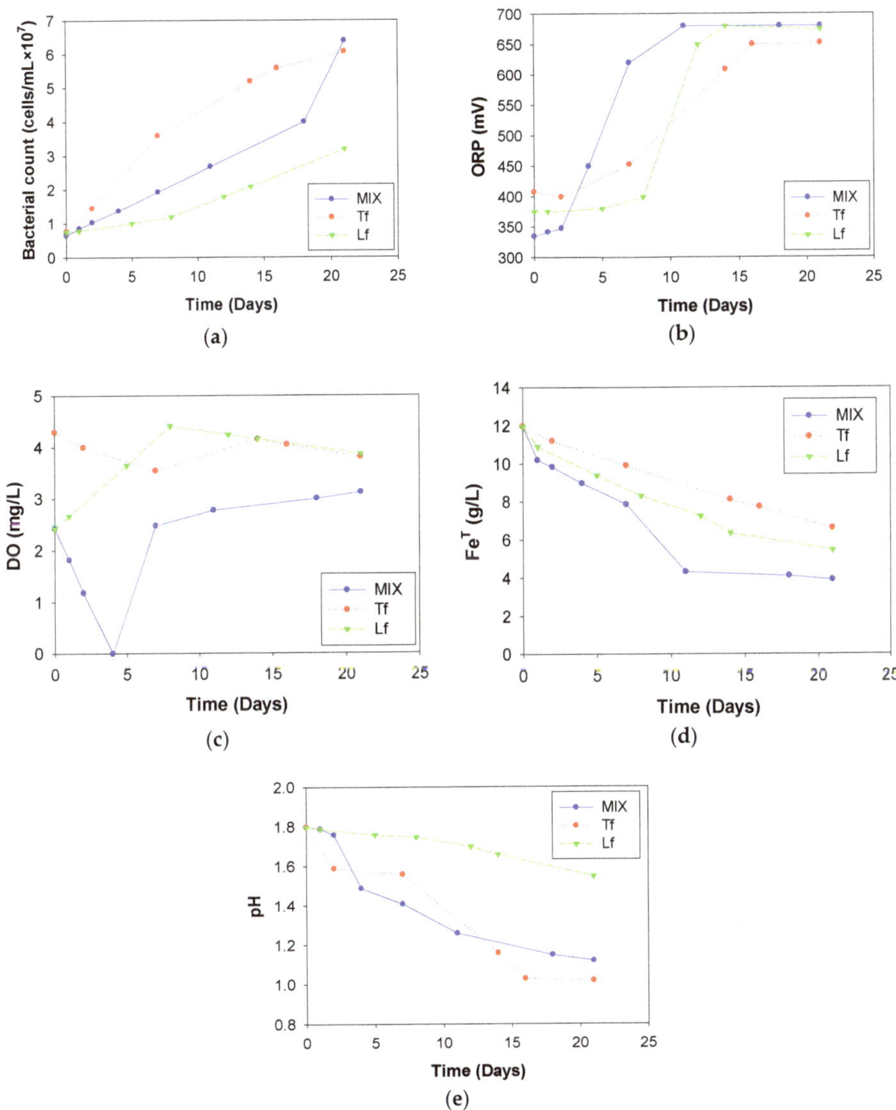

Figure 1. Exploring variation of process parameters in three different cultures for control tests during 21 days monitoring. (**a**) Bacterial count; (**b**) ORP; (**c**) DO; (**d**) Fe^T; (**e**) pH.

Feature Selection

Exploring linear relationships by Pearson correlation through various measured parameters in three different cultures over 21 days indicate that the pH and Fe^T have the highest negative "r" value with MC (Figure 2) while DO shows an insignificant correlation. In other words, when pH and Fe^T decrease, MC increases (Figure 2). Moreover, Pearson correlation shows a high relationship between ORP and Fe^T. MI was used to explore nonlinear relationships between parameters and rank them based on their importance. If the MI was close to 1, it means there is a high correlation between X and Y, and a value close to 0 means there is no relationship. MI can be used to rank variables based on their effectiveness on a dependent variable and rank independent variables based on their importance (VIM) [52]. In this study, MI was used to rank the measured parameters (pH, ORP, Fe^T, and DO) in the control tests and rank them based on their effects that may receive from the MC value (VIM). Using VIM by MI and Pearson correlation together provides a direct determination to decide whether to add an additional variable for assessments or not. MI results (Figure 3) illustrate that pH and Fe^T receive the highest effectiveness from MC among all measured parameters. Thus, these two parameters are selected to study the effect of the conventional flotation frothers (MIBC and PO) on the different cultures.

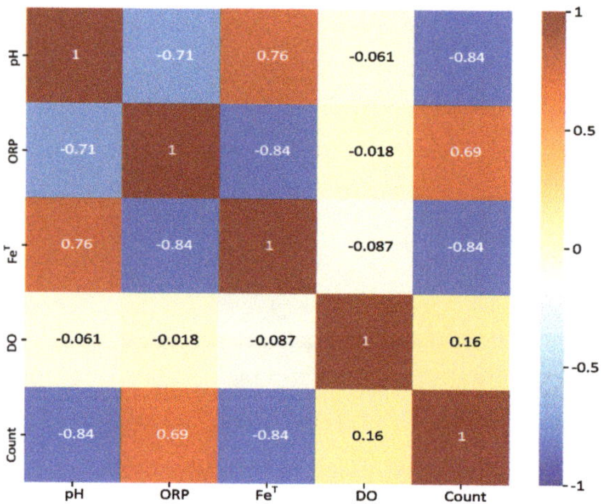

Figure 2. Pearson correlations between the measured parameters in the control tests.

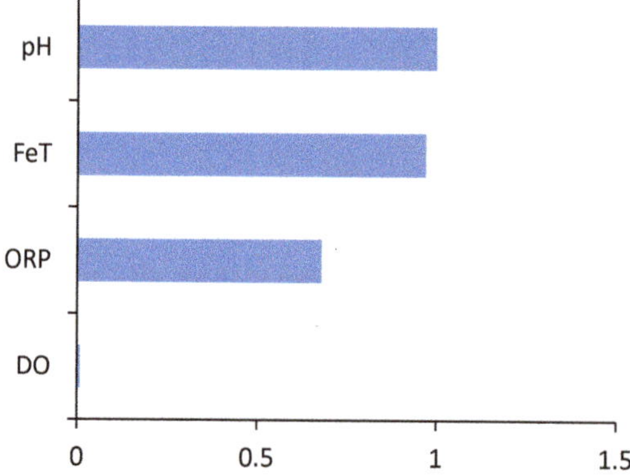

Figure 3. Ranking effectiveness of the measured parameters on microorganism population by Mutual Information.

3.2. Frothers

3.2.1. Population of Microorganisms

A comparison between the population ratio of microorganisms ($\frac{MC\ after\ 21\ days}{Initial\ MC}$) in the absence (control tests) with the presence of frothers indicates (Table 5) that MIBC and PO reduce the MC during the process. In other words, generally by increasing the frother dosages the MC is decreased. This decrease in the highest examined dosage (1 g/L) was considerably higher than the other dosages while in the case of L.f and mixed culture the population is even lower than the initial day (MC ratio < 1) (Table 5). In the presence of frothers and their different dosages, the MC ratio has the following order: T.f > mixed > L.f. In general, the MC ratio is higher in the presence of PO compering with MIBC.

Table 5. The $\frac{MC\ after\ 21\ days}{Initial\ MC}$ ratio in various conditions.

		MIBC (g/L)		
Culture	Control	0.01	0.1	1
T.f	7.8	5.4	6.3	3.1
L.f	4.2	2.4	2.1	0.5
Mixed	9.7	5.9	1.2	0.6
		PO (g/L)		
Culture	Control	0.01	0.1	1
T.f	7.8	6.2	7.6	3.9
L.f	4.2	5.3	3.7	0.8
Mixed	9.7	6	8.5	1.8

3.2.2. Fe Total

Figure 4 shows the negative relationship between Fe^T and MC for three different cultures in all tests where by increasing bacteria population the Fe^T is decreasing. In general, by increasing the dosages of frothers, by stopping the growth MC, the amount of Fe^T in the solution remains high through the process (Figure 4). Furthermore, these results illustrate that, after 21-day measurement, the amount of Fe^T in the solution for T.f culture is higher than two other cultures since there is a moderate slope of reduction between MC and Fe^T for T.f culture in all experiments. The mixed culture generally shows the highest Fe^T reduction in the solution compared with two other cultures and the Fe^T reduction ratio during the process has the following order: mixed > T.f > L.f. The Fe^T in the solution is approximately lower in the presence of MC.

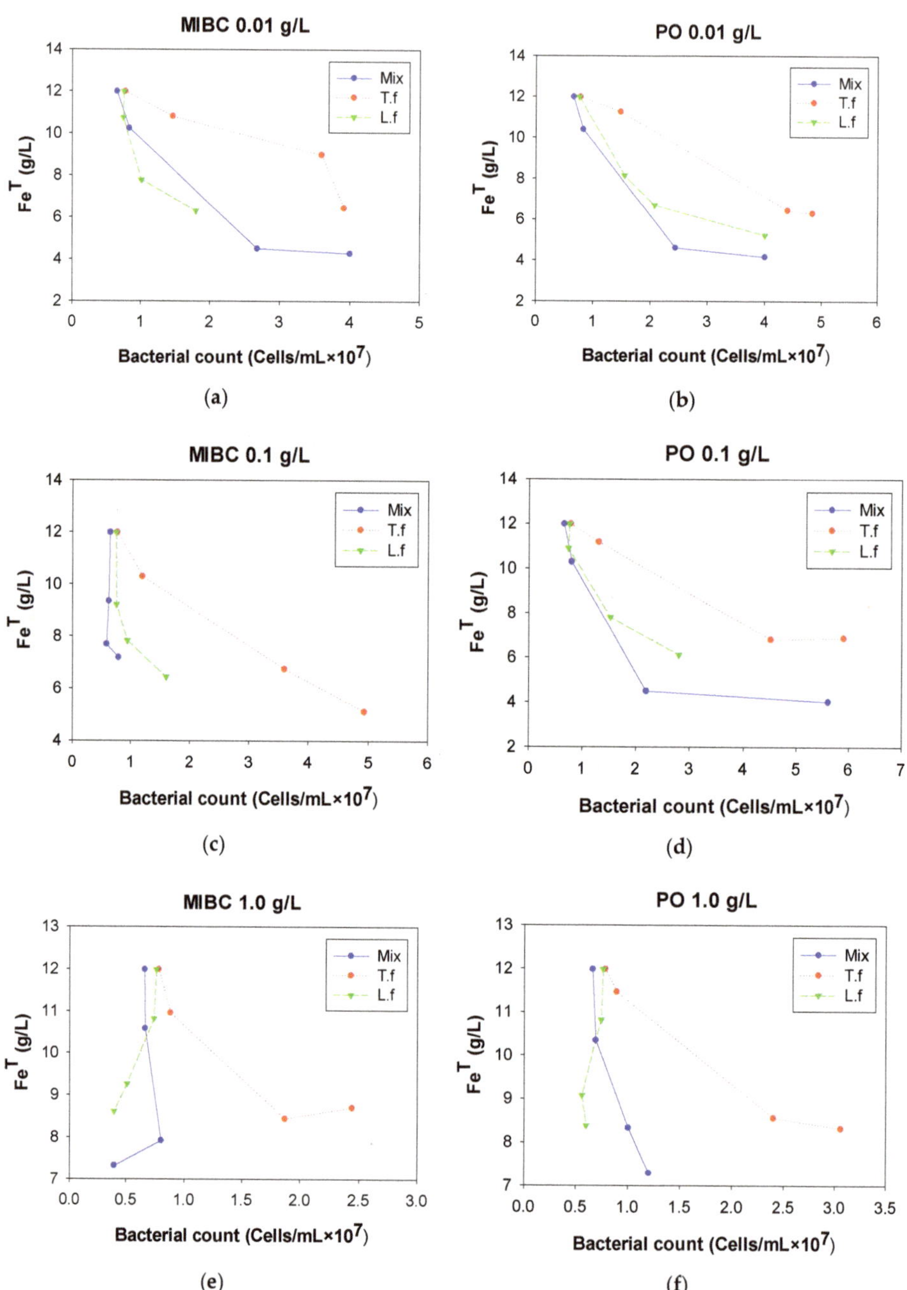

Figure 4. *Cont.*

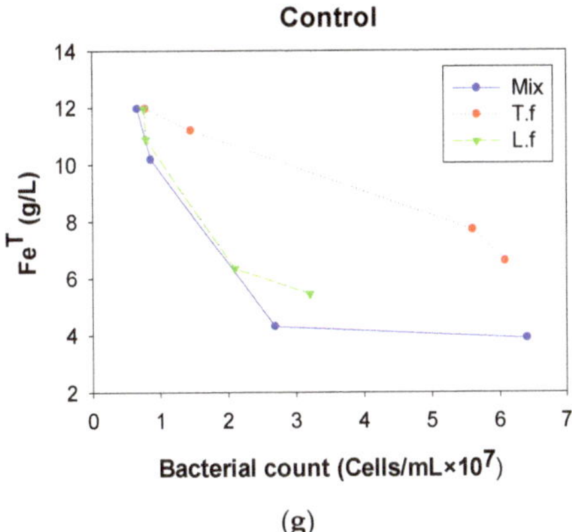

(**g**)

Figure 4. Relationship between population of bacteria and Fe total in different conditions. (**a**) 0.01 g/L MIBC; (**b**) 0.01 g/L PO; (**c**) 0.1 g/L MIBC; (**d**) 0.1 g/L PO; (**e**) 1 g/L MIBC; (**f**) 1 g/L PO; (**g**) Control test.

3.2.3. pH

Figure 5 illustrates the negative relationship between pH and MC for three different cultures in all conditions where by increasing bacteria population (MC) the pH value is decreasing. In other words, by increasing MC and as a result of their activities, the pH value is reducing. In general, the rate of pH reduction is decreased by increasing the dosages of frothers (Figure 5). These results indicate that L.f has the highest and T.f has the lowest pH value during the process monitoring (L.f > mixed > T.f). In the presence of PO, the pH reduction is moderately continuous for all cultures while in the presence of MIBC (above 0.01 g/L), the pH reduction is only detectable for the T.f culture (Figure 5).

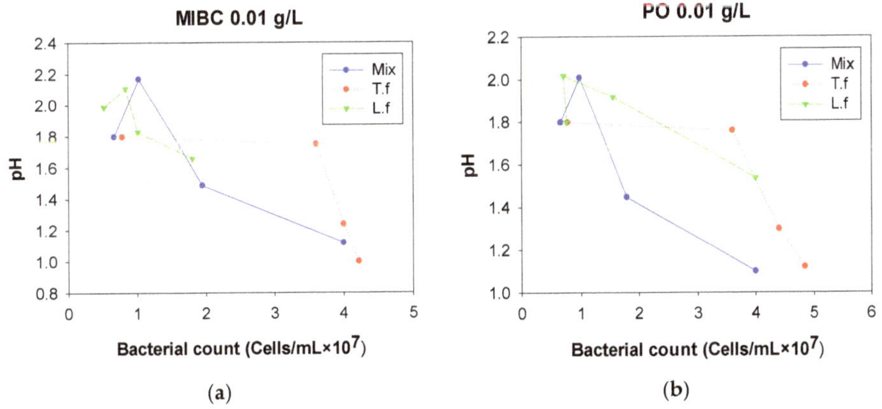

(**a**)　　　　　　　　　　　　　　(**b**)

Figure 5. *Cont.*

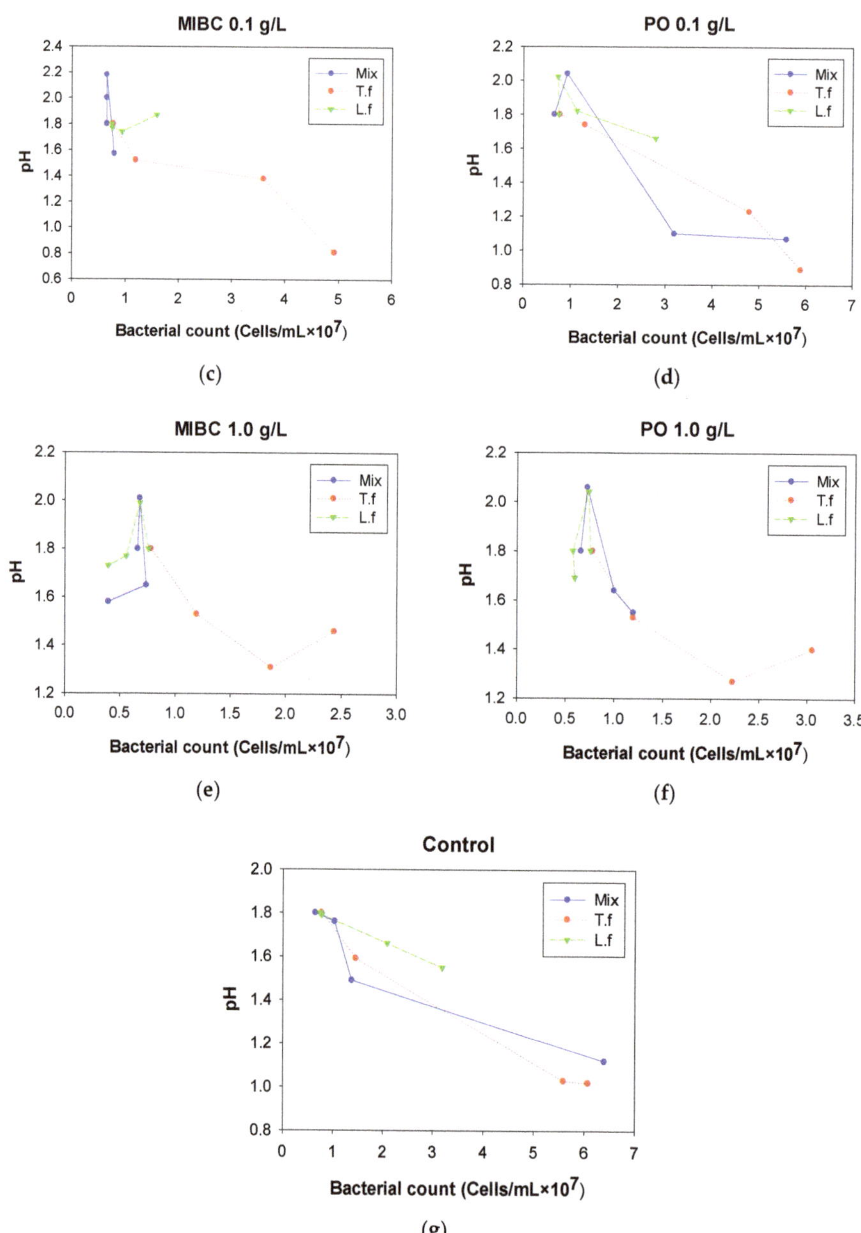

Figure 5. Relationship between population of bacteria and pH in different conditions. (**a**) 0.01 g/L MIBC; (**b**) 0.01 g/L PO; (**c**) 0.1 g/L MIBC; (**d**) 0.1 g/L PO; (**e**) 1 g/L MIBC; (**f**) 1 g/L PO; (**g**) Control test.

4. Discussion

Oxidizing metal sulfides to sulfate via contact between bacteria and mineral (direct) and oxidizing Fe^{2+} to Fe^{3+} or/and S^{o} to SO_4 (without contact: indirect) are the main mechanisms of metal extraction in the bioleaching process [53–59]. Moreover, it was well understood that oxidation of Fe^{2+} to Fe^{3+}

during bioprocess decreases pH values and Fe^T in the solution (precipitation of iron as jarosite and other iron oxides/hydroxides: Equations (2)–(4)) [60–62]. These phenomena lead to both direct and indirect bioleaching mechanisms [63–69]. Moreover, increasing bacteria population (MC) plays a fundamental role in the bioleaching process, mostly affecting pH and Fe^T. Thus, there should be negative correlations between MC-pH, and MC-Fe^T during bio-activities (Figures 2 and 3). On the other hand, there should be also a positive relationship between pH and Fe^T (Figure 2). Akinci et al. demonstrated that the rate of pH reduction during bioleaching in different bacterial cultures have a decreasing order as follows: *A. thiooxidans* > mixed culture > *A. ferooxidans* [70] that supports outcomes presented in Table 5.

$$Fe^{3+} + 2H_2O \rightarrow FeOOH \downarrow +3H^+, \tag{2}$$

$$Fe^{3+} + 3H_2O \rightarrow Fe(OH)_3 \downarrow +3H^+, \tag{3}$$

$$3Fe^{3+} + M^+ + 2HSO_4^- + 6H_2O \rightarrow MFe_3(SO_4)_2(OH)_6 \downarrow +8H^+. \tag{4}$$

The toxicity of flotation reagents is well documented and the presence of their substrates in the flotation products indicated many environmental issues which may inactivate bacteria metabolism [71–73]. Thus, frothers can change the surface properties of energy resources in the culture, limit the surface tension of the media, and inhibit microorganism activities [74,75]. The toxicity of flotation reagents depends on their chemical composition, and their dosages [25–27,32,76]. When flotation concentrate of sulfides is subjected for metal extraction via bioleaching, presence of frothers in the solution can increase the pH at the initial stage of the process [28,29,43,77]. Loon and Madgwick reported that flotation reagents reduced the bacteria growth and limited the formation of soluble iron in the bioleaching process. Since MIBC and PO are unstable at pH below 3, therefore, they may consume H^+ from the solution, decompose, and increase the pH. Thus, presented results in Figures 4 and 5 are in good agreement with the literature where by increasing the dosages of these frothers, the rate of decreasing in pH and Fe^T value into the solutions are slowing down [29].

It was reported that the growth rate of L.f is lower than T.f (around half of T.f) [58]. This can translate as the rate of its activities also lower than two other examined cultures in a certain period (21 days) of the process (Figure 1 and Table 5). These mean that the rate of the negative effect of reagents can be related to the bacteria sensitivity. Okibe and Johnson reported that L.f is more sensitive than T.f in the presence of flotation reagents which comprises the presented results in Figures 4 and 5 [45]. In general, in the mixed culture, bacteria show a better activity and lower sensitivity than other cultures to the toxicity of frothers while the sensitivity of T.f to the frothers in their highest dosages (1 g/L) is lower than two other bacteria (Figures 4 and 5). This can be as a result of the simultaneous presence of iron and sulfur-oxidizing bacteria in the mixed culture that positively improves microorganism activities. Zhang et al. reported that the oxidation activity of the mixed culture (*Acidithiobacillus ferrooxidans*, *Acidithiobacillus thiooxidans*, and *Leptospirillum*) is higher than that of the pure culture, and the mixed culture has the highest adaptability to the bioleaching conditions [78]. Furthermore, meanwhile, the solubility of MIBC is six times higher than PO (at the same condition) [47], its inhibitory effect on bacteria activity can be higher than PO. This also is in a good agreement with the outcome of analyses (Figures 4 and 5). Thus, although PO produces larger and less stable bubbles than MIBC within flotation separation, its toxicity in terms of bioleaching and environmental issues is lower than MIBC.

5. Conclusions

A comparison between the mixed and pure cultures during 21 days of monitoring indicated that bacteria concentration of the mixed culture is higher than pure ones. These results indicated that the insensitivity of the mixed culture to the toxicity of MIBC and PO as conventional flotation frothers in low dosages (0.001 and 0.01 g/L) is more than pure cultures. MC showed the highest population in the presence of frothers (0.001 and 0.01 g/L). Mutual information and Pearson correlation assessments released that pH value and total iron in the solution are the main parameters during bacteria activities.

There is a significant negative correlation between bacteria population and pH (as the most important factor of bioleaching). Presence of frothers disrupted bacteria activities; thus, the rate of pH reduction and oxidation-reduction of iron were decreasing by increasing the dosage of frothers. In the absence and presence (0.001 g/L) of the flotation frothers the rate of pH reduction during the process has the following order for the examined cultures: mixed > T.f > L.f. In general, the mixed culture has the highest Fe oxidation-reduction ratio in both the absence and presence of frothers. Results demonstrated that although during flotation MIBC can produce smaller and more stable bubbles than PO, its toxicity is higher than PO for various microorganisms.

Author Contributions: M.J., S.Z.S., M.G. (Mahdi Gharabaghi), and H.A. conceived and designed the experiments; M.J. performed the experiments; M.G. (Mehdi Golzadeh) and S.C.C. analyzed the data; S.C.C. and M.J. wrote the paper.

Acknowledgments: Authors would like to appreciate all support and provision of microorganisms for this investigation from the Center of Research and Development in Sarcheshmeh mine, Kerman, Iran. Additionally, we would like to thank Ali Rezai, and Hossayni for their assistance in the mineral processing laboratory at the University of Tehran, and Ghanbar Zad for the AAS analyses in the geochemistry laboratory at the University of Tehran.

Conflicts of Interest: The authors declare no conflict of interest.

References

1. Fugleberg, S.; Järvinen, A. Method for leaching zinc concentrate in atmospheric conditions. Google Patents: U.S. Patent No. 6,340,450, 22 January 2002.
2. Baláž, P.; Achimovičová, M.; Bastl, Z.; Ohtani, T.; Sanchez, M. Influence of mechanical activation on the alkaline leaching of enargite concentrate. *Hydrometallurgy* **2000**, *54*, 205–216. [CrossRef]
3. Babu, M.N.; Sahu, K.; Pandey, B. Zinc recovery from sphalerite concentrate by direct oxidative leaching with ammonium, sodium and potassium persulphates. *Hydrometallurgy* **2002**, *64*, 119–129. [CrossRef]
4. Rawlings, D.E. Microbially-assisted dissolution of minerals and its use in the mining industry. *Pure Appl. Chem.* **2004**, *76*, 847–859. [CrossRef]
5. Pradhan, N.; Das, B.; Gahan, C.S.; Kar, R.N.; Sukla, L.B. Beneficiation of iron ore slime using Aspergillus niger and Bacillus circulans. *Bioresour. Technol.* **2006**, *97*, 1876–1879. [CrossRef] [PubMed]
6. Brierley, C. Biohydrometallurgical prospects. *Hydrometallurgy* **2010**, *104*, 324–328. [CrossRef]
7. Panda, S.; Mishra, S.; Rao, D.S.; Pradhan, N.; Mohapatra, U.; Angadi, S.; Mishra, B.K. Extraction of copper from copper slag: Mineralogical insights, physical beneficiation and bioleaching studies. *Korean J. Chem. Eng.* **2015**, *32*, 667–676. [CrossRef]
8. Akcil, A. Potential bioleaching developments towards commercial reality: Turkish metal mining's future. *Miner. Eng.* **2004**, *17*, 477–480. [CrossRef]
9. Anjum, F.; Bhatti, H.N.; Asgher, M.; Shahid, M. Leaching of metal ions from black shale by organic acids produced by Aspergillus niger. *Appl. Clay Sci.* **2010**, *47*, 356–361. [CrossRef]
10. D'Hugues, P.; Spolaore, P.; Consortium, B. Biohydrometallurgy applied to exploitation of black shale resources: Overview of Bioshale FP6 European project. *Trans. Nonferrous Met. Soc. China* **2008**, *18*, 1485–1490. [CrossRef]
11. Zhao, H.; Zhang, Y.; Zhang, X.; Qian, L.; Sun, M.; Yang, Y.; Zhang, Y.; Wang, J.; Kim, H.; Qiu, G. The dissolution and passivation mechanism of chalcopyrite in bioleaching: An overview. *Miner. Eng.* **2019**, *136*, 140–154. [CrossRef]
12. Borja, D.; Nguyen, K.; Silva, R.; Park, J.; Gupta, V.; Han, Y.; Lee, Y.; Kim, H. Experiences and future challenges of bioleaching research in South Korea. *Minerals* **2016**, *6*, 128. [CrossRef]
13. Borja, D.; Nguyen, K.A.; Silva, R.A.; Ngoma, E.; Petersen, J.; Harrison, S.T.; Park, J.H.; Kim, H. Continuous bioleaching of arsenopyrite from mine tailings using an adapted mesophilic microbial culture. *Hydrometallurgy* **2019**, *187*, 187–194. [CrossRef]
14. Ngoma, E.; Borja, D.; Smart, M.; Shaik, K.; Kim, H.; Petersen, J.; Harrison, S.T. Bioleaching of arsenopyrite from Janggun mine tailings (South Korea) using an adapted mixed mesophilic culture. *Hydrometallurgy* **2018**, *181*, 21–28. [CrossRef]

15. Hong, J.; Silva, R.A.; Park, J.; Lee, E.; Park, J.; Kim, H. Adaptation of a mixed culture of acidophiles for a tank biooxidation of refractory gold concentrates containing a high concentration of arsenic. *J. Biosci. Bioeng.* **2016**, *121*, 536–542. [CrossRef] [PubMed]

16. Park, J.; Han, Y.; Lee, E.; Choi, U.; Yoo, K.; Song, Y.; Kim, H. Bioleaching of highly concentrated arsenic mine tailings by Acidithiobacillus ferrooxidans. *Sep. Purif. Technol.* **2014**, *133*, 291–296. [CrossRef]

17. Brierley, C. How will biomining be applied in future? *Trans. Nonferrous Met. Soc. China* **2008**, *18*, 1302–1310. [CrossRef]

18. Panda, S.; Akcil, A.; Pradhan, N.; Deveci, H. Current scenario of chalcopyrite bioleaching: a review on the recent advances to its heap-leach technology. *Bioresour. Technol.* **2015**, *196*, 694–706. [CrossRef]

19. Torma, A.E.; Walden, C.; Branion, R. Microbiological leaching of a zinc sulfide concentrate. *Biotechnol. Bioeng.* **1970**, *12*, 501–517. [CrossRef]

20. Pourhossein, F.; Mousavi, S.M. A novel step-wise indirect bioleaching using biogenic ferric agent for enhancement recovery of valuable metals from waste light emitting diode (WLED). *J. Hazard. Mater.* **2019**, *378*, 120648. [CrossRef]

21. Wang, S.; Zheng, Y.; Yan, W.; Chen, L.; Mahadevan, G.D.; Zhao, F. Enhanced bioleaching efficiency of metals from E-wastes driven by biochar. *J. Hazard. Mater.* **2016**, *320*, 393–400. [CrossRef]

22. Gahan, C.S.; Sundkvist, J.E.; Sandström, Å. A study on the toxic effects of chloride on the biooxidation efficiency of pyrite. *J. Hazard. Mater.* **2009**, *172*, 1273–1281. [CrossRef] [PubMed]

23. Mahmood, Q.; Zheng, P.; Cai, J.; Wu, D.; Hu, B.; Li, J. Anoxic sulfide biooxidation using nitrite as electron acceptor. *J. Hazard. Mater.* **2007**, *147*, 249–256. [CrossRef] [PubMed]

24. Lee, E.; Han, Y.; Park, J.; Hong, J.; Silva, R.A.; Kim, S.; Kim, H. Bioleaching of arsenic from highly contaminated mine tailings using Acidithiobacillus thiooxidans. *J. Environ. Manag.* **2015**, *147*, 124–131. [CrossRef] [PubMed]

25. Jafari, M.; Abdollahi, H.; Shafaei, S.Z.; Gharabaghi, M.; Jafari, H.; Akcil, A.; Panda, S. Acidophilic bioleaching: a review on the process and effect of organic–inorganic reagents and materials on its efficiency. *Miner. Process. Extr. Met. Rev.* **2018**, *40*, 87–107. [CrossRef]

26. Jafari, M.; Shafaei, S.; Abdollahi, H.; Gharabaghi, M.; Chehreh Chelgani, S. Effect of flotation reagents on the activity of L. ferrooxidans. *Miner. Process. Extr. Met. Rev.* **2018**, *39*, 34–43. [CrossRef]

27. Jafari, M.; Shafaei, S.Z.; Abdollahi, H.; Gharabaghi, M.; Chehreh Chelgani, S.; Ghassa, S. Examining the effects of typical reagents for sulfide flotation on bio-oxidation activity of ferrous iron oxidizing microorganisms. In *Proceedings of Solid State Phenomena*; Trans Tech Publications: freiberg, Germany, 2017; Volume 262, pp. 84–87.

28. Dehghan, R.; Dianati, M. The effects of Pb-Zn flotation reagents on the bioleaching process by mesophilic bacteria. *Int. J. Mineral. Process.* **2015**, *143*, 80–86. [CrossRef]

29. Loon, H.Y.; Madgwick, J. The effect of xanthate floatation reagents on bacterial leaching of chalcopyrite by Thiobacillus ferrooxidans. *Biotechnol. Lett.* **1995**, *17*, 997–1000. [CrossRef]

30. Puhakka, J.; Tuovinen, O.H. Effect of organic compounds on the microbiological leaching of a complex sulphide ore material. *MIRCEN J. Appl. Microbiol. Biotechnol.* **1987**, *3*, 429–436. [CrossRef]

31. Zhang, C.G.; Xia, J.L.; Zhang, R.Y.; Peng, A.A.; Nie, Z.Y.; Qiu, G.Z. Comparative study on effects of Tween-80 and sodium isobutyl-xanthate on growth and sulfur-oxidizing activities of Acidithiobacillus albertensis BY-05. *Trans. Nonferrous Met. Soc. China* **2008**, *18*, 1003–1007. [CrossRef]

32. Jafari, M.; Shafaie, S.; Abdollahi, H.; Gharabaghi, M.; Chehreh Chelgani, S. Study of the effects of conventional reagents for sulfide flotation on bio-oxidation activity of Acidithiobacillus ferrooxidans. *Chem. Eng. Commun.* **2019**, *206*, 365–377. [CrossRef]

33. Rawlings, D.E. *Biomining: Theory, Microbes And Industrial Processes*; Springer Science & Business Media: Berlin, Germany, 2013.

34. Battaglia-Brunet, F.; d'Hugues, P.; Cabral, T.; Cezac, P.; Garcia, J.-L.; Morin, D. The mutual effect of mixed thiobacilli and leptospirilli populations on pyrite bioleaching. *Miner. Eng.* **1998**, *11*, 195–205. [CrossRef]

35. Norris, P. Iron and mineral oxidation with Leptospirillum-like bacteria. In *Proceedings of International symposium on biohydrometallurgy*; Recent Progress in Biohydrometalurgy: Cagliari, Canada, 1983; pp. 83–96.

36. Goebel, B.M.; Stackebrandt, E. Cultural and phylogenetic analysis of mixed microbial populations found in natural and commercial bioleaching environments. *Appl. Environ. Microbiol.* **1994**, *60*, 1614–1621. [PubMed]

37. Gomez, E.; Ballester, A.; Blázquez, M.; Gonzalez, F. Silver-catalysed bioleaching of a chalcopyrite concentrate with mixed cultures of moderately thermophilic microorganisms. *Hydrometallurgy* **1999**, *51*, 37–46. [CrossRef]

38. Tuovinen, O.; Kelley, B.; Groudev, S. Mixed cultures in biological leaching processes and mineral biotechnology. In *Mixed Cultures in Biotechnology*; McGraw-Hill Book Co.: New York, NY, USA, 1991; pp. 373–427.

39. Akcil, A.; Ciftci, H.; Deveci, H. Role and contribution of pure and mixed cultures of mesophiles in bioleaching of a pyritic chalcopyrite concentrate. *Miner. Eng.* **2007**, *20*, 310–318. [CrossRef]

40. Akcil, A.; Ciftci, H. Bacterial leaching of Kure copper ore. *J. Chamb. Min. Eng. Turk.* **2003**, *42*, 15–25.

41. Dong, Y.; Lin, H. Influences of flotation reagents on bioleaching of chalcopyrite by Acidthiobacillus ferrooxidans. *Miner. Eng.* **2012**, *32*, 27–29. [CrossRef]

42. Huerta, G.; Escobar, B.; Rubio, J.; Badilla-Ohlbaum, R. Adverse effect of surface-active reagents on the bioleaching of pyrite and chalcopyrite by Thiobacillus ferrooxidans. *World J. Microbiol. Biotechnol.* **1995**, *11*, 599–600. [CrossRef]

43. Dopson, M.; Sundkvist, J.-E.; Lindström, E.B. Toxicity of metal extraction and flotation chemicals to Sulfolobus metallicus and chalcopyrite bioleaching. *Hydrometallurgy* **2006**, *81*, 205–213. [CrossRef]

44. Zeng, G.-M.; Shi, J.-G.; Yuan, X.-Z.; Liu, J.; Zhang, Z.-B.; Huang, G.-H.; Li, J.-B.; Xi, B.-D.; Liu, H.-L. Effects of Tween 80 and rhamnolipid on the extracellular enzymes of Penicillium simplicissimum isolated from compost. *Enzym. Microb. Technol.* **2006**, *39*, 1451–1456. [CrossRef]

45. Okibe, N.; Johnson, D.B. Toxicity of flotation reagents to moderately thermophilic bioleaching microorganisms. *Biotechnol. Lett.* **2002**, *24*, 2011–2016. [CrossRef]

46. Foucher, S.; Battaglia-Brunet, F.; d'Hugues, P.; Clarens, M.; Godon, J.; Morin, D. Evolution of the bacterial population during the batch bioleaching of a cobaltiferous pyrite in a suspended-solids bubble column and comparison with a mechanically agitated reactor. *Hydrometallurgy* **2003**, *71*, 5–12. [CrossRef]

47. Nazari, S.; Chelgani, S.C.; Shafaei, S.; Shahbazi, B.; Matin, S.; Gharabaghi, M. Flotation of coarse particles by hydrodynamic cavitation generated in the presence of conventional reagents. *Sep. Purif. Technol.* **2019**, *220*, 61–68. [CrossRef]

48. Matin, S.; Farahzadi, L.; Makaremi, S.; Chelgani, S.C.; Sattari, G. Variable selection and prediction of uniaxial compressive strength and modulus of elasticity by random forest. *Appl. Soft Comput.* **2018**, *70*, 980–987. [CrossRef]

49. Hadavandi, E.; Chelgani, S.C. Estimation of coking indexes based on parental coal properties by variable importance measurement and boosted-support vector regression method. *Measurement* **2019**, *135*, 306–311. [CrossRef]

50. Nazari, S.; Shafaei, S.; Shahbazi, B.; Chelgani, S.C. Study relationships between flotation variables and recovery of coarse particles in the absence and presence of nanobubble. *Colloids Surf. A Physicochem. Eng. Asp.* **2018**, *559*, 284–288. [CrossRef]

51. Chelgani, S.C.; Matin, S.S. Study the relationship between coal properties with Gieseler plasticity parameters by random forest. *Int. J. OilGas. Coal Technol.* **2018**, *17*, 113–127. [CrossRef]

52. Chelgani, S.C.; Shahbazi, B.; Hadavandi, E. Support vector regression modeling of coal flotation based on variable importance measurements by mutual information method. *Measurement* **2018**, *114*, 102–108. [CrossRef]

53. Schippers, A.; Sand, W. Bacterial leaching of metal sulfides proceeds by two indirect mechanisms via thiosulfate or via polysulfides and sulfur. *Appl. Environ. Microbiol.* **1999**, *65*, 319–321.

54. Bosecker, K. Bioleaching: metal solubilization by microorganisms. *FEMS Microbiol. Rev.* **1997**, *20*, 591–604. [CrossRef]

55. Stott, M.; Watling, H.; Franzmann, P.; Sutton, D. The role of iron-hydroxy precipitates in the passivation of chalcopyrite during bioleaching. *Miner. Eng.* **2000**, *13*, 1117–1127. [CrossRef]

56. Rohwerder, T.; Sand, W. The sulfane sulfur of persulfides is the actual substrate of the sulfur-oxidizing enzymes from Acidithiobacillus and Acidiphilium spp. *Microbiology* **2003**, *149*, 1699–1710. [CrossRef] [PubMed]

57. Keeling, S.; Palmer, M.; Caracatsanis, F.; Johnson, J.; Watling, H. Leaching of chalcopyrite and sphalerite using bacteria enriched from a spent chalcocite heap. *Miner. Eng.* **2005**, *18*, 1289–1296. [CrossRef]

58. Brandl, H. Microbial leaching of metals. In *Biotechnology Set*, 2nd ed.; WILEY: Zürich, Switzerland, 2008; pp. 191–224.

59. Silva, R.A.; Park, J.; Lee, E.; Park, J.; Choi, S.Q.; Kim, H. Influence of bacterial adhesion on copper extraction from printed circuit boards. *Sep. Purif. Technol.* **2015**, *143*, 169–176. [CrossRef]

60. Dutrizac, J. Factors affecting alkali jarosite precipitation. *Metall. Trans. B* **1983**, *14*, 531–539. [CrossRef]

61. Chowdhury, F.; Ojumu, T. Investigation of ferrous-iron biooxidation kinetics by Leptospirillum ferriphilum in a novel packed-column bioreactor: Effects of temperature and jarosite accumulation. *Hydrometallurgy* **2014**, *141*, 36–42. [CrossRef]

62. Silva, R.A.; Borja, D.; Hwang, G.; Hong, G.; Gupta, V.; Bradford, S.A.; Zhang, Y.; Kim, H. Analysis of the effects of natural organic matter in zinc beneficiation. *J. Clean. Prod.* **2017**, *168*, 814–822. [CrossRef]

63. Deo, N.; Natarajan, K. Biological removal of some flotation collector reagents from aqueous solutions and mineral surfaces. *Miner. Eng.* **1998**, *11*, 717–738. [CrossRef]

64. Pomianowski, A.; Leja, J. Spectrophotometric study of xanthate and dixanthogen solutions. *Can. J. Chem.* **1963**, *41*, 2219–2230. [CrossRef]

65. Jones, M.; Woodcock, J. Decomposition of alkyl dixanthogens in aqueous solutions. *Int. J. Mineral. Process.* **1983**, *10*, 1–24. [CrossRef]

66. Khoshdast, H.; Sam, A. Flotation frothers: review of their classifications, properties and preparation. *Open Mineral. Process. J.* **2011**, *4*, 25–44. [CrossRef]

67. Bulatovic, S.M. *Handbook of Flotation Reagents: Chemistry, Theory and Practice: Volume 1: Flotation of Sulfide Ores*; Elsevier: Amsterdam, The Netherlands, 2007.

68. Sun, Z.; Forsling, W. The degradation kinetics of ethyl-xanthate as a function of pH in aqueous solution. *Miner. Eng.* **1997**, *10*, 389–400. [CrossRef]

69. Boxall, N.J.; Cheng, K.Y.; Bruckard, W.; Kaksonen, A.H. Application of indirect non-contact bioleaching for extracting metals from waste lithium-ion batteries. *J. Hazard. Mater.* **2018**, *360*, 504–511. [CrossRef] [PubMed]

70. Akinci, G.; Guven, D.E. Bioleaching of heavy metals contaminated sediment by pure and mixed cultures of Acidithiobacillus spp. *Desalination* **2011**, *268*, 221–226. [CrossRef]

71. Liu, G.; Yin, J.; Cong, W. Effect of fluid shear and particles collision on the oxidation of ferrous iron by Acidithiobacillus ferrooxidans. *Miner. Eng.* **2007**, *20*, 1227–1231. [CrossRef]

72. Smith, R. Liquid and solid wastes from mineral processing plants. *Mineral. Process. Extr. Metullargy Rev.* **1996**, *16*, 1–22. [CrossRef]

73. Pacholewska, M.; Cwalina, B.; Steindor, K. The influence of flotation reagents on sulfur-oxidizing bacteria Acidithiobacillus thiooxidans. *Physicochem. Probl. Mineral. Process.* **2008**, *42*, 37–46.

74. Tuovinen, O.H. Inhibition of Thiobacillus ferrooxidans by mineral flotation reagents. *Appl. Microbiol. Biotechnol.* **1978**, *5*, 301–304. [CrossRef]

75. Torma, A.E.; Gabra, G.; Guay, R.; Silver, M. Effects of surface active agents on the oxidation of chalcopyrite by Thiobacillus ferrooxidans. *Hydrometallurgy* **1976**, *1*, 301–309. [CrossRef]

76. Jafari, M.; Shafaei, S.; Abdollahi, H.; Gharabaghi, M.; Chehreh Chelgani, S. A comparative study on the effect of flotation reagents on growth and iron oxidation activities of Leptospirillum ferrooxidans and Acidithiobacillus ferrooxidans. *Minerals* **2016**, *7*, 2. [CrossRef]

77. Iwasaki, I.; Cooke, S. Dissociation constant of xanthic acid as determined by spectrophotometric method. *J. Phys. Chem.* **1959**, *63*, 1321–1322. [CrossRef]

78. Zhang, Y.-S.; Qin, W.-Q.; Jun, W.; Zhen, S.-J.; Yang, C.-R.; Zhang, J.-W.; Nai, S.-S.; Qiu, G.-Z. Bioleaching of chalcopyrite by pure and mixed culture. *Trans. Nonferrous Met. Soc. China* **2008**, *18*, 1491–1496. [CrossRef]

Article

Combining Mechanistic Modeling and Raman Spectroscopy for Monitoring Antibody Chromatographic Purification

Fabian Feidl [1], Simone Garbellini [1], Martin F. Luna [1], Sebastian Vogg [1], Jonathan Souquet [2], Hervé Broly [2], Massimo Morbidelli [1] and Alessandro Butté [1,*]

[1] Institute for Chemical and Bioengineering, ETH Zurich, 8093 Zurich, Switzerland; Fabian.feidl@chem.ethz.ch (F.F.); simon.garbellini@outlook.com (S.G.); lunam@ethz.ch (M.F.L.); sebastian.vogg@chem.ethz.ch (S.V.); massimo.morbidelli@chem.ethz.ch (M.M.)
[2] Merck Serono S.A. Biotech Process Sciences, 1809 Corsier-sur-Vevey, Switzerland; jonathan.souquet@merckgroup.com (J.S.); herve.broly@merckgroup.com (H.B.)
* Correspondence: alessandro.butte@chem.ethz.ch; Tel.: +41-44-633-38-98

Received: 30 August 2019; Accepted: 23 September 2019; Published: 1 October 2019

Abstract: Chromatography is widely used in biotherapeutics manufacturing, and the corresponding underlying mechanisms are well understood. To enable process control and automation, spectroscopic techniques are very convenient as on-line sensors, but their application is often limited by their sensitivity. In this work, we investigate the implementation of Raman spectroscopy to monitor monoclonal antibody (mAb) breakthrough (BT) curves in chromatographic operations with a low titer harvest. A state estimation procedure is developed by combining information coming from a lumped kinetic model (LKM) and a Raman analyzer in the frame of an extended Kalman filter approach (EKF). A comparison with suitable experimental data shows that this approach allows for the obtainment of reliable estimates of antibody concentrations with reduced noise and increased robustness.

Keywords: Raman spectroscopy; downstream processing; chromatography; flow cell; extended Kalman filter

---

1. Introduction

The application of spectroscopic techniques to monitor chromatographic processes in the frame of the so-called process analytical technology (PAT) initiative is very promising due to its potential for gathering important on-line process information in a non-invasive way [1–5]. Available spectroscopic techniques range from UV/vis and Fourier transform infrared spectroscopy to dynamic light scattering [6]. Several applications of Raman spectroscopy have been reported in upstream processing [7–9], showing the potential of this technology, which often requires specific modeling techniques, such as partial least squares (PLS) regression, to extract the desired information from the measured spectra. Recently, a successful implementation of Raman spectroscopy for the on-line monitoring of monoclonal antibody (mAb) concentrations in downstream processing was reported by Feidl et al [10]. An ad hoc developed flow cell enabled the integration of the Raman technology into the capture (protein A) step of a mAb manufacturing process, providing accurate on-line estimates of mAb concentration. However, in spite of these results, the use of this technology remains limited due to the intrinsic weakness of the Raman signal [11–13].

In this work, we explore the possibility of overcoming these difficulties by combining estimates from the Raman signal with the predictions of a mechanistic model. This is particularly convenient in the chromatographic purification of mAbs because these processes are well understood and reliable mechanistic simulation models are available [14]. It has been already shown in many other areas that the combination of deterministic knowledge and on-line measurements can lead to more accurate and reliable estimates [15–18], e.g., by using extended Kalman filtering (EKF) [19]. While a general introduction to Kalman filtering is given in [20], a more detailed description is provided in [21].

In this work, the implementation of an EKF for a chromatographic capture step to estimate antibody concentration is shown by combining the information of a deterministic chromatographic model with on-line information derived from Raman-based PLS estimates. In particular, the objective is to monitor the chromatographic breakthrough curves of a low-concentrated monoclonal antibody harvest. The beneficial effect of the combination of the two approaches with respect to the stand-alone Raman and chromatographic model is discussed.

2. Materials and Methods

2.1. Raman Spectral Acquisition and Flow Cell

Raman spectra were acquired with a Kaiser RamanRxn2 analyzer (Kaiser Optical Systems, Inc., Ann Arbor, MI, USA), including a 785 nm laser at 400 mW and a cooled charged-coupled device (CCD) detector, measuring inelastic photon scattering across a 150–3425 cm^{-1} wavenumber range. A laser exposure time of 30 s was chosen to collect the single scan spectra. A flow cell with an optimized flow characteristic, signal enhancement, pressure tolerance, and a single use potential was developed. A schematic illustration of the flow cell is shown in Figure 1. It includes four main modules: (A) An analyzer adapter, which connects the flow cell to the Raman analyzer via a fiber cable; (B) a non-contact objective to focus the laser beam within the flow path; (C) a flow path, which guides the sample longitudinally to the laser beam; and (D) a reflector, reflecting scattered and unscattered light to the analyzer via the flow path. In the application to chromatographic purifications, the inlet connection is coupled to the elution stream, and the outlet connection is linked to the sample fractionator.

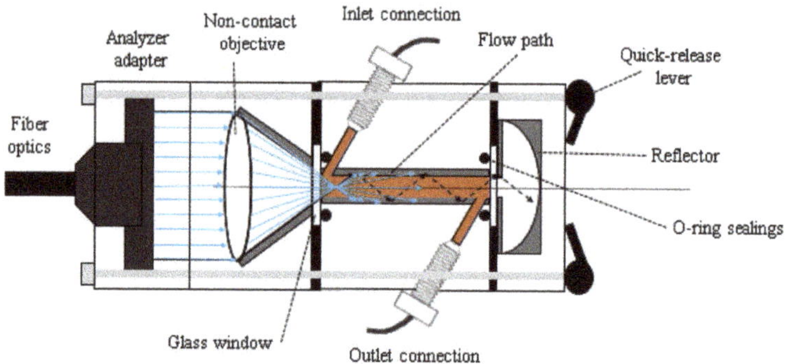

Figure 1. Schematic illustration of the developed flow cell.

2.2. Cell Culture Supernatant

Two cell culture supernatant pools containing a recombinant mAb with product concentrations between 0.30 and 0.60 mg/mL were obtained from a CHO cell perfusion process, as reported in [22]. Besides cell filtering through the perfusion hollow fiber module (0.5 µm pore size, Spectrum Laboratories, Netherlands), no other treatment was applied to the supernatant, which therefore contained a large quantity of impurities, e.g., media components, host cell proteins (HCP, 3×10^5 ppm), DNA (4×10^4 ppm) and high molecular weight (HMW) species (1.1%).

2.3. Breakthrough Runs and Reference Analytic

Fifteen breakthrough (BT) runs were performed on MabSelect SuRe columns (GE Healthcare, Uppsala, Sweden), prepacked by Repligen GmbH (Ravensburg, Germany, 0.5×5 cm), as described in [10]. The feed concentration, flow rate, and the fraction duration were changed between the different BT runs, as described in Table 1. Since several Raman measurements were acquired while collecting the sample of a single fraction (between 4 and 16 spectra per fraction), a spline approach was applied to interpolate missing reference measurements for each Raman spectrum.

Table 1. Chromatographic settings and Raman measurements availability of performed breakthrough curves.

	Breakthrough Curve ID	Feed Conc. (mg/mL)	Feed Flow Rate (mL/Min)	Run Duration (Min)	Fraction Duration (Min)	Raman Measurements
Used for PLS and EKF	BT#1	0.34	1.0	200	8	+
	BT#2	0.34	1.5	230	4	+
	BT#3	0.34	1.0	250	6	+
	BT#4	0.42	1.5	130	2	+
	BT#5	0.42	1.0	120	3.5	+
	BT#6	0.42	1.0	170	3.5	+
Used for LKM fitting	BT#7	0.43	1.0	240	4	-
	BT#8	0.43	1.5	160	4	-
	BT#9	0.43	0.5	480	4	-
	BT#10	0.30	0.5	690	10	-
	BT#11	0.30	1.5	230	10	-
	BT#12	0.30	1.0	340	10	-
	BT#13	0.60	0.5	200	3	-
	BT#14	0.60	1.5	80	3.5	-
	BT#15	0.60	1.0	120	3.5	-

The mAb concentrations were determined off-line by HPLC with an analytical standard deviation of 0.01 mg/mL as described in [23]. The HMW, HCP, and DNA content of the harvest were determined as described in [10]. A schematic illustration of the experimental set-up is shown in Figure 2.

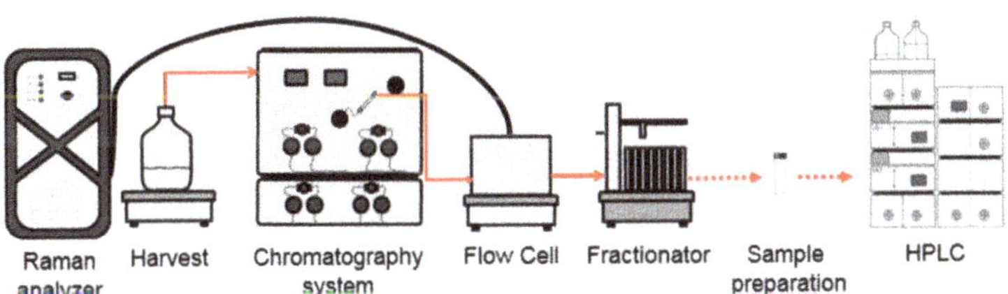

Figure 2. Schematic illustration of the experimental set up to perform chromatographic breakthrough runs and collecting synchronized Raman measurements and breakthrough fractions.

2.4. Chemometric Modeling Procedure

All calculations were performed with MATLAB R2018a (Mathworks, Natick, MA, USA) using in-house developed routines, if not stated otherwise. The modeling procedure included Savitzky–Golay smoothing with a second polynomial order and a frame size of 51 Raman shift wavenumbers [24], spectrum wise standard normal variate (SNV) processing [25], and Raman shift wavenumber wise mean centering on spectra and reference values [26]. The removal of spectral regions based on the bioprocess modeling experience resulted in spectral ranges of 450–1820 cm^{-1}, 1880–2530 cm^{-1} and 2590–3100 cm^{-1} to eliminate interferences with the window material and water as well as non-informative regions. No derivative, Raman shift wavenumber selection, or automated outlier removal tools were applied. The nonlinear iterative partial least squares (NIPALS) algorithm [27] was used to calibrate predictive PLS models, which regressed spectral data on HPLC reference values, including a threefold cross validation (CV) [28]. The optimal number of latent variables (LV) was determined based on the minimum CV error.

After calibrating the model on five different BT runs, the model was tested on an external BT run, i.e., not included in the calibration. In order to evaluate the model performance, the root mean square error was calculated for both cross validation (RMSECV) and external prediction (RMSEP), using Equation (1), where $\hat{y}_i$ is the predicted value of the i-th observation, y_i is the corresponding measured value, and n is the total number of observations:

$$RMSE = \sqrt{\frac{1}{n}\sum_{i=1}^{n}(y_i - \hat{y}_i)^2}$$ (1)

Furthermore, the coefficient of determination (R^2) was calculated for the external prediction as follows:

$$R^2 = 1 - \frac{\sum_{i=1}^{n}(y_i - \hat{y}_i)^2}{\sum_{i=1}^{n}(y_i - \overline{y})^2}$$ (2)

A rotational approach was used to judge upon the model transferability and data set similarity. Hence, sets were rotated in order to have every BT run in the external prediction set once, as shown in Table 2. As an example, to build the model for rotation 1 (ROT1), the data of BT#2–6 were used for model calibration and tested on BT#1 (for further details, see [10]).

Table 2. Modeling procedure including the Raman-PLS (partial least squares) calibration, mechanistic model fitting, as well as the tuning, validation and perturbation of the extended Kalman filtering (EKF).

For a given Rot i

1. **Raman-PLS calibration:**
   $PLS_{Rot\ i} \leftarrow$ calibrate Raman-PLS on BT#1–6 except BT#i
2. **Mechanistic model fitting:**
   $LKM \leftarrow$ fit LKM on BT#7–15 with respective process inputs ($PI_{BT\#7-15}$)
3. **EKF tuning:**
   $EKF_{Rot\ i} \leftarrow$ tune EKF (Q, R, $k_Q^{Rot\ i}$) on BT#1–6 except BT#i
      for n = BT#1–6 except BT#i
         $k_{Q_{best}}^{n} = min_{k_Q}RMSEP_{EKF_n}$
      end for
      $k_Q^{Rot\ i} =$ average $k_{Q_{best}}^{n}$ for all n
4. **EKF validation:**
   Run $EKF_{Rot\ i}$ with inputs LKM, $PI_{Rot\ i}$, $PLS_{Rot\ i}$ and $k_Q^{Rot\ i}$
5. **EKF perturbation:**
      for n = 1–200 simulations
         PI_{RS} = random sampling of ε, c_{in}, Q_{flow} from Gaussian probability distribution
         Run $EKF_{Rot\ i}$ with inputs LKM, PI_{RS}, $PLS_{Rot\ i}$ and $k_Q^{Rot\ i}$
      end for

2.5. Deterministic Modeling Procedure

The chromatographic process was modelled using the lumped kinetic model (LKM) [14,29]:

$$\frac{\partial c}{\partial t} = -v\frac{\partial c}{\partial x} + D_L\frac{\partial^2 c}{\partial x^2} - \varphi\frac{\partial q}{\partial t} \quad t \in [0, t_{end}], \ x \in [0, L_{Col}]$$ (3)

$$\frac{\partial q}{\partial t} = k_m(q^* - q)$$ (4)

where c is the liquid phase concentration of the protein, t is the time, t_{end} is the end of the BT run, v is the interstitial velocity ($v = \frac{Q_{flow}}{A_{col}\ \varepsilon}$), Q_{flow} is the volumetric flow rate (see Table 1), A_{col} is the column cross sectional area ($A_{col} = 0.196$ cm^2), ε is the bed porosity ($\varepsilon = 0.36$ [30]), x is the coordinate along the column longitudinal axis, L_{Col} is the column length ($L_{Col} = 5$ cm), D_L is the apparent axial dispersion coefficient, $\varphi = (1 - \varepsilon)/\varepsilon$ is the phase ratio of the column, q is the solid phase concentration of the protein, k_m is the mass transfer coefficient, and q^* is the equilibrium solid phase concentration of the protein.

The following initial conditions were applied:

$$c(t = 0, x) = c_0(x) \tag{5}$$

$$q(t = 0, x) = q_0(x) \tag{6}$$

They were then combined with the classical Danckwerts' boundary conditions:

$$c(t > 0, x = 0) = c_{in}(t) + \frac{D_L}{v} \frac{\partial c}{\partial x}\bigg|_{x=0} \tag{7}$$

$$\frac{\partial c}{\partial x}\bigg|_{x=L_{col}} = 0 \tag{8}$$

where $c_{in}(t)$ is the feed concentration (see Table 1). Both $c_0(x)$ and $q_0(x)$ are zero for all values of x. The apparent axial dispersion coefficient D_L can be estimated from the reduced van Deemter equation [31]:

$$D_L = A\frac{d_p}{2}v \tag{9}$$

where A is the intercept of the reduced van Deemter equation, d_p the average particle diameter (d_p = 85 µm). The van Deemter eddy diffusion coefficient A was experimentally determined from pulse injection experiments at different flow rates with mAb under non-adsorbing conditions (A = 17.15; data not shown). An empirical correlation was used for the mass transfer coefficient k_m, approximating hindered mass transfer due to pore blockage and other effects [32]:

$$k_m = k_m^{max}\left(S_1 + (1 - S_1)\left(1 - \frac{q}{q_{sat}}\right)^{S_2}\right) \tag{10}$$

where k_m^{max} is the maximum mass transfer coefficient, q_{sat} is the saturation capacity of the resin, and S_1 is a maximum hindrance coefficient ($0 < S_1 \leq 1$). The coefficient S_2 (with $S_2 > 0$) accounts for the nonlinear increase of the hindrance. The protein adsorption process was described using a Langmuir isotherm, where H is the Henry coefficient:

$$q^* = \frac{H\,c}{1 + \frac{H\,c}{q_{sat}}} \tag{11}$$

Coefficients k_m, S_1, S_2, q_{sat} and H were fitted on BT#7–15 using the corresponding process inputs (PI), such as ε, c_{in} and Q_{flow}, as shown in Table 2. The partial differential equations were discretized along the x coordinate using a first order central finite difference method, and the resulting system of ordinary differential equations was solved using 100 grid points.

2.6. Extended Kalman Filter Tuning, Validation and Perturbation

A prerequisite for this technique is a general nonlinear time-invariant system in continuous time, which generates measurements at discrete time steps $t_k = k\Delta t$ [33,34]:

$$\frac{\partial x}{\partial t} = f(x(t), u(t), p) + w(t) \tag{12}$$

$$y(t_k) = h(x(t_k)) + v(t_k) \tag{13}$$

where x denotes the states, u is the deterministic inputs, p is the time-invariant parameters, and y is the measurements of the system. The nonlinear function $f()$ describes the state dynamics, and $h()$ is the measurement function that relates state x with measurement y. The process noise w and the measurement noise v are assumed to be uncorrelated zero-mean Gaussian random processes with covariances $Q(t)$ and $R(t)$, respectively ($E \triangleq$ expectation operator):

$$E[w(t)] = E[v(t_k)] = 0 \tag{14}$$

$$E\left[w(t)w^T(\tau)\right] = Q(t)\delta(t - \tau) \tag{15}$$

$$E\left[v(t_k)v^T(t_k)\right] = R(t_k) \tag{16}$$

$$E\left[w(\tau)v^T(t_k)\right] = 0 \tag{17}$$

Given such a system, the EKF can estimate states from noisy measurements through a recursive procedure, including two main steps [34,35]. In the first step (prediction step), the a posteriori state estimate $\hat{x}\left(t_{k-1}^+\right)$ and covariance matrix $P\left(t_{k-1}^+\right)$ are propagated from t_{k-1}^+ to t_k^-, leading to the a priori state estimate and covariance matrix (superscripts indicate values before (-) and (+) after measurement update):

$$\hat{x}\left(t_k^-\right) = \hat{x}\left(t_{k-1}^+\right) + \int_{t_{k-1}^+}^{t_k^-} f(\hat{x}(\tau), u(\tau), p)d\tau \tag{18}$$

$$\hat{y}\left(t_k^-\right) = h\left(\hat{x}\left(t_k^-\right)\right) \tag{19}$$

As well as the state covariance matrix:

$$P\left(t_k^-\right) = P\left(t_{k-1}^+\right) + \int_{t_{k-1}^+}^{t_k^-} \left(Z(\tau)P(\tau) + P(\tau)Z^T(\tau) + Q(\tau)\right)d\tau \tag{20}$$

The EKF formulation uses linearized models of the nonlinear system for state estimation. Hence, the system is linearized at each time t_k to obtain local state–space matrices:

$$Z(t) = \left(\frac{\partial f}{\partial x}\right)_{\hat{x}(t), u(t), p} \tag{21}$$

$$C(t_k) = \left(\frac{\partial h}{\partial x}\right)_{\hat{x}(t_k^-)} \tag{22}$$

In the second step (update step), conducted as soon as a new measurement $y(t_k)$ becomes available, the Kalman filter gain $K(t_k)$ is calculated and used to update the a priori state estimates and covariance matrix to the a posteriori values:

$$K(t_k) = P\left(t_k^-\right)C(t_k)^T\left(C(t_k)P\left(t_k^-\right)C(t_k)^T + R(t_k)\right)^{-1} \tag{23}$$

$$\hat{x}\left(t_k^+\right) = \hat{x}\left(t_k^-\right) + K(t_k)\left(y(t_k) - h\left(\hat{x}\left(t_k^-\right)\right)\right) \tag{24}$$

$$P\left(t_k^+\right) = (I - K(t_k)C(t_k))P\left(t_k^-\right) \tag{25}$$

The aim of the procedure is to obtain improved state estimates $\hat{x}\left(t_k^+\right)$, characterized by small values of the covariance matrix $P\left(t_k^+\right)$. In order to achieve this for a specific application, the design parameters of the EKF, such as the measurement noise covariance R, the initial state estimates $\hat{x}\left(t_0^+\right)$ the corresponding covariance P_0, and the process noise covariance Q need to be carefully selected. To initialize the filter, a consistent pair of $\hat{x}_0$ and P_0 needs to be selected to enable a fast convergence to the correct estimate [36].

In this work, the in-built KF toolbox of MATLAB was used. The discretized lumped kinetic model served as state transition function $f()$, and the Raman-PLS results were used as physical measurements of the outlet concentration of the column. The corresponding variance of the rotation specific RMSEP multiplied by the identity matrix was used as the error covariance matrix R. The time-varying process noise covariance Q was computed on-line at any given time t_k using a Monte Carlo approach, as reported in [33]. This takes the knowledge from the LKM parameter identification step into account by using the nominal parameter values and its parameter covariance matrix, resulting in Q_{MC}. Additionally, a model mismatch factor k_Q [34] was used and tuned for each rotation.

$$Q = Q_{MC}\, k_Q \tag{26}$$

For a certain rotation, as shown in Table 2, the EKF approach was separately applied to all n BT curves of the calibration set. For each BT curve, k_Q was optimized ($k_{Q_{best}}^n$) based on the respective prediction error ($RMSEP_{EKF_n}$). Subsequently, the received $k_{Q_{best}}^n$ of all n BT curves of the calibration set were averaged, resulting in the rotational specific model mismatch factor ($k_Q^{Rot\ i}$). The EKF was applied to the external BT curve, which was included in neither the Raman-PLS calibration nor the EKF tuning, to externally validate its effect. In a perturbation study, 200 simulations of the mechanistic model were ran with random process input values (PI_{RS}) for bed porosity ε, the feed concentration c_{in} as well as the volumetric flow rate Q_{flow} sampled from a Gaussian probability distribution with a standard deviation of 5% to compare the robustness of the LKM and EKF.

3. Results and Discussion

3.1. Partial Least Squares Raman Modeling

The acquired Raman spectra of the BT curves are comparable with the spectra described in [10] and are shown in Figure S1A. Due to the high impurity content in the harvest, the spectral features of different species (i.e., target mAb, media components, HCPs, DNA and HMW) overlapped, leading to broad bands and no distinct peak profiles within single spectra. Hence, only small variations between different spectra could be observed, and a suitable data pretreatment and multivariate model calibration were needed to extract useful information, such as the target protein titer. This had an obvious influence on the spectral appearance shown in Figure S1B. The variable importance in projection (VIP), shown in Figure S1C, indicates the regions between 2300 and 2700 cm^{-1} as well as between 2900 and 3000 cm^{-1} as very important. This is in line with the fact that proteins exhibit several Raman bands in the region between 2500–4000 cm^{-1} [37]. Results of the PLS modeling for different rotations are shown in Table 3. Though the number of observations in the calibration set varied slightly between rotations, the optimal number of selected latent variables (LVs) was consistent among rotations and was either 11 or 12. The RMSECV was constant around 0.040 mg/mL on a calibration range from 0 to 0.42 mg/mL and was thus almost independent of the rotation scheme. However, variations in the RMSEP between 0.045 and 0.072 mg/mL as well as varying values of R^2 between 0.70 and 0.86 indicated slight differences between the BT runs.

Table 3. Data set information and PLS modeling results of all rotations.

	Calibr.set (# Obs)	Pred.set (# Obs)	Opt. num. of LV	RMSECV (mg/mL)	RMSEP (mg/mL)	R^2
ROT1	1755	398	12	0.042	0.051	0.80
ROT2	1711	442	12	0.042	0.047	0.86
ROT3	1656	497	12	0.040	0.061	0.78
ROT4	1906	247	12	0.041	0.072	0.70
ROT5	1918	235	11	0.042	0.045	0.84
ROT6	1819	334	12	0.041	0.055	0.82

The prediction of titer as a function of time for ROT1 is exemplarily shown in Figure 3. The red dots represent the off-line HPLC titer measurements for each fraction, whereas the continuous blue line represents the Raman-PLS-based prediction. It can be seen that the trend of the BT curve was generally well captured by Raman. However, the predictions were clearly scattered around the reference values. It is worth mentioning that increasing the number of scans per Raman measurement could improve the signal-to-noise ratio. However, this would extend the measurement duration, which is critical in most of the downstream processing applications.

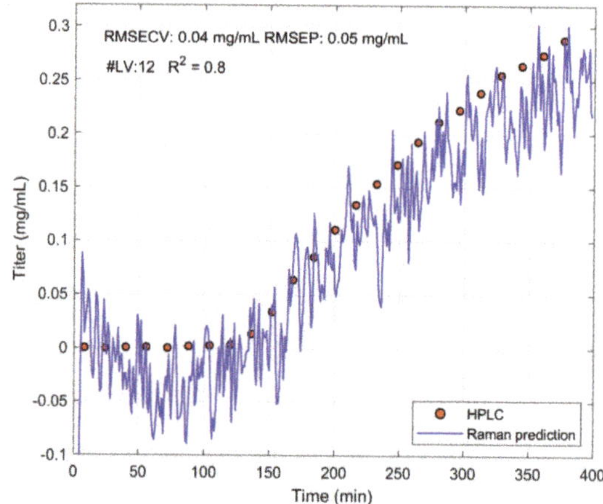

Figure 3. Time evolutions of Raman-PLS predictions (blue) for rotation 1 (ROT1) compared to off-line HPLC measurements (red).

As for other rotations, shown in Figure 4, one can observe a clear offset of the predictions at the initial phase (i.e., before breakthrough started). In spite of an optimized laser exposure time, signal enhancement through the flow cell and spectral pretreatment methods, the signal-to-noise ratio was rather small and probably close to the detection limit. Though the obtained averaged RMSEP of 0.05 mg/mL and averaged R^2 of 0.8 is remarkable, one must probably conclude that Raman-PLS is insufficient for a precise monitoring of a breakthrough at such small concentrations.

3.2. Mechanistic Modeling

As a next step, an additional set of nine BT runs (BT#7–15) were used to fit the LKM parameters by minimizing the RMSE with the measured concentration values in the breakthrough. The corresponding estimated values (along with 95% confidence intervals) are reported in Table 4 and contain the Henry coefficient H, saturation capacity q_{sat}, the maximum mass transfer coefficient k_m^{max}, maximum hindrance coefficient S_1, and the nonlinearity increase of hindrance coefficient S_2.

Table 4. Fitting parameter of the lumped kinetic model (LKM) optimized on breakthrough (BT)#7–BT#15 (ε = 0.36; A = 17.15; d_p = 85 μm).

$H(-)$	q_{sat}(mg/mL)	k_m^{max}(min^{-1})	$S_1(-)$	$S_2(-)$
449.3 ± 31.5	109.7 ± 4.7	8.37 ×10^{-4} ± 0.94 ×10^{-4}	0.36 ± 0.24	1.76 ± 1.38

The corresponding model predictions of the breakthrough together with the fitting data sets are shown in Figure 5. The red dots represent the HPLC titer measurements, whereas the continuous blue lines represent the predictions of the LKM. Additionally, the RMSE in fitting (RMSEF) is indicated.

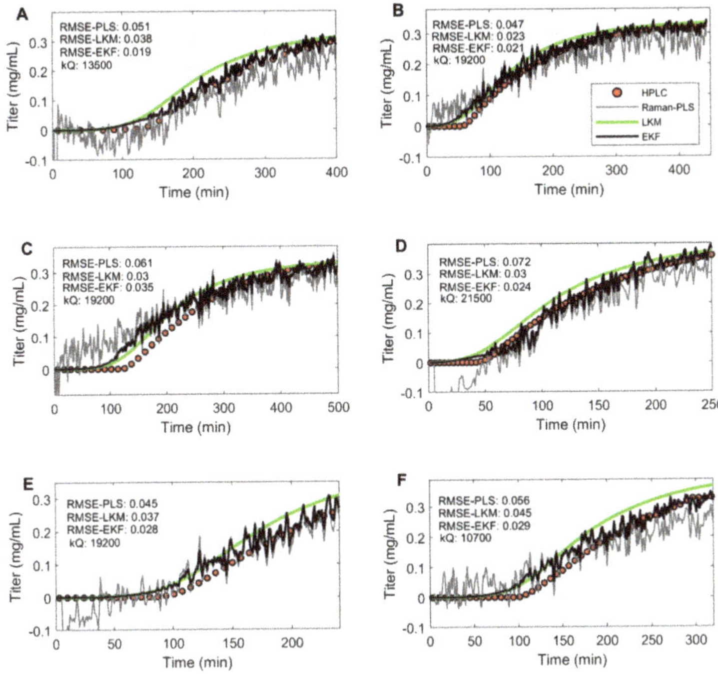

Figure 4. Titer predictions of Raman-PLS (grey), the LKM (green), the EKF model with tuned k_Q (black) and HPLC off-line measurements (red) for rotations 1–6 (**A–F**).

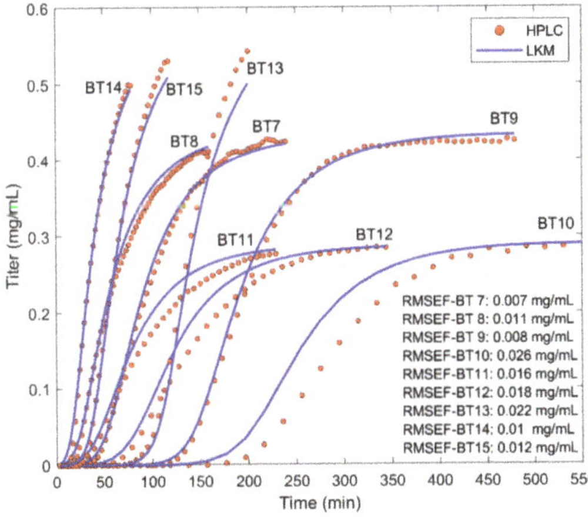

Figure 5. Internal mechanistic model (LKM) predictions of titer for BT#7–15 (blue) and corresponding HPLC off-line measurements (red).

It can be seen that the shapes of BT curves vary in steepness, inflection point and saturation level. This is due to the differences in feed concentration and loading flow rate. At complete column saturation, the asymptotic value of the outlet concentration in the BT tended to the feed concentration. Moreover, higher feed concentrations generally produced earlier breakthroughs. Similarly, larger flow rates not only reduced the residence time in the column but also increased the convection rate along the column with respect to the diffusion rate to the resin, thus producing faster and flatter BT curves. In most cases, the LKM was able to closely predict the trend of the reference measurements. The worst results were obtained in the case of BT#10–12, where RMSEF ranged from 0.016 to 0.026 mg/mL. For such runs, which correspond to the smaller feed concentrations, the error was mostly due to a shift of the predicted BT time with respect to the measured one. In spite of this, it appears that the slope of the predicted BT curves in the inflection point was very similar to the measured one. This result could be explained by an underestimation of the effective column capacity. However, it is difficult to identify the exact origin of this disagreement.

The model ability of predicting new runs was tested by applying the LKM to the first six BT curves (BT#1–6), which were not included in the fitting process of the LKM. The results of the predictions are shown in Figure 6, where the red dots represent HPLC measurements and the blue line the LKM predictions.

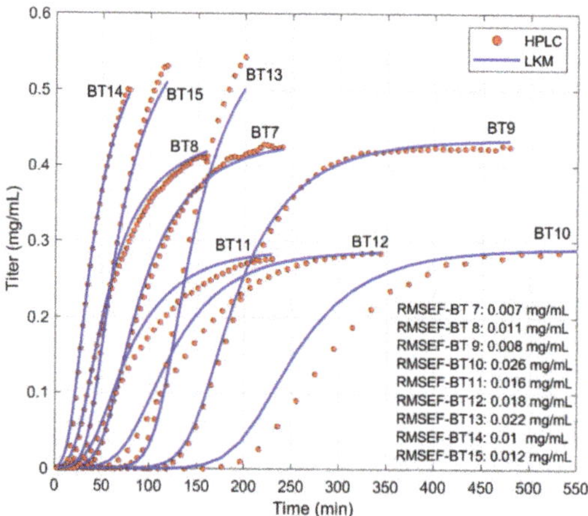

Figure 6. Mechanistic model (LKM) predictions of titer for BT# 1–6 with parameter values fitted on BT# 7–15 (blue) compared to HPLC off-line measurements (red).

It can be observed that the model was able to predict the shape of the BT curves and, in particular, the steepness, indicating a good estimation of the mass transport properties. Additionally, the saturation level seemed to be well predicted, since there is no significant mismatch between estimated and measured times for reaching saturation conditions. However, BT#1–4 showed a significant offset, leading to RMSEP values up to 0.039 mg/mL. This may be related to the fact that the model might not have precisely captured the adsorption mechanism at lower feed concentrations, which could also explain the offsets in Figure 5. Moreover, one can also observe a different behavior in the curves at early BT times. The measured BT curves seem sharper than what was predicted by the model. Again, this might be due to unaccounted differences in the feed composition, resin aging, column packing quality or a more complex behavior of the system than described by the model. Of course, more complex models could be introduced to improve the description of, particularly, the mass transfer process [31,38]. On the contrary, the good ability of the model to predict the shape of the BT curve well in spite of its generality and simplicity makes it a good candidate for its application in the frame of the EKF, where such inaccuracies could be corrected in real-time by experimental measurements.

3.3. Extended Kalman Filter Tuning

Before applying the EKF concept, the filter design parameters R, $\hat{x}(t_0^+)$, Q and k_Q needed to be carefully selected and tuned. For this, the variance of the Raman-PLS model ($RMSECV_{PLS}^2$) was used as the measurement noise R, while the process noise Q was computed on-line at any given t_k, as described in Section 2.6. Since mismatches between the LKM predictions and external data set were expected, a rotation specific mismatch factor k_Q was applied. In the following, the determination of k_Q for ROT1 ($k_Q^{Rot\ 1}$) is described: The EKF was applied for each BT curve of the calibration set of ROT1, i.e., BT#2–6, using the Raman-PLS calibrated for ROT1, LKM predicting the distinct BT curve and varying k_Q in a range between 10 and 1×10^6. The resulting $RMSEP_{EKF}$ as a function of k_Q for each BT curve of the training set is shown in Figure 7.

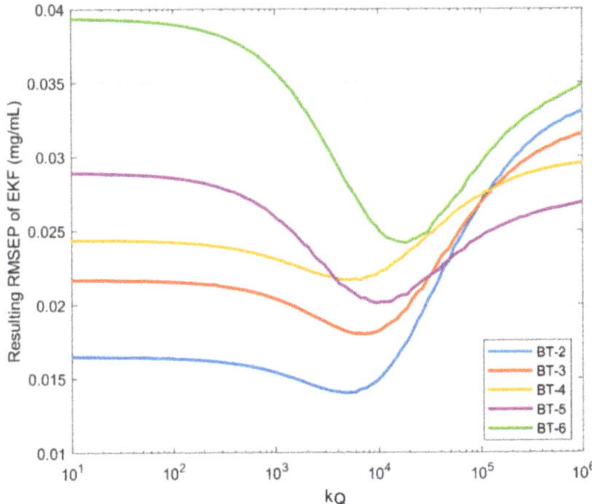

Figure 7. Error in the EKF prediction of the monoclonal antibody (mAb) concentration as a function of the model mismatch factor k_Q for each of the BT curves of the calibration set of ROT1.

It can be seen that a minimal $RMSEP_{EKF}$ could be obtained by selecting k_Q around 1×10^4 for all BT curves. Note that k_Q can be regarded as a measure of the confidence in Raman measurements versus the confidence in the mechanistic model. The higher k_Q, the more the EKF relied on Raman-PLS, while for small k_Q, the LKM was considered as more trustworthy. It is also important to note that the absolute value of k_Q depended strongly on the absolute values of the estimation of both measurement and process noise. The presence of the minimum in the middle of the investigated k_Q range indicates the beneficial effect of considering both types of information in producing the estimates, thus indicating that this approach is better than using only the LKM (small k_Q) or the Raman-PLS model (large k_Q). It can be assumed that by further increasing k_Q above 1×10^6, even higher RMSEP could be obtained, since the filter would singly rely on Raman-PLS. In contrast, when k_Q tended towards 10, it singly relied on the LKM. This procedure was repeated for all rotations, and the resulting rotation-specific model mismatch factors ($k_Q^{Rot\ i}$) are summarized in Table 5. It can be seen that the values of $k_Q^{Rot\ i}$ were similar for all rotations, thus indicating a robust confidence balancing between Raman-PLS and LKM predictions.

Table 5. Rotational specific model mismatch factor $k_Q^{Rot\ i}$ for all rotations.

	ROT1	ROT2	ROT3	ROT4	ROT5	ROT6
$k_Q^{Rot\ i}$	1.35×10^4	1.92×10^4	1.92×10^4	2.15×10^4	1.92×10^4	1.07×10^4

3.4. Extended Kalman Filter Validation

To externally validate the EKF, the rotational approach explained in Section 2.6 and Table 2, was applied for all rotations. In Figure 4A–F, the final results of Raman-PLS, the LKM and EKF for ROT1-6 are shown, respectively.

The red dots represent the off-line HPLC measurements; the continuous grey and green lines represent the Raman-PLS and LKM predictions, respectively; the black line represents the results of the EKF. As mentioned above, the Raman-PLS predictions showed significant noise, although they captured the trend of the actual BT curve. A significant prediction offset can be seen in ranges at the beginning of the breakthrough. Here, the Raman-PLS prediction was rather untrustworthy, which might be explained through difficulties in training the model on samples which did not contain the target molecule or contained a very small amount of it, close to the limit of detection. On the other hand, the LKM was able to smoothly predict the shape of the BT curve, although it showed a constant and significant error. As also noted before, the LKM tended to anticipate the onset of breakthrough, which appeared sharper in the experiments. The line of the EKF seemed, as expected, to be a combination of the curves above, with the beneficial effect of eliminating the mechanistic model offsets on one hand and smoothening the high noise of the Raman predictions on the other. The EKF was particularly reliable in the region of the incipient breakthrough, where its predictions relied mostly on the LKM results, thus eliminating the negative concentration values predicted by the Raman-PLS. On the other hand, at concentration values of 0.1 mg/mL, the offset with respect to the off-line measurements was eliminated and simultaneously reduced the noise. This same pattern exists in all other rotations, as seen by the data summarized in Table 6, where the errors in terms of RMSEP for Raman-PLS, LKM and EKF predictions are compared for all considered rotations.

Table 6. Prediction errors of the Raman-PLS, the LKM and EKF model for all rotations.

	Raman-PLS RMSEP (mg/mL)	LKM RMSEP (mg/mL)	EKF RMSEP (mg/mL)
ROT1	0.051	0.038	0.019
ROT2	0.047	0.023	0.021
ROT3	0.061	0.030	0.035
ROT4	0.072	0.030	0.024
ROT5	0.045	0.037	0.028
ROT6	0.055	0.045	0.029
Mean:	0.055	0.034	0.026

As can be observed in Table 6, the $RMSEP_{PLS}$ and $RMSEP_{LKM}$ could be reduced by applying the EKF. The only exception to this trend was ROT3, where the LKM was slightly better. In this case, both the Raman-PLS and the LKM both largely anticipated the BT time. Nevertheless, the advantage of using an EKF estimator appears very clear from this table, especially in significantly reducing the error of those rotations exhibiting a large Raman-PLS model error.

3.5. Extended Kalman Filter Perturbation

One of the major drawbacks of the LKM approach is its sensitivity to the input process parameters. This is illustrated in Figure 8 with reference to ROT1, where the effect of a 5% Gaussian distributed perturbation in feed concentration and flow rate, as well as bed porosity on the predictions of the LKM and the EKF, is compared for 200 simulations. The selected perturbed process parameters are representative of the variables that are actually subject to perturbations in real applications.

The solid line indicates the mean of the predictions of all 200 simulations using the LKM (red) and EKF (blue), whereas the shaded areas show the corresponding 68% confidence intervals, respectively. While the red shaded area is broad and clearly deviates from the reference off-line HPLC measurements, the blue shaded area is rather narrow and distributed around the reference values. It can be concluded that the LKM was strongly affected by perturbations of its input parameters, while this sensitivity was reduced for the EKF predictions due to the influence of the Raman-based estimates.

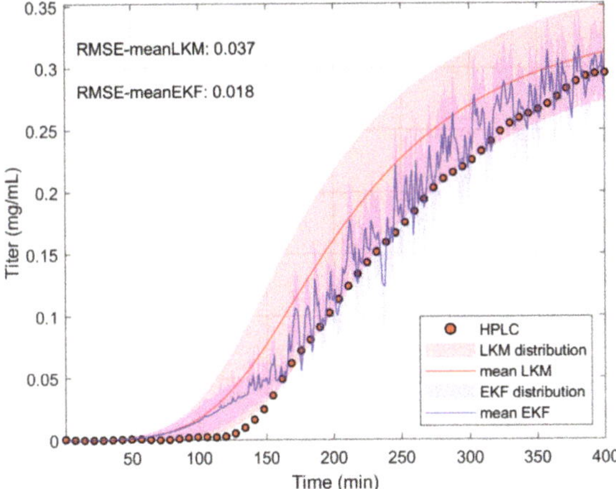

Figure 8. Effect of 5% Gaussian distributed perturbations in feed concentration and flow rate, as well as bed porosity on the predictions of the LKM (mean = dark red; standard deviation interval = light red) and the EKF (mean = dark blue; standard deviation interval = light blue) of ROT1 in 200 simulations.

4. Conclusions

In this work, an EKF estimator of the monoclonal antibody concentration at the outlet of a chromatographic column was developed. Its predictions considerably improved with respect to the use of the single Raman-PLS or the LKM. This was demonstrated for the case of a low titer harvest (mAb conc. < 0.42 mg/mL), which is typical for perfusion bioreactors. In general, the PLS-based predictive Raman models were able to capture the shape of the breakthrough curves, but the obtained results were too noisy for practical applications—for example, in the direction of process control. On the other hand, the mechanistic LKM, properly tuned on an external data set, was able to capture the qualitative shape of the breakthrough curves, but it exhibited deviations that were too large with respect to the off-line reference measurements. The proper application of the EKF requires the preliminary estimation of the parameter k_Q, which is responsible for weighting the contribution of the Raman-PLS and the LKM predictions in the final estimate of the filter. By applying the tuned EKF to an external data set, its superiority as compared to Raman-PLS and LKM became obvious. While the LKM predictions served as a solid backbone for the EKF, the Raman-PLS real-time information updated the state estimates and significantly reduced the LKM offset. Though the LKM showed, in some cases, comparable prediction errors, the perturbation analysis showed the additional benefit of EKF through the increased robustness with respect to the model input parameter values. It is worth noting that the RMSE-EKF of 0.026 mg/mL on a range of 0–0.42 mg/mL is very close to the analytical standard deviation of the reference off-line HPLC measurements (0.01 mg/mL). However, it needs to be mentioned that the low detection limit of Raman spectroscopy becomes critical at very low protein concentrations. Here, the LKM could of course be of help, and the EKF predictions should rely mostly on these values. To fully benefit from the LKM, it should be specifically tuned in the region of low concentrations, which is at the incipient breakthrough. Nevertheless, the EKF performance reported in this work is already sufficient for the implementation in the frame of the control of a capture chromatographic step, where the range of interest is around 70% of the breakthrough value [39]. It can be concluded that the EKF is a powerful tool for smart sensors and should be considered more often for monitoring and control within bioprocesses. In the future, this approach might be extended to other applications where deterministic knowledge is available, like in the monitoring of protein aggregation [40], crystallization [41] or in-line buffer preparation. At the same time, research activities on the different components of a Raman analyzer towards increased signal intensities should be continued to further increase the prediction accuracy and reduce the measurement duration.

Supplementary Materials: The following are available online at http://www.mdpi.com/2227-9717/7/10/683/s1, Figure S1: Acquired Raman spectra of the calibration set of ROT1.

Author Contributions: Conceptualization, F.F.; formal analysis, M.F.L.; investigation, S.G. and S.V.; validation, J.S. and H.B.; supervision, M.M. and A.B.

Funding: This research was funded by the KTI (CTI)-Program of the Swiss Economic Ministry Project 19190.2 PFIW-IW.

Acknowledgments: We would like to acknowledge Merck-Serono SA for providing the cell line and nutrition media, ChromaCon AG and Kaiser Optical Systems Inc. for the discussions and partnership as well as Dr. Moritz Wolf for producing the harvest.

Conflicts of Interest: The authors declare no conflicts of interest.

References

1. Glassey, J.; Gernaey, K.V.; Clemens, C.; Schulz, T.W.; Oliveira, R.; Striedner, G.; Mandenius, C.-F. Process Analytical Technology (PAT) for Biopharmaceuticals. *Biotechnol. J.* **2011**, *6*, 369–377. [CrossRef] [PubMed]

2. Rathore, A.S.; Bhambure, R.; Ghare, V. Process Analytical Technology (PAT) for Biopharmaceutical Products. *Anal. Bioanal. Chem.* **2010**, *398*, 137–154. [CrossRef] [PubMed]

3. Simon, L.L.; Pataki, H.; Marosi, G.; Meemken, F.; Hungerbühler, K.; Baiker, A.; Tummala, S.; Glennon, B.; Kuentz, M.; Steele, G.; et al. Assessment of Recent Process Analytical Technology (PAT) Trends: A Multiauthor Review. *Org. Process Res. Dev.* **2015**, *19*, 3–62. [CrossRef]

4. Read, E.K.; Park, J.T.; Shah, R.B.; Riley, B.S.; Brorson, K.A.; Rathore, A.S. Process Analytical Technology (PAT) for Biopharmaceutical Products: Part I. Concepts and Applications. *Biotechnol. Bioeng.* **2010**, *105*, 276–284. [CrossRef] [PubMed]

5. Sommeregger, W.; Sissolak, B.; Kandra, K.; von Stosch, M.; Mayer, M.; Striedner, G. Quality by control: Towards model predictive control of mammalian cell culture bioprocesses. *Biotechnol. J.* **2017**, *12*, 1600546. [CrossRef] [PubMed]

6. Rüdt, M.; Briskot, T.; Hubbuch, J. Advances in Downstream Processing of Biologics—Spectroscopy: An Emerging Process Analytical Technology. *J. Chromatogr. A* **2017**, *1490*, 2–9. [CrossRef] [PubMed]

7. Santos, R.M.; Kessler, J.-M.; Salou, P.; Menezes, J.C.; Peinado, A. Monitoring MAb cultivations with in-situ raman spectroscopy: The influence of spectral selectivity on calibration models and industrial use as reliable PAT tool. *Biotechnol. Prog.* **2018**, *34*, 659–670. [CrossRef] [PubMed]

8. Berry, B.; Moretto, J.; Matthews, T.; Smelko, J.; Wiltberger, K. Cross-scale predictive modeling of CHO cell culture growth and metabolites using raman spectroscopy and multivariate analysis. *Biotechnol. Prog.* **2015**, *31*, 566–577. [CrossRef]

9. Craven, S.; Whelan, J.; Glennon, B. Glucose concentration control of a fed-batch mammalian cell bioprocess using a nonlinear model predictive controller. *J. Process Control* **2014**, *24*, 344–357. [CrossRef]

10. Feidl, F.; Garbellini, S.; Vogg, S.; Sokolov, M.; Souquet, J.; Broly, H.; Butté, A.; Morbidelli, M. A new flow cell and chemometric protocol for implementing in-line raman spectroscopy in chromatography. *Biotechnol. Prog.* **2019**. [CrossRef]

11. Buckley, K.; Ryder, A.G. Applications of Raman spectroscopy in biopharmaceutical manufacturing: A short review. *Appl. Spectrosc.* **2017**, *71*, 1085–1116. [CrossRef] [PubMed]

12. Esmonde-White, K.A.; Cuellar, M.; Uerpmann, C.; Lenain, B.; Lewis, I.R. Raman spectroscopy as a process analytical technology for pharmaceutical manufacturing and bioprocessing. *Anal. Bioanal. Chem.* **2017**, *409*, 637–649. [CrossRef] [PubMed]

13. Lewis, I.R. *Handbook of Raman Spectroscopy*; CRC Press: Boca Raton, FL, USA, 2001. [CrossRef]

14. Guiochon, G. Preparative liquid chromatography. *J. Chromatogr. A* **2002**, *965*, 129–161. [CrossRef]

15. Auger, F.; Hilairet, M.; Guerrero, J.M.; Monmasson, E.; Orlowska-Kowalska, T.; Katsura, S. Industrial applications of the Kalman Filter: A Review. *IEEE Trans. Ind. Electron.* **2013**, *60*, 5458–5471. [CrossRef]

16. Krämer, D.; King, R. On-line monitoring of substrates and biomass using near-infrared spectroscopy and model-based state estimation for enzyme production by S. cerevisiae. *IFAC PapersOnLine* **2016**, *49*, 609–614. [CrossRef]

17. Stelzer, I.V.; Kager, J.; Herwig, C. Comparison of particle filter and extended kalman filter algorithms for monitoring of bioprocesses. In *Computer Aided Chemical Engineering*; Elsevier: Amsterdam, The Netherlands, 2017; Volume 40, pp. 1483–1488. [CrossRef]

18. Arndt, M.; Hitzmann, B. Kalman Filter Based Glucose Control at Small Set Points during Fed-Batch Cultivation of Saccharomyces Cerevisiae. *Biotechnol. Prog.* **2008**, *20*, 377–383. [CrossRef]

19. Kalman, R.E. A new approach to linear filtering and prediction problems. *J. Basic Eng.* **1960**, *82*, 35. [CrossRef]

20. Faragher, R. Understanding the Basis of the Kalman Filter Via a Simple and Intuitive Derivation [Lecture Notes]. *IEEE Signal Process. Mag.* **2012**, *29*, 128–132. [CrossRef]

21. Simon, D. THE H∞ FILTER. In *Optimal State Estimation*; John Wiley & Sons, Inc.: Hoboken, NJ, USA, 2006; pp. 331–371. [CrossRef]

22. Wolf, M.K.F.; Müller, A.; Souquet, J.; Broly, H.; Morbidelli, M. Process design and development of a mammalian cell perfusion culture in shake-tube and benchtop bioreactors. *Biotechnol. Bioeng.* **2019**, *116*, 1973–1985. [CrossRef]

23. Feidl, F.; Vogg, S.; Wolf, M.K.F.; Podobnik, M.; Ruggeri, C.; Ulmer, N.; Wälchli, R.; Souquet, J.; Broly, H.; Butte, A.; et al. Process-Wide Control and Automation of an Integrated Continuous Manufacturing Platform for Therapeutic Proteins. *Submiss* **2019**. submitted for publication.

24. Savitzky, A.; Golay, M.J.E. Smoothing and Differentiation of Data by Simplified Least Squares Procedures. *Anal. Chem.* **1964**, *36*, 1627–1639. [CrossRef]

25. Barnes, R.J.; Dhanoa, M.S.; Lister, S.J. Standard normal variate transformation and de-trending of near-infrared diffuse reflectance spectra. *Appl. Spectrosc.* **1989**, *43*, 772–777. [CrossRef]

26. Van den Berg, R.A.; Hoefsloot, H.C.; Westerhuis, J.A.; Smilde, A.K.; van der Werf, M.J. Centering, scaling, and transformations: Improving the biological information content of metabolomics data. *BMC Genom.* **2006**, *15*, 1–15. [CrossRef]

27. Wold, S.; Sjöström, M.; Eriksson, L. PLS-Regression: A Basic Tool of Chemometrics. *Chemom. Intell. Lab. Syst.* **2001**, *58*, 109–130. [CrossRef]

28. Kohavi, R. A study of cross-validation and bootstrap for accuracy estimation and model selection. *IJCAI* **1995**, *14*, 1137–1145.

29. Baur, D.; Angarita, M.; Müller-Späth, T.; Morbidelli, M. Optimal model-based design of the twin-column capturesmb process improves capacity utilization and productivity in protein A affinity capture. *Biotechnol. J.* **2016**, *11*, 135–145. [CrossRef] [PubMed]

30. Hahn, R.; Schlegel, R.; Jungbauer, A. Comparison of protein A affinity sorbents. *J. Chromatogr. B* **2003**, *790*, 35–51. [CrossRef]

31. Guiochon, G.; Shirazi, D.G.; Felinger, A.; Katti, A.M. *Fundamentals of Preparative and Nonlinear Chromatography*; Academic Press: Cambridge, MA, USA, 2006.

32. Ng, C.K.S.; Osuna-Sanchez, H.; Valéry, E.; Sørensen, E.; Bracewell, D.G. Design of high productivity antibody capture by protein A chromatography using an integrated experimental and modeling approach. *J. Chromatogr. B* **2012**, *899*, 116–126. [CrossRef]

33. Valappil, J.; Georgakis, C. Systematic estimation of state noise statistics for extended kalman filters. *AIChE J.* **2000**, *46*, 292–308. [CrossRef]

34. Schneider, R.; Georgakis, C. How to not make the extended Kalman filter fail. *Ind. Eng. Chem. Res.* **2013**, *52*, 3354–3362. [CrossRef]

35. Kuo, A.D. An optimal state estimation model of sensory integration in human postural balance. *J. Neural Eng.* **2005**, *2*, S235–S249. [CrossRef] [PubMed]

36. Jazwinski, A.H. *Stochastic Processes and Filtering Theory*; Courier Corporation: North Chelmsford, MA, USA, 2007.

37. Wen, Z. Raman Spectroscopy of Protein Pharmaceuticals. *J. Pharm. Sci.* **2007**, *96*, 2861–2878. [CrossRef] [PubMed]

38. Carta, G.; Jungbauer, A. *Protein Chromatography*; Wiley: Weinheim, Germany, 2010. [CrossRef]

39. Steinebach, F.; Angarita, M.; Karst, D.J.; Müller-Späth, T.; Morbidelli, M. Model based adaptive control of a continuous capture process for monoclonal antibodies production. *J. Chromatogr. A* **2016**, *1444*, 50–56. [CrossRef] [PubMed]

40. Nicoud, L.; Owczarz, M.; Arosio, P.; Morbidelli, M. A multiscale view of therapeutic protein aggregation: A colloid science perspective. *Biotechnol. J.* **2015**, *10*, 367–378. [CrossRef] [PubMed]
41. Rupp, B. Predictive models for protein crystallization. *Methods* **2004**, *34*, 390–407. [CrossRef] [PubMed]

 *processes*

Article

Bioenvironmental Zonal Controlling of Incubated Avian Embryo Using Localised Infrared Heating

Ali Youssef, Tomas Norton * and Daniel Berckmans

Department of Biosystems, Animal and Human Health Engineering Division, M3-BIORES: Measure, Model & Manage of Bioresponses Laboratory, KU Leuven, Kasteelpark Arenberg 30, 3001 Heverlee, Belgium; ali.youssef@kuleuven.be (A.Y.); daniel.berckmans@kuleuven.be (D.B.)
* Correspondence: tomas.norton@kuleuven.be

Received: 25 August 2019; Accepted: 17 September 2019; Published: 23 September 2019

Abstract: The main objective of any bioenvironmental controller is to create favourable bioenvironmental conditions around the living-system. In industrial incubation practice of chicken embryo, it is sometimes difficult to fill large incubators with uniform eggs, which leads to suboptimal results. The ideal incubation solution is a machine that is capable of coping with all sorts of variabilities in eggs. This can be realised in practice by creating different zones of different environmental conditions within the same machine. In the present study, a two-levels controller was designed and implemented to combine both convective and radiative heating to incubate eggs. On the higher level, three model-predictive-control (MPC) constrained controllers were developed to regulate the power applied to nine IR-radiators divided into three zones based on continuous feedback of the eggshell temperatures in each zone. On the lower level, a PID controller was used to maintain the air temperature within an experimental incubator at a fixed level (34 °C) lower than the standard incubation temperature. Four full incubation trials were carried out to test and implement the developed zonal controllers. The implementation results showed that the developed controllers were able to follow the reference trajectory defined for each zone. It was possible to keep the eggshell temperatures within the middle region (zone) different from the sidelong regions (zones) while the air temperature kept constant at 34 °C. The average hatching result (HOF) of the four full incubation trial was 84.0% (±0.5). The developed two-levels control system is a promising technique for demand-based climate controller and to optimizing energy use by using multi-objectives MPCs with constraint on total energy consumption.

Keywords: bioenvironmental control; model-predictive controller; zonal controlling; dynamic modelling

---

1. Introduction

The production of meat and eggs worldwide is increasing because of the growing population and the high demand of animal proteins [1]. In the poultry industry, meat and egg production sectors require large-scale incubation of eggs at a hatchery and a well-controlled and monitored environment [2].

Eggs of different origins and with different pre-incubation treatments are put together in an incubator, resulting in a non-uniformity between the hatching times of the different eggs [3]. This non-synchronised time of hatch, referred to as a large hatch window, is negative in terms of animal welfare and post-hatching performance, as the chicks are deprived from food and water (until the rest of the eggs are hatched and transferred to the farm) [4–7]. Hatcheries therefore have a strong objective to synchronise the hatch time.

In practice, the incubation process takes place in two different machines, namely, (i) the setter (from incubation day 0 to day 16–18) where the embryos stay during the largest part of their development and (ii) the hatcher (day 17–19 to day 20) where the embryos hatch.

In practice, today's tendency is that incubators are becoming larger. This has however implications on hatchery management. In practice, as hen houses are not increasing in size, it becomes more difficult to fill such larger incubators with uniform eggs from a single flock and storage time. Therefore, it regularly happens that eggs from two or three different flocks are combined in one machine, which can lead to sub-optimal results. For instance, putting eggs from a flock with an age of 30 weeks with those from another flock of 60 weeks of age in the same incubator will certainly lead to a decrease in the number of hatched chicks, poorer chick quality, higher post-hatch mortality, etc., under present conditions where all eggs in the machine are treated equally [8–10].

The ideal incubation solution is a machine that is capable of coping with all sorts of variabilities in eggs (e.g., flock age, strain, storage time, etc.,). This can be realised in practice by creating different zones with different environmental conditions within the same machine. Localised manipulation of the environmental variables inside the incubator (mainly temperature) is a potential solution to create different zones within the incubator space. The creation of different temperature zones by means of air-heat flow control is a very complex and expensive process [11–14]. The best option, according to many scientists, is the radiation (radiant) method of heating. In practice, for the creation of localized zones with a higher temperature, heating systems based on electric radiation (infrared) heaters have been widely used [11]. Radiation (infrared) heating systems radiantly heat surfaces rather than air volumes, which allows them to be used to heat individual zones of eggs, which cannot be achieved with conventional forced convective heating.

During incubation, the thermoregulatory system of the chicken embryo evolves through different stages from a poikilothermic to a homeothermic system [15,16]. The incubated egg is considered as a complex, individually different, time varying and dynamic (CITD) system as introduced by Berckmans et al. (e.g., [15,17,18]). Hence, the dynamic thermal response of the fertile egg to changes in ambient temperature is different from one day to another during the embryonic development [15]. As such, modelling and controlling a biological system such as the fertile incubated eggs is more complex than modelling of non-living physical systems (such as electric circuits). Most of the biological responses including heat production of incubated eggs are results of a complex network of interactions among many components inside the egg. We have ([12]) successfully implemented the multi-zonal controlling in an empty ventilated chamber using a multi-objective proportional-integral-plus (PIP). However, implementing such multi-zonal controller in a bio-environment around living systems (e.g., incubated embryo) is a great challenge because of the inherent non-linearity of the system. One of the challenges that are engaged with controlling different thermal zones simultaneously (multi-zonal control) is the controllability of the system under question. The controllability property of the system plays a crucial role in many control problems, such as stabilization of unstable systems by feedback, or optimal control [19,20]. The system's controllability can be roughly defined as an ability to do whatever we want with our system, or in more technical terms, the ability to transfer our system from any initial state to any desired final state in a finite time [18]. The challenge facing us to control multi-thermal-zones was to control the temperature in a certain number of zones inside the test chamber with a minimal number of control variables. Model-based control techniques, such as model-predictive-control (MPC) are suitable approaches to handle the inherent nonlinearity of the living systems and the interaction of the different thermal zones. The MPC is well-known and frequently used in the industry for optimal control of time-varying systems with constraints [21]. MPC benefits from simple and intuitive tuning and the ability to control a range of simple and complex phenomena, including systems with time delays, non-minimum phase dynamics, and instability [22]. Additionally, the framework of MPC incorporates straightforwardly system's constraints and multiple operating conditions, exhibits an intrinsic compensation for dead time, and provides the flexibility to formulate and tailor a control objective [21,22].

The main objective of this paper is to investigate the possibility of zonal controlling the bio-environment of incubated chicken embryo by combining forced convection and localised infrared heat using adaptive predictive-controlling approach.

2. Materials and Methods

2.1. Experimental Setup

2.1.1. Experimental Incubator

Experiments were carried out in a small-scale experimental incubator (see Figure 1) at the division of Measure, Model and Manage Bio-responses (M3-BIORES), Leuven University (KU Leuven), Belgium. The experimental incubator is composed of the main chamber and the air preparation chamber whose inner dimension is 0.8 × 0.6 × 0.4 m and 0.8 × 0.25 × 0.4 m (l × w × h), respectively (Figure 1). Both chambers have 0.04 m thick surrounding walls made of propylene. The air preparation chamber is equipped with an air re-circulation, one inlet opening for fresh air, four re-circulation openings, and a pipe system (Figure 1) with a three-way valve to regulate the flow rate of fresh air (refreshment) and re-circulated air coming from the main chamber. The fresh air and re-circulated air were mixed within the air preparation chamber to be pumped into the main chamber through two inlet pipes (Figure 1). Volumetric flow rate, of fresh air versus that of the re-circulated air, into the main chamber is conditioned by the open degree of the three-way valve. With 100% valve open, maximum ventilation rate of 0.17 m^3 h^{-1} (0.885 volume refreshment per hour) was achieved with no air re-circulation. The main chamber is equipped with two heater fans connected to the two inlet pipes, which create forced air ventilation and supply maximum total heat of 200 W in the system (Figure 1). The prepared air (in the air preparation chamber) is pumped through the heater fan to be heated up, if the heaters were on, before flowing into the main chamber (Figure 1). The air inside the main chamber is exhausted through the four openings on the top sidewall. Portion of the exhausted air (based on the control suggestion) was mixed with the fresh inlet air and the rest was removed out of the incubator through the main outlet opening. A mixing fan was positioned in the middle of the main chamber to accelerate the mixing of the inlet air with the air inside the chamber. The main chamber was designed to hold a standard incubation egg tray above the mixing fan and the air inlet pipes. The used egg tray was a Petersime N.V. standard setter tray (B00568) for chicken eggs with 150 egg places arranged in a 10 × 15 matrix and made of polypropylene. Air temperature, eggshell temperature and relative humidity inside the test chamber were automatically controlled using the Petersime Focus^{TM} controller. Control actions are calculated based on continuous feedback from a temperature/humidity probe, which is placed inside the main chamber. The controller compares the measurements from these sensors with the set points to make the decision of heating up by turning on the heater fan, cooling down by cool air ventilation or humidifying by supply steam from the steam generator or dehumidifying by dry air ventilation. The Petersime Focus^{TM} control box is connected to the data acquisition and control PC where the incubation process and control actions were programmed using the Petersime Focus Software (v.2.0).

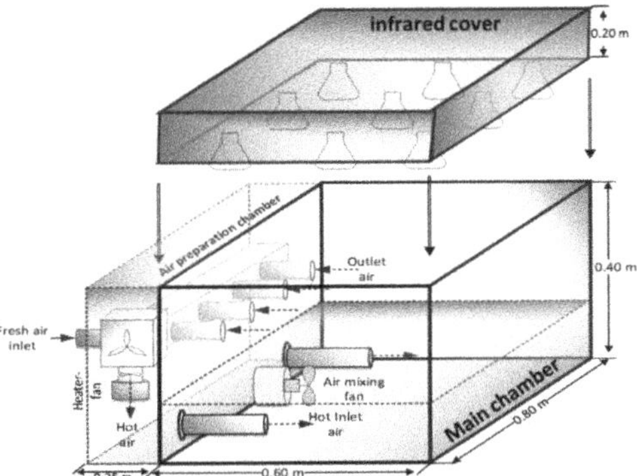

Figure 1. Schematic representation of the experimental incubator showing the main chamber, air preparation chamber, and the infrared cover.

2.1.2. Infrared Cover

The cover of the experimental incubator was designed to provide infrared heating in the top of the incubated eggs. The infrared cover was equipped of nine ceramic infrared (IR) radiators (lamps), of the type Elstein® IOT/75. The used Elstein IOT/75 lamps were ceramic infrared dark radiators with maximum power of 100 Watts and operating infrared wavelength range of 3–10 μm. The IR lamps were equipped with E27 threads that can be screwed-in like bulbs into porcelain sockets. Nine porcelain sockets were fixed on three metal bars with three sockets each (Figure 2). The three metal bars were fixed to the inner side of a wooden frame (with an inner dimension 0.8 × 0.60 × 0.20 m) in such a way that the IR lamps were facing downward (Figure 2). The distance (*h*) between the IR lamps and the eggs was adjustable, within the range h = 0.1 and 0.3 m, using four adjustment knobs (Figure 2, top view). A plexiglass cover was placed on the top of the wooden frame. A rubber washer was placed between the wooden frame and the incubator chamber to prevent air leakage. The nine IR lamps were divided into three groups, each consisting of three IR lamps. The power applied to each of the three groups were individually controlled via a model-predictive-controller (MPC) designed for this purpose (see Section 2.4).

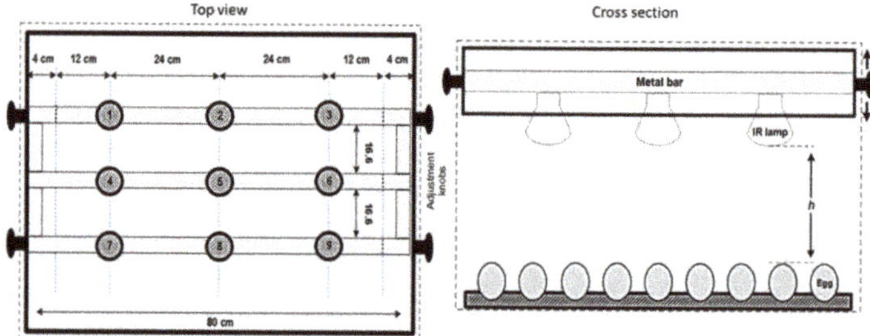

Figure 2. Schematic representation of the infrared cover showing the nine IR lamps (top view graph) and the adjustable distance from the eggs (h) using the adjustment knob (cross section graph).

2.2. Measurements and Data Acquisition

Measurements from incubator's built-in air temperature/humidity probe and CO_2 were recorded every two minutes and saved together with the controller set points in the data acquisition PC. The egg tray was spatially divided (through the short side) into three regions, each of which consisted of 50 eggs and facing one corresponding group of three IR lamps (Figure 3). In each region, two thermocouples (type-T) were placed on the equators of two eggs to represent the average eggshell temperature within each region (Figure 3). All the sensors were covered with aluminium foil for protection from overheating caused by the direct exposition to radiation heating.

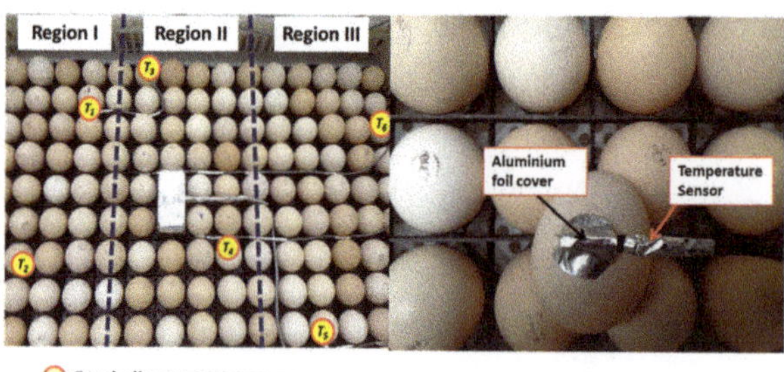

Figure 3. The location of the eggshell temperature sensors within each region (I, II and III) in the experimental incubator (left picture) and the temperature sensor placed on the equator of the eggshell covered with an aluminium foil to be protected from the IR heating (right picture).

2.3. Experiments

2.3.1. Pilot Experiments

In total, 12 pilot experiments were conducted to investigate the thermal profile (i.e., the main effective diameter of thermal radiation over the eggs surface, which is a function of the distance h between the lamp and the eggs) of the IR lamps. Additionally, these pilot experiments were conducted to investigate the maximal allowable operating power applied to the IR lamps to avoid over heating of the incubated eggs (to define the controller constraints). To investigate the thermal profile of the IR lamps over the incubated eggs a thermal camera (VarioCAM®, *InfraTech*) with 640×480 thermal resolution, is used. Two infertile eggs were placed on a small tray and one IR lamp was suspended from the top with an adjustable distance (h) from the surface of the two eggs.

Two infertile eggs ("equipped eggs") were equipped with two thermocouples. One thermocouple was fixed to the eggshell on the equator to measure the eggshell temperature (T_{egg}) and another was inserted 2 cm inside the egg, through a drilled hole in the narrow end of the egg (Figure 4), to measure the egg core temperature. The equipped eggs (Figure 4) were used to investigate how much the egg's internal temperature differs from the eggshell temperature and how fast the heat transferred from the eggshell to the internal parts of the eggs when using the IR heating. This enables the definition of the optimal constraints necessary for designing the IR controller to avoid any harms to the living embryo during incubation.

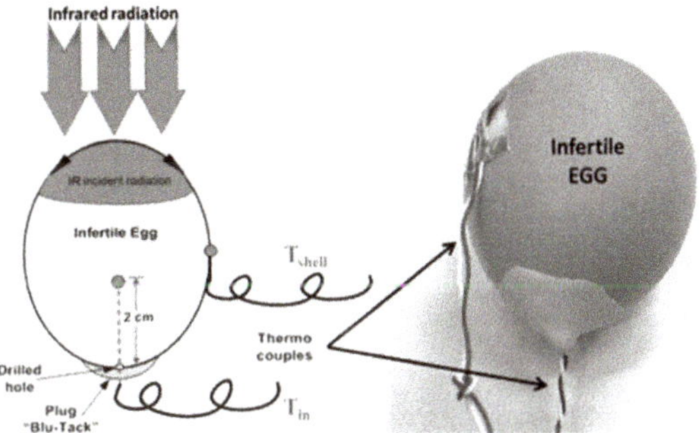

Figure 4. The "equipped egg" is an infertile egg, which was equipped with two thermocouples (type-T), one being placed on the equator of the egg, and the other one placed inside the egg through a drilled hole in the narrow end of the egg.

2.3.2. Control and System Identification Experiments

During the course of the research work reported in this paper, a set of five-step experiments were conducted to model the dynamic responses of eggshell temperatures to changes in the power (pulse width modulation, or PWM) applied to the IR radiators. The main goal was to develop a model predictive controller (MPC) to regulate, locally, the eggshell temperature of the incubated eggs within the incubator. The power applied to the IR radiators was manipulated by changing the duty cycle (percentage) of the PWM signals. The step experiments were carried out by applying step changes in the PWM duty cycle (percentage) to the IR radiators over the range 10–20%, while maintaining a fixed distance ($h = 0.15$ m, obtained from the pilot experiments) between the radiators and the eggs and constant air temperature ($T_{air} \approx 34$ °C). Figure 5 shows an example of the applied step changes in the power applied to the IR radiators and the corresponding dynamic response of the eggshell temperature for two incubated eggs in region II.

2.3.3. Full Incubation Trials and Controller Implementation

Four full incubation trials were carried out to implement the developed MPCs. In order to test the performance of the developed predictive controllers in regulating the eggshell temperatures within the three regions (I, II and III) of the experimental incubator, 150 eggs were incubated with 50 eggs per region until hatching. During each incubation trial, reference trajectories of different set points were applied to evaluate the performance of each controller. To test the capability of the developed MPCs to create different thermal zones of eggshell temperature within the IRinc1 the reference trajectories to the three MPC's were defined in such a way that the eggshell temperatures within the middle regions (region II) was kept different (from day 10 until day 6) from the two sidelong regions (regions I and III).

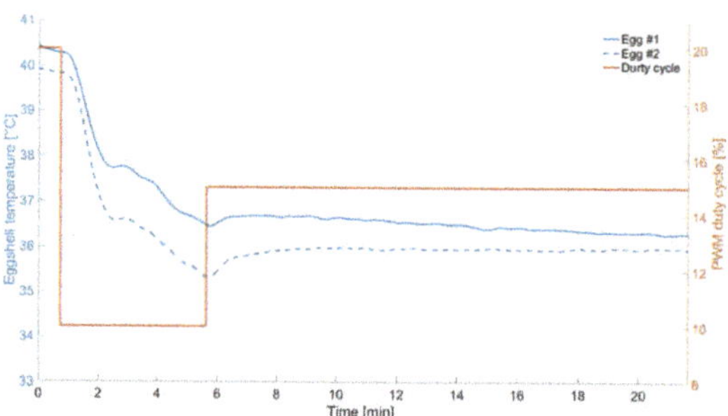

Figure 5. Step changes in the pulse width modulation (PWM) duty cycle (%) and the corresponding responses of the eggshell temperatures of two eggs in region II.

2.4. Model Predictive Controller (MPC) and System Identification

The proposed control strategy in this paper is based on controlling the temperature of the incubated eggs by combining the convective and radiative heating mechanisms (Figure 6). The forced convective heating/cooling was controlled using the Petersime Focus^{TM} controller and set to maintain the air temperature (T_{air}), within the main chamber, at a lower value (set point $\approx$ 34 °C) than the standard temperature for egg incubation (T_{std} = 37.8 °C). The radiative heating, using the IR radiators, was used to bring the eggshell temperature (T_{egg}) to the desired reference value $R_{Tegg}(k)$ at time k. A model predictive controller (MPC) was developed with the objective of maintaining the eggshell temperature (T_{egg}) around a certain predefined desired reference trajectory (R_{Tegg}) along the incubation period, which was achieved by actively manipulating the power applied to the IR radiators. The input power applied to the IR radiators was implemented through the pulse-widths modulated (PWM) signals generated using NI USB6251 interface.

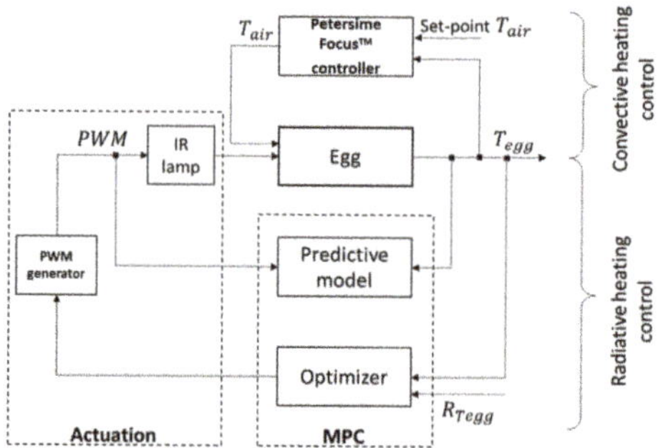

Figure 6. Block diagram representing the control strategy to combine both convective and radiative heating to control the eggshell temperature using model-predictive-controller (MPC) to regulate the eggshell temperature of incubated eggs.

2.4.1. System Identification and Parameter Estimation

A single-input, single-output (SISO) discrete transfer function (DTF) model was used to describe the static and dynamic responses of the eggshell temperature (T_{egg}) to step changes in the power (PWM duty cycle '%') applied to the IR radiators. The model has the following general structure [23]:

$$T_{egg}(k) = \frac{B\left(z^{-1}\right)}{A\left(z^{-1}\right)} u(k - \delta) - \xi(k) \tag{1}$$

where: $T_{egg}(k)$ is the output (average eggshell temperature per region) at time k; $u(k)$ is the input (PWM duty cycle to the IR radiators) at time k (min); $\xi(k)$ is additive noise, assumed to be a zero mean, serially uncorrelated sequence of random variables with variance σ^2 accounting for measurement noise, modelling errors and effects of unmeasured inputs to the process.

The two polynomials $A\left(z^{-1}\right)$ and $B\left(z^{-1}\right)$ are given by:

$$\begin{aligned}
A\left(z^{-1}\right) &= 1 + a_1 z^{-1} + a_2 z^{-2} + \ldots + a_n z^{-na} \\
B\left(z^{-1}\right) &= b_0 + b_1 z^{-1} + b_2 z^{-2} + \ldots + b_m z^{-nb}
\end{aligned} \tag{2}$$

where a_i and b_i are the model parameters to be estimated; z^{-i} is the backward shift operator, $z^{-1} \cdot y(k) = y(k-1)$; and n, m are the orders of the respective polynomials. In the present paper, the simplified refined Instrumental variable (SRIV) algorithm was utilised in the identification and estimation of the models [24]. The appropriate model structure was identified, i.e., the most appropriate values for the triad $[n, m, \delta]$ (see Equation (1)). Two main statistical measures were employed to determine the most appropriate values of this triad. Namely, the coefficient of determination R_2^T, based on the response error; and YIC (Young's information criterion), which provides a combined measure of model fit and parametric efficiency, with large negative values indicating a model which explains the output data well and yet avoids over-parameterisation [25,26].

2.4.2. MPC and Cost Function Formulation

The general idea behind any MPC design is to select a sequence of N_c future control moves to minimise a cost function J (Equation (3)) over a prediction horizon of N_p sample times [27]. In this paper, a quadratic programming cost function with quadratic objective function and linear constraints was used. The quadratic programming form leads to smoother control actions in comparison to the linear form. The model predictive controller uses the model (Equation (1)) to predict the response of the system based on the past measured inputs and outputs. This predicted output ($\hat{T}_{egg}$) was then used to calculate the optimal input by mathematical optimisation techniques in order to reduce the difference between this output and the desired one (R_{Tegg}). This optimal input was calculated by minimizing the following cost function [21]:

$$J\left(N_1, N_p, N_c\right) = \sum_{j=N_1}^{N_p} \alpha_j \left[\hat{T}_{egg}\left(k + j | k\right) - R_{Tegg}[k + j]\right]^2 + \sum_{j=1}^{N_c} \lambda_j [\Delta u[k + j - 1]]^2 \tag{3}$$

where, $\Delta u(k)$ is the change in input (power applied to the IR radiators), $\hat{T}_{egg}(k + j | k)$ is the predicted output (eggshell temperature) sequence, $R_{Tegg}(k)$ is the desired value of the output, N_1 is the minimum of the prediction horizon, α and λ are the weighting factors. The block diagram depicted in Figure 7. shows the basic structure of the designed MPC system in the present work to control the eggshell temperature using localised IR heating.

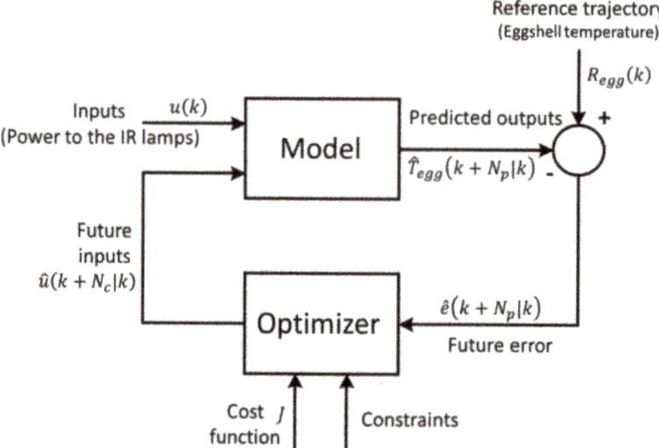

Figure 7. Block diagram representing the basic structure of the proposed MPC system to control the eggshell temperature of incubated eggs.

The model (Equation (1)) is the cornerstone of the MPC system and should be robust enough to fully capture the process dynamics [21]. In other words, the identified model should be able to describe the dynamic responses of the eggshell temperature to changes in the control input (i.e., power applied to the IR radiators). There is a wide family of MPC algorithms, each member of which is defined by the choice of the prediction model, the cost function and obtaining the control law [21]. In the present paper, the dynamic matrix control (DMC) algorithm was considered. The DMC formulation uses the step response to model the process [21,28]. The process model employed in this formulation is the step response of the eggshell temperature to step increase in the input, while the disturbance was considered constant along the prediction horizon (N_p). The procedure to obtain the predictions is as follows. As a *step response* model (Equation (1)) was employed:

$$T_{egg}(k+j) = \sum_{j=1}^{\infty} g_j \Delta u(k+j-1) \tag{4}$$

where g_j is the *step response* coefficient.

The predicted eggshell temperature along the prediction horizon is:

$$\hat{T}_{egg}(k+j|k) = \sum_{j=1}^{N_p} g_j \Delta u(k+j-1) + f(k+j) \tag{5}$$

where the $f(k+j)$ is the *free response* of the system.

Equation (5) can be written as follows:

$$\begin{bmatrix} \hat{T}_{egg}[k+1|k] \\ \hat{T}_{egg}[k+2|k] \\ \cdots \\ \hat{T}_{egg}[k+N_p|k] \end{bmatrix} = \begin{bmatrix} g_1 & 0 & \cdots & 0 \\ g_2 & g_1 & \cdots & 0 \\ \vdots & \vdots & \ddots & \vdots \\ g_{N_p} & g_{N_p-1} & \cdots & g_1 \end{bmatrix} \cdot \begin{bmatrix} \Delta u[k] \\ \Delta u[k+1] \\ \cdots \\ \Delta u[k+N_c-1] \end{bmatrix} + \begin{bmatrix} f[k] \\ f[k+1] \\ \cdots \\ f[k+N_p-1] \end{bmatrix} \tag{6}$$

or

$$\hat{T} = \mathbf{G}\mathbf{u} + \mathbf{f} \tag{7}$$

where $\mathbf{G}$ is the $N_p \times N_c$ *dynamic matrix*, $\hat{T}$ is the N_p-dimension vector contains the predicted eggshell temperatures along the prediction horizon, $\mathbf{u}$ represents the N_c-dimension vector of the control inputs and $\mathbf{f}$ is the *free response* vector. Hence, using Equation (7) the cost Function (3) can be represented in the following form [21]:

$$J = (\mathbf{G}\mathbf{u} + \mathbf{f} - \mathbf{R})^{\mathrm{T}}(\mathbf{G}\mathbf{u} + \mathbf{f} - \mathbf{R}) + \lambda \mathbf{u}^{\mathrm{T}}\mathbf{u} \tag{8}$$

$$J = \frac{1}{2}\mathbf{u}^T\mathbf{H}\mathbf{u} + \mathbf{b}^T\mathbf{u} + \mathbf{f}_0 \tag{9}$$

with $\mathbf{H} = 2\big(\mathbf{G}^T\mathbf{G} + \lambda\mathbf{I}\big)$, $\mathbf{b}^T = 2(\mathbf{f} - \mathbf{R})^T\mathbf{G}$ and $\mathbf{f}_0 = (\mathbf{f} - \mathbf{R})^T(\mathbf{f} - \mathbf{R})$.

The cost Function (9) is quadratic, therefore, the minimum is unique. The optimal input change can be calculated by setting the derivative equal to zero:

$$\frac{dJ}{d\mathbf{u}} = 2\big(\mathbf{G}^T\mathbf{G} + \lambda\mathbf{I}\big)\mathbf{u} + 2\mathbf{G}^T(\mathbf{f} - \mathbf{R}) = 0 \tag{10}$$

Then the optimal input change is given by:

$$\mathbf{u} = \big(\mathbf{G}^T\mathbf{G} + \lambda\mathbf{I}\big)^{-1}\mathbf{G}^T(\mathbf{R} - \mathbf{f}) \tag{11}$$

For the control law, as such, only the first element of vector $\mathbf{u}$ was implemented.

3. Results

3.1. Pilot Experiments

The results of the conducted experiments have shown that the optimal thermal profile of the IR lamps was achieved at distance (h) of 0.15 m, which corresponds to an effective diameter of 0.30 m.

Figure 8 shows a comparison between the temperature responses of the equipped eggs under convective heating and radiative heating (at PWM of 10%). The results (an example is shown in Figure 8) have shown that the temperature difference between the egg-core and the eggshell was vanishing (quasi zero) faster (5.2 ± 1.5 min) in case of radiative heating in comparison to convective heating (9.4 ± 1.8 min). The average time constants of eggshell temperature in case of radiative and convective heating were 7.6 ± 1.2 min and 7.45 ± 1.23 min, respectively. The steady-state eggshell temperature at different power levels (PWM) applied to the radiative lamps are shown in Figure 9. A linear regression model was fit the relation between the eggshell temperature (°C) and PWM (%) with a slope of 0.42.

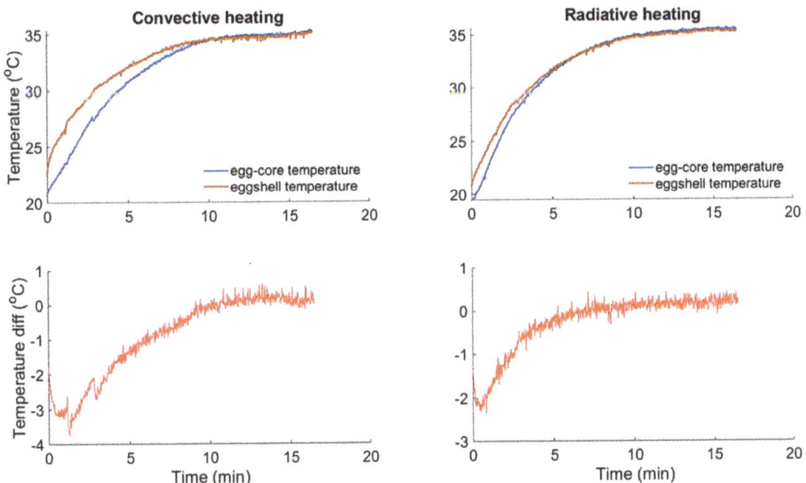

Figure 8. Step responses of eggshell and egg core temperatures to step up increases in convective-heating (left graphs) and radiative-heating (right graphs), with PWM = 10%, inside the experimental incubator. The 'temperature diff' is the difference between the egg-core and the eggshell temperatures.

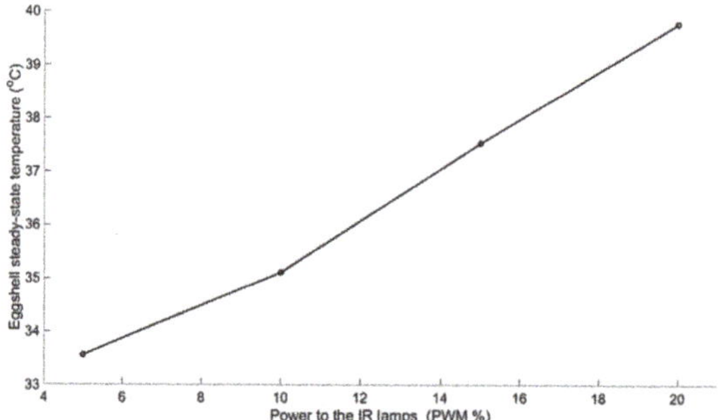

Figure 9. The resulted steady-state temperature of the eggshell at different PWM (%) to the IR lamps at distance $h = 0.15$ m.

3.2. System Identification and Predictive Model Generation

It should be stated here that the objective of this stage was to identify an approximation mode of the incubated fertile-egg system. The identification step in the present paper was tuned towards the main objective of control-oriented model design.

The SRIV algorithm, combined with the YIC and R_2^T, suggested that a second-order (number of poles, $n = 2$) DTF model with one minute pure delay ($\delta = 1$ min) was most suitable (i.e., $R_2^T = 0.98 \pm 0.01$ and $YIC = -13.00 \pm 1.45$) to describe the dynamic responses of the eggshell temperature ($T_{egg}(k)$) to step changes in the power applied to the IR radiators ($u(k)$). More specifically, the SRIV algorithm identified the following general DTF model structure (denoted by the triad [2 2 2], i.e., $n = 2$, $m = 2$ and $\delta = 2$),

$$T_{egg}(k) = \frac{b_1 z^{-1} + b_2 z^{-2}}{1 + a_1 z^{-1} + a_2 z^{-2}} u(k - \delta)$$

(12)

or in the following difference equation form,

$$T_{egg}(k) = -a_1.T_{egg}(k-1) - a_2.T_{egg}(k-2) + b_1.u(k - \delta - 1) + b_2.u(k - \delta - 2)$$

(13)

Table 1 shows the average parameter estimates for the identified model structure for the three eggs.

Table 1. The resulted model parameter estimates (average and ± standard error) obtained from 15 incubated eggs at embryonic day (ED) 15.

$A(z^{-1})$		$B(z^{-1})$		R_T^2	YIC
a_1	a_2	b_1	b_2		
$-1.997\ (\pm 0.003)$	$0.997\ (\pm 0.003)$	$0.0010\ (\pm 0.0017)$	$-0.0010\ (\pm 0.0017)$	$0.98\ (\pm 0.01)$	$-13.00\ (\pm 1.45)$

3.3. Model Predictive Control Design

3.3.1. MPC Cost Function and Constraints

During the incubation process, the thermoregulatory system of the chicken embryo evolves through different stages from a poikilothermic to a homeothermic system. Hence, the thermal response of the fertile egg to changes in ambient temperature is different from one day to another during the embryonic development [15]. Such a complex and sensitive process is subject to a number of limitations and constraints pertaining to the ranges of tolerable eggshell temperatures and acceptable IR operating power range. Table 2 shows the applied constraints on the controlled, T_{egg}, and the manipulated, $u(k)$, variables during designing the MPC system. Based on the results of pilot experiments, the control signal, $u(k)$, was constrained within the allowable PWM range between 0 and 20%. In order to prevent large increments in the rate of change in the manipulated variable $\Delta u(k)$, the maximum boundary was set as small as 0.5%. On the other hand, to prevent overheating a fast decrease in the IR heating was allowed by setting the minimum boundary to −2%. Additionally, the controlled variable, $T_{egg}(k)$, was constrained within the allowable incubation range between 36 and 40 °C.

Table 2. The predefined constraints on both the controlled and manipulated variables, which should be considered for designing the MPC system.

	Constraints
Manipulated variable (power to the IR, PWM, %)	$u(k) = (0, 20)$
Change in the Manipulated variable (%)	$\Delta u(k) = (-2, 0.5)$
Controlled variable (average eggshell temperature, °C)	$T_{egg}(k) = (34, 40)$

The developed MPC system should anticipate constraint violations and correct them in an appropriate way. Therefore, the minimization of the cost Function (9) is subject to constraints on the output ($T_{egg}(k)$, eggshell temperature), input ($u(k)$, power applied to the IR radiators) and changes in the input, $\Delta u(k)$, as the quadratic programing (QP) formulation:

$$\min_{\mathbf{u}} \frac{1}{2}\mathbf{u}^{\mathrm{T}}\mathbf{H}\mathbf{u} + \mathbf{b}^{\mathrm{T}}\mathbf{u}, \text{ subject to} \begin{cases} 0 \leq u\{k \leq 20 \\ -2 \leq \Delta u\{k \leq 0.5 \\ 34 \leq T_{egg}\{k \leq 40 \end{cases} \tag{14}$$

3.3.2. MPC Simulation

To simulate the closed-loop MPC of the eggshell temperature (controlled variable, $T_{egg}(k)$) using the power applied to the IR radiators (manipulated variable, $u(k)$) the following design parameters were defined:

1. Sampling time (or period) t_s, which determines the rate at which the control algorithm was executed. The shorter the sampling period the better controller performance to deal with fast disturbances. On the other hand, diminishing t_s can increase the computational burden. In any case, t_s should not exceed the maximal expected time necessary for running one iteration of the MPC algorithm [29]. In the current study the sampling time was chosen to be one sample ($t_s = 1$ min), which corresponds to a value ten times smaller than the average observed rise time ($t_r = 10.12 \pm 0.22$ min).

2. Defining the prediction and control horizons, which should be at least same or larger than the settling time of the system. The control horizon in general should be less than the defined prediction horizon and based on many applications an optimal control horizon should be between 20–30% of the prediction horizon to ensure smooth control actions and yet low computational costs. In the current study, the prediction N_p and control N_c horizons were set to 20 samples (N_p > average settling time = 16.20 ± 1.13 min) and five samples (25% of the prediction horizon).

3. To achieve a balanced performance between the competing control objectives (i.e., a close tracking of the set-point together with smooth control moves), the weighting factors α and λ were set to 0.9 and 0.8, respectively [30], to avoid any conflict between the control objectives.

The controller algorithm is initialized by computing the *dynamic matrix* **G** for the system (13), which is defined, based on (6), as follows:

$$\mathbf{G} = \begin{bmatrix} g_1 & 0 & 0 \\ g_2 & g_1 & 0 \\ g_3 & g_2 & g_1 \end{bmatrix} \tag{15}$$

where,

$$g_1 = -a_1.T_{egg}(k-1) - a_2.T_{egg}(k-2) + b_1.u(k-\delta-1) + b_2.u(k-\delta-2)$$
$$g_2 = -a_1.T_{egg}(k) - a_2.T_{egg}(k-1) + b_1.u(k-\delta) + b_2.u(k-\delta-1)$$
$$g_3 = -a_1.T_{egg}(k+1) - a_2.T_{egg}(k) + b_1.u(k-\delta+1) + b_2.u(k-\delta)$$

and computing the control *gain matrix* **K**, which is defined as follows:

$$\mathbf{K} = \left(\mathbf{G}^\mathrm{T}\mathbf{G} + \lambda\mathbf{I}\right)^{-1}\mathbf{G}^\mathrm{T}$$

A zero mean ($\mu = 0$) white noise term with variance (σ) of 0.1 was added to both output $T_{egg}(k)$ and input $u(k)$ signals to simulate the measurement and actuator noises (unmeasured disturbances), respectively. A simulation example of the closed-loop response of the designed MPC controller based on the DTF model (12) is depicted in Figure 10. Despite the added disturbances in both input and output signals, the simulation of the MPC closed-loop response was able to follow the reference signal (set point) efficiently with max error ($T_{egg} - R_{Tegg}$) of ±0.4 °C.

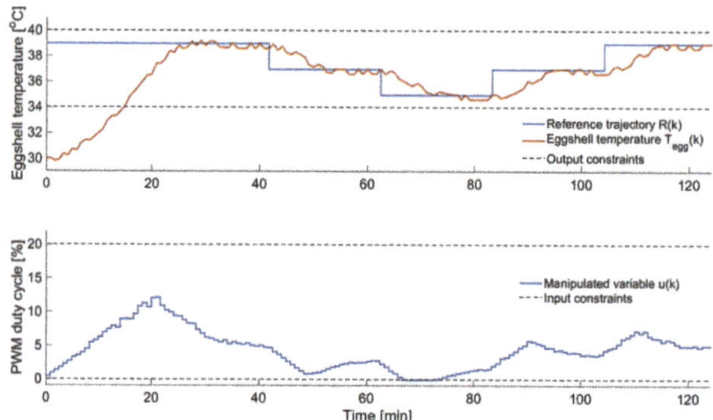

Figure 10. A simulation example of the closed-loop step response $T_{egg}(k)$ (upper graph) and unmeasured disturbances in both input and output using the designed constrained MPC controller based on the general TF model structure (12), showing the control signal $u(k)$ (lower graph). The closed-loop simulation of the developed MPC was implemented on MATLAB on a computer with Intel® 8 core i7 (2.7 GHz) processor and 16GB RAM. The average computational time for one iteration on this computer was 15.6 s.

3.3.3. MPC Implementation and Full Incubation Experiments

A two-level zonal control system was developed combining both convective and radiative heating to regulate the eggshell temperatures within three different zones (region I, II and III) simultaneously. On the higher-level three MPC systems were used to regulate the eggshell temperatures within the three regions. Additionally, a PID controller (Petersime Focus™) was employed in the lower level to regulate the incubation air temperature within the experimental incubator.

Four full incubation trials were carried out to implement and tune the developed MPC system to regulate the eggshell temperatures of incubated eggs in three different zones inside the experimental incubator (see Figure 11). To investigate the possibility of the controllers to regulate the eggshell temperatures within the three regions, the reference trajectory for region II (middle region) was different from those for sidelong regions (i.e., region I and III). The programmed reference trajectories were including some extreme set points (e.g., 34 °C), which do not follow the standard eggshell temperature (around 38 °C). Therefore, it was expected that such treatments might affect the final hatching results. Figure 11 shows an example of the implementation results of the developed MPCs to regulate the eggshell temperatures inside the experimental incubator IRin1. By employing different set points, it was possible to create two thermal zones, between days 0 and 6, (at region I and III) of incubated eggs with more or less same eggshell temperature sandwiching another with different eggshell temperature.

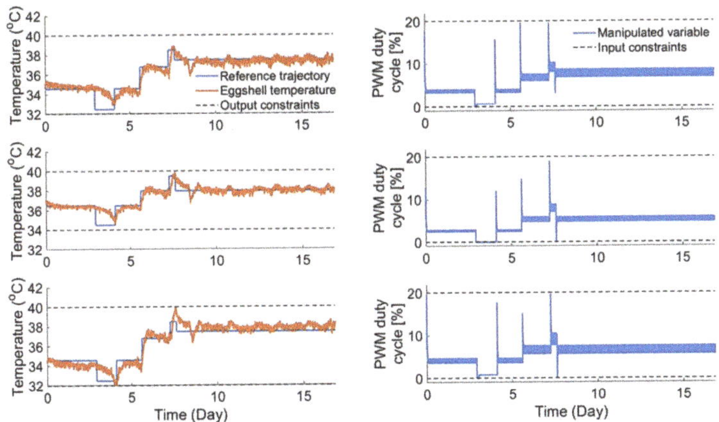

Figure 11. Example of the implementation results of the developed MPCs to regulate the eggshell temperatures at three different regions, I (upper graphs), II (middle graphs) and II (lower graphs) simultaneously.

Although the controllers were able to follow the reference trajectories in each zone (region), it was noticed (Figure 11) that the responses of the eggshell temperatures in each zone exhibited an oscillation around the set point values with an average error of ±0.5 °C. This can be attributed to the unmeasured disturbances resulting from the interaction between adjacent zones with different temperatures and to the control actions of the PID (Petersime Focus^{TM}) controller, which regulates the incubation air temperature around the eggs in the whole incubator. Another possible reason for such oscillated deviation between the actual eggshell temperature and the set point is the fact that, during this study, we have designed the MPCs based on one predictive model (12). That with the assumption that the DTF model (12) is representative of the controlled dynamic system (incubated embryo). However, in reality as shown in previous studies (e.g., [15]) the incubated embryo is inherently a nonlinear system, which exhibits different dynamics and responses almost every embryonic day. Therefore, we are proposing for future work an adaptive control approach, in which a linear model (with a fixed model structure as (12)) is estimated on the fly as the operation conditions are changing, hence the internal system-model of the MPC is updated at each scheduled time period (e.g., each day).

Previous studies (e.g., [31–34]) showed that the incubated chicken embryos are evolving at early stage of development (between incubation days ED 5–7) from an ectothermic (gaining its required heat from the surrounding environment) organism to an endothermic (produces its own heat) organism. Hence, in practice of industrial incubation, most of the energy is used to cool down the incubated embryo to the standard eggshell temperature (~38 °C). Therefore, the proposed two-level control system, which combine convective and radiative heating, is believed to be a promising technique to use the energy more efficiently during incubation. This can be achieved by locally heating up the required zones using localized IR heating (i.e., demand-based climate controlling). Additionally, a multi-objective cost function can be used to optimize the energy used by including an extra constraint on total energy consumption.

The results of the four full incubation trials showed that combining both convective and radiative heating mechanisms was successful to hatch the incubated eggs with average hatch-of-fertile (HOF = (hatched chicks/number of true fertile eggs) × 100) of 84.0% (±0.5). A breakout of the unhatched eggs was performed [35,36] in the end of each incubation trial. The average breakout results of the unhatched eggs are shown in Figure 12.

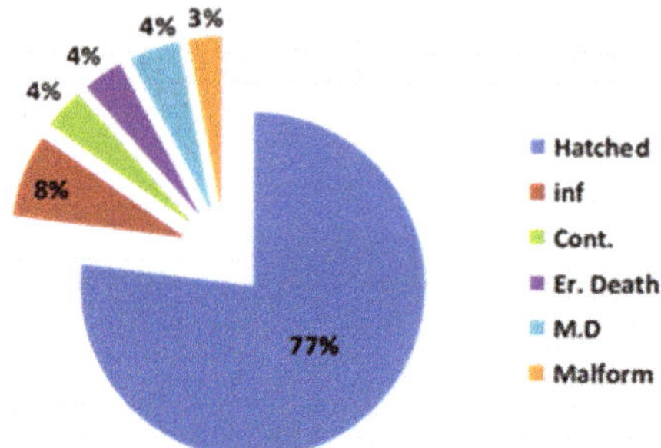

Figure 12. Average breakout results of the unhatched eggs during the four full incubation trials showing the percentage of hatched, infertile (inf), contaminated (Cont.), early death (Er. Death), middle death (M.D) and malformed embryo.

4. Conclusions

During the present study, a two-levels controller was designed and implemented to combine both convective and radiative heating to incubate eggs. On the higher level, three MPC constrained controllers were developed to regulate the power applied to nine IR-radiators divided into three zones based on continuous feedback of the eggshell temperatures in each zone. On the lower level, a PID controller (Petersime Focus^{TM}) was used to maintain the air temperature within an experimental incubator at a fixed level (34 °C) lower than the standard incubation temperature. Four full incubation trials were carried out to test and implement the developed zonal controllers. The implementation results showed that the developed controllers were able to follow the reference trajectory defined for each zone. It was possible to keep the eggshell temperatures within the middle region (zone) different from the sidelong regions (zones) while the air temperature kept fixed at 34 °C. The average hatching result (HOF) of the four full incubation trial was 84.0% (±0.5). The developed two-levels control system is a promising technique for demand-based climate controller and to optimize energy use by using multi-objectives MPCs with constraint on total energy consumption.

Author Contributions: Conceptualization, A.Y. and D.B.; methodology, A.Y.; software, A.Y.; validation, A.Y.; formal analysis, A.Y.; investigation, A.Y.; resources, D.B. and T.N.; data curation, A.Y.; writing—original draft preparation, A.Y.; writing—review and editing, A.Y. and T.N.; visualization, A.Y.; supervision, A.Y., T.N. and D.B.; project administration, T.N. and D.B.; funding acquisition, D.B.

Funding: This research was funded by the Belgian government agency for Innovation by Science and Technology (IWT), grant number IWT110404: 2011–2014.

Acknowledgments: The authors gratefully thank to the technical and financial support of Petrsime N.V. (Belgium) and the support of Petersime R&D team, namely, Luc Gabrial, Rudy Verhelst, Pascal Garain and Paul Degraeve.

Conflicts of Interest: The authors declare no conflict of interest.

References

1. OECD/FAO. *OECD-FAO Agricultural Outlook 2019–2028*; OECD: Paris, France, 2017.
2. Fasenko, G.M. Egg Storage and the Embryo. *Poult. Sci.* **2013**, *86*, 1020–1024. [CrossRef] [PubMed]
3. Bergoug, H.; Burel, C.; Guinebretiere, M.; Tong, Q.; Roulston, N.; Romanini, C.E.; Exadaktylos, V.; McGonnell, I.M.; Demmers, T.G.; Verhelst, R.; et al. Effect of pre-incubation and incubation conditions on hatchability, hatch time and hatch window, and effect of post-hatch handling on chick quality at placement. *World's. Poult. Sci. J.* **2013**, *69*, 313–334. [CrossRef]
4. Uni, Z.; Noy, Y.; Sklan, D. Posthatch development of small intestinal function in the poult. *Poult. Sci.* **1999**, *78*, 215–222. [CrossRef] [PubMed]
5. Noy, Y.; Sklan, D. Yolk and Exogenous Feed Utilization in the Posthatch Chick. *Poult. Sci.* **2001**, *80*, 1490–1495. [CrossRef] [PubMed]

6. Romanini, C.E.; Exadaktylos, V.; Tong, Q.; McGonnel, I.; Demmers, T.G.; Bergoug, H.; Eterradossi, N.; Roulston, N.; Garain, P.; Bahr, C.; et al. Monitoring the hatch time of individual chicken embryos. *Poult. Sci.* **2013**, *92*, 303–309. [CrossRef] [PubMed]

7. Tong, Q.; McGonnell, I.M.; Roulston, N.; Bergoug, H.; Romanini, C.E.; Garain, P.; Eterradossi, N.; Exadaktylos, V.; Bahr, C.; Berckmans, D.; et al. Higher levels of CO_2 during late incubation alter the hatch time of chicken embryos. *Br. Poult. Sci.* **2015**, *56*, 503–509. [CrossRef] [PubMed]

8. Gucbilmez, M.; Ozlu, S.; Shiranjang, R.; Elibol, O.; Brake, J. Effects of preincubation heating of broiler hatching eggs during storage, flock age, and length of storage period on hatchability. *Poult. Sci.* **2013**, *92*, 3310–3313. [CrossRef]

9. Hulet, R.; Gladys, G.; Hill, D.; Meijerhof, R.; El-Shiekh, T. Influence of egg shell embryonic incubation temperature and broiler breeder flock age on posthatch growth performance and carcass characteristics. *Poult. Sci.* **2007**, *86*, 408–412. [CrossRef]

10. Lourens, A. *Embryo Temperature during Incubation: Practice and Theory*; Wageningen University: Wageningen, The Netherlands, 2008.

11. Shelehov, I.Y.; Smirnov, E.I.; Inozemsev, V.P. Localized Electrical Heating System for Various Types of Buildings. *IOP Conf. Ser. Mater. Sci. Eng.* **2017**, *262*, 012083. [CrossRef]

12. Youssef, A.; Exadaktylos, V.; Ozcan, S.E.; Berckmans, D. Proportional-Integral-Plus (PIP) Control System for Individual Thermal Zones in a Small Ventilated Space. *ASHRAE Trans.* **2011**, *117*, 48–56.

13. Youssef, A.; Yen, H.; Özcan, S.E.; Berckmans, D. Data-based mechanistic modelling of indoor temperature distributions based on energy input. *Energy Build.* **2011**, *43*, 2965–2972. [CrossRef]

14. Janssens, K.; van Brecht, A.; Desta, T.Z.; Boonen, C.; Berckmans, D. Modeling the internal dynamics of energy and mass transfer in an imperfectly mixed ventilated airspace. *Indoor Air* **2004**, *14*, 146–153. [CrossRef] [PubMed]

15. Youssef, A.; Exadaktylos, V.; Berckmans, D. Modelling and quantification of the thermoregulatory responses of the developing avian embryo: Electrical analogies of a physiological system. *J. Therm. Biol.* **2014**, *44*, 14–19. [CrossRef] [PubMed]

16. Tzschentke, B.; Rumpf, M. Embryonic development of endothermy. *Respir. Physiol. Neurobiol.* **2011**, *178*, 97–107. [CrossRef] [PubMed]

17. Quanten, S.; de Valck, E.; Cluydts, R.; Aerts, J.-M.; Berckmans, D. Individualized and time-variant model for the functional link between thermoregulation and sleep onset. *J. Sleep Res.* **2006**, *15*, 183–198. [CrossRef] [PubMed]

18. Youssef, A. *Model-Based Control of Micro-Environment with Real-Time Feedback of Bioresponses*; KU Leuven: Leuven, Belgium, 2014.

19. Trentelman, H.L.; Stoorvogel, A.A.; Hautus, M. *Control Theory for Linear Systems*; Springer: London, UK, 2001.

20. Leigh, J.R. *Applied Control Theory*; Rev Sub. P. Peregrinus Ltd. on behalf of the Institution of Electrical Engineers: London, UK, 1987.

21. Camacho, E.F.; Bordons, C. *Model Predictive Control*; Springer: London, UK, 2007.

22. Kaiser, E.; Kutz, J.N.; Brunton, S.L. Sparse identification of nonlinear dynamics for model predictive control in the low-data limit. *Proc. R. Soc. A Math. Phys. Eng. Sci.* **2018**, *474*, 20180335. [CrossRef] [PubMed]

23. Godina, R.; Rodrigues, E.M.G.; Pouresmaeil, E.; Catalão, J.P.S. Optimal residential model predictive control energy management performance with PV microgeneration. *Comput. Oper. Res.* **2018**, *96*, 143–156. [CrossRef]

24. Young, P.C.; Chotai, A.; Tych, W. *Identification, Estimation and Control of Continuous-Time Systems Described by Delta Operator Models*; Kluwer Academic Publishers: Dordrecht, The Netherlands, 1991.

25. Young, P.C.; Jakeman, A. Refined instrumental variable methods of recursive time-series analysis Part III. Extensions. *Int. J. Control* **1980**, *31*, 741–764. [CrossRef]

26. Young, P.C. *Recursive Estimation and Time-Series Analysis: An Introduction for the Student and Practitioner*; Springer: Berlin/Heidelberg, Germany, 2011.

27. Camacho, E.F.; Bordons, C. Model Predictive Control. *Int. J. Robust Nonlinear Control* **1999**, *13*, 280.

28. Cutler, C.R.; Ramaker, B.C. Dynamic Matrix Control? A Computer Control Algorithm. *Autom. Control Conf.* **1980**, *17*, 72. [CrossRef]

29. Wysocki, A.; Ławryńczuk, M. On Choice of the Sampling Period and the Horizons in Generalized Predictive Control. In *Recent Advances in Automation, Robotics and Measuring Techniques*; Springer: Cham, Switzerland, 2014; pp. 329–339.

30. Borrelli, F.; Bemporad, A.; Morari, M. *Predictive Control for Linear and Hybrid Systems*; Cambridge University Press: New York, NY, USA, 2017.

31. Mueller, C.A.; Burggren, W.W.; Tazawa, H. The Physiology of the Avian Embryo Chapter 32. In *Sturkie's Avian Physiology*; Academic Press: Cambridge, MA, USA, 2015; pp. 739–766.

32. Whittow, G.C.; Tazawa, H. The early development of thermoregulation in birds. *Physiol. Zool.* **2002**, *64*, 1371–1390. [CrossRef]

33. Janke, O.; Tzschentke, B.; Höchel, J.; Nichelmann, M. Metabolic responses of chicken and muscovy duck embryos to high incubation temperatures. *Comp. Biochem. Physiol. Part A Mol. Integr. Physiol.* **2002**, *131*, 741–750. [CrossRef]

34. Tzschentke, B. Monitoring the development of thermoregulation in poultry embryos and its influence by incubation temperature. *Comput. Electron. Agric.* **2008**, *64*, 61–71. [CrossRef]

35. Fayeye, T.; Olapade, A. Hatch-Out Analysis and Repeatability Estimates of Common Hatchability Problems in Issa-Brown Breeder Stock. *Agrosearch* **2013**, *13*, 51. [CrossRef]

36. Mauldin, J.M. Breakout Analyses Guide for Hatcheries|The Poultry Site. 2003. Available online: https://thepoultrysite.com/articles/breakout-analyses-guide-for-hatcheries-1 (accessed on 25 August 2019).

Article

Model-Based Monitoring of Occupant's Thermal State for Adaptive HVAC Predictive Controlling

Ali Youssef, Nicolás Caballero and Jean-Marie Aerts *

Department of Biosystems, Animal and Human Health Engineering Division, M3-BIORES: Measure, Model & Manage of Bioresponses Laboratory, KU Leuven, Kasteelpark Arenberg 30, 3001 Heverlee, Belgium; ali.youssef@kuleuven.be (A.Y.); nicolas.caballero@gmail.com (N.C.)
* Correspondence: jean-marie.aerts@kuleuven.be

Received: 31 August 2019; Accepted: 5 October 2019; Published: 10 October 2019

Abstract: Conventional indoor climate design and control approaches are based on static thermal comfort/sensation models that view the building occupants as passive recipients of their thermal environment. Recent advances in wearable sensing technologies and their generated streaming data are providing a unique opportunity to understand the user's behaviour and to predict future needs. Estimation of thermal comfort is a challenging task given the subjectivity of human perception; this subjectivity is reflected in the statistical nature of comfort models, as well as the plethora of comfort models available. Additionally, such models are using not-easily or invasively measured variables (e.g., core temperatures and metabolic rate), which are often not practical and undesirable measurements. The main goal of this paper was to develop dynamic model-based monitoring system of the occupant's thermal state and their thermoregulation responses under two different activity levels. In total, 25 participants were subjected to three different environmental temperatures at two different activity levels. The results have shown that a reduced-ordered (second-order) multi-inputs-single-output discrete-time transfer function (MISO-DTF), including three input variables (wearables), namely, aural temperature, heart rate, and average skin heat-flux, is best to estimate the individual's metabolic rate (non-wearable) with a mean absolute percentage error of 8.7%. A general classification model based on a least squares support vector machine (LS-SVM) technique is developed to predict the individual's thermal sensation. For a seven-class classification problem, the results have shown that the overall model accuracy of the developed classifier is 76% with an F1-score value of 84%. The developed LS-SVM classification model for prediction of occupant's thermal sensation can be integrated in the heating, ventilation and air conditioning (HVAC) system to provide an occupant thermal state-based climate controller. In this paper, we introduced an adaptive occupant-based HVAC predictive controller using the developed LS-SVM predictive classification model.

Keywords: thermal sensation; thermal comfort; machine-learning; prediction; adaptive controlling

---

1. Introduction

Thermal comfort (TC) is an ergonomic aspect determining the satisfaction about the surrounding environment and is defined as "that condition of mind which expresses satisfaction with the thermal environment and is assessed by subjective evaluation" [1]. The effect of thermal environments on occupants might also be assessed in terms of thermal sensation (T_S), which can be defined as "a conscious feeling commonly graded into the categories cold, cool, slightly cool, neutral, slightly warm, warm, and hot" [1]. Thermal sensation and thermal comfort are both subjective judgments, however, thermal sensation is related to the perception of one's thermal state, and thermal comfort to the evaluation of this perception [2]. The assessment of thermal sensation has been regarded as more reliable and. as such. is often used to estimate thermal comfort [3].

Human thermal sensation mainly depends on the human body temperature (core body temperature), which is a function of sets of comfort factors [4,5]. These comfort factors include indoor environmental factors, namely the mean air temperature around the body, relative air velocity around the body, humidity, and the mean radiant temperature to the body [5]. Additionally, some personal (individual-related) factors, namely, metabolic rate or internal heat production in the body, which vary with the activity level and clothing thermal-physical properties (such as clothing insulation and vapour clothing resistance), are included. It should be mentioned that the individual thermal perception is deepening, as well, on psychological factors, including naturalness (an environment where the people tolerate wide changes of the physical environment), expectations and short/long-term experience, which directly affect individuals' perceptions, time of exposure, perceived control and environmental stimulation [6]. The most considered method to have an accurate assessment of T_S is to ask the individuals directly about their thermal sensation perception [4,5]. Thermal sensation mathematical models have been developed in order to overcome the difficulties of direct enquiry of subjects. The development of such models is mostly dependent on statistical approaches that correlates experimental conditions (i.e., environmental and person-related variables) data to thermal sensation votes obtained from human subjects [3,5]. Most of these models (e.g., predicted mean vote, PMV) are static in the sense that they predict the average vote of a large group of people based on the seven-point thermal sensation scale. Instead of individual thermal comfort, they only describe the overall thermal sensation of multiple occupants in a shared thermal environment. To overcome the disadvantages of static models, adaptive thermal comfort models aim to provide insights in increasing opportunities for personal and responsive control, thermal comfort enhancement, energy consumption reduction and climatically responsive and environmentally responsible building design [7,8]. The idea behind adaptive model is that occupants and individuals are no longer regarded as passive recipients of the thermal environment but, rather, play an active role in creating their own thermal preferences [8]. In addition to regression analysis, thermal sensation prediction can also be seen as a classification problem where various classification algorithms can be implemented [7]. Recently, a number of studies (e.g., [9–13]) have demonstrated the possibility of using machine learning techniques, such as a support vector machine (SVM), to assess and predict human thermal sensation. It can be concluded based on the published work (see the recent literature review [7] showing that classification-based models have performed so well as regression models).

Recent advances in mobile technologies in healthcare, in particular wearable technologies (m-health) and smart clothing, have positively contributed to new possibilities in controlling and monitoring health conditions and human wellbeing in daily life applications. The wearable sensing technologies and their generated streaming data are providing a unique opportunity to understand the user's behaviour and to predict future needs [14]. The generated streaming data is unique due to the personal nature of the wearable devices. However, the generated streaming data is forming a challenge pertaining to the need of personalized adaptive models that can handle newly arrived personal data.

Current heating, ventilation, and air conditioning (HVAC) control systems can be divided into two types: air temperature regulator (ATR) and thermal comfort regulator (TCR). Most TCR controllers use static models, mainly PMV, as a performance criterion.

This paper aims to develop an adaptive model for real-time monitoring of human thermal states using personal non-intrusive sensing techniques. The developed model should be suitable for real-time adaptive control of indoor climate systems and smart wearable applications.

2. Materials and Methods

2.1. Experiments and Experimental Setup

2.1.1. Climate Chambers (Body and Mind Room)

The *"Body and Mind Room"* consists of three climate-controlled chambers (A, B and C) designed and built to investigate the dynamic mental and physiological responses of humans to specific indoor climate conditions. The Body and Mind rooms are experimental facilities at the M3-BIORES laboratory (Animal and Human Health Engineering Division, KU Leuven, Leuven, Belgium). The three rooms are dimensionally identical; however, each room is designed to provide different ranges of climate conditions, as shown in Table 1.

Table 1. The different temperature and relative humidity ranges that can be provided by the different Body and Mind rooms (A, B and C).

Room	Air Temperature Range (°C)	Relative Humidity Range (%)
A	+23–+37	50–80
B	+10–+25	50–80
C	−5–+10	40–60

The three rooms are equipped with axial fans to simulate wind velocities between 2.5 and 50 km·h^{-1}.

2.1.2. Experimental Protocol

The experimental protocol used in the present study is designed in such way to investigate the subjects' thermal and physiological responses to predefined three different temperature (*low, normal* and *high*) that under two levels of physical activities (seated = *low* and cycling = *high*). The three predefined temperatures (*low* = 5 °C, *normal* = 24 °C and *high* = 37 °C) are chosen based on the thermal comfort chart from [15] and the effects on health according to the Wind Chill Chart for cold exposure (National Weather Service of the US) and for hot temperatures exposure according to [16]. The conducted experiments consist of two phases (Figure 1, upper graph), namely, low activity and high activity phases. During the first experimental phase, low activity phase, the test subjects (while being seated = low activity) are exposed, for 55 min, to three levels of temperatures in the following order: normal, low, high and normal again (Figure 1). During the high activity phase, the test subjects is exposed to a 15 min of light physical stress (80 W of cycling on a fastened racing bicycle). During the course (75 min) of the active phase, each test subject is exposed to the predefined three temperature levels (Figure 1, lower graph). During each temperature level, starting from the normal level (24 °C), the test subjects are performed 15 min of cycling (with 80 W power) and followed 4 min of resting (seated). During the course of conducted experiments, the clothing insulation factor (*Col*) is kept constant at *Col* = 0.34, which accounts for a cotton short and t-shirt as standard clothing for all test subjects. The experimental protocol is approved by the SMEC (Sociaal-Maatschappelijke Ethische Comissie) on 16 January 2019 with number G-2018-12-1464.

2.1.3. Test Subjects

In total, 25 healthy participants (six females and 19 males were asked before the experiments if they had been diagnosed with any cardiac problems, diabetic or any other health problems), between the age of 25 and 35 (average age 26 ± 4.2) years, with average weight and height of 70.90 (±12.70) kg and 1.74 (±0.10) m, respectively, volunteered to perform the aforementioned experimental protocol.

2.1.4. Measurements and Gold Standards

During the course of the experiments, participants' heart rate, metabolic rate, average skin temperature, heat flux between the skin and the ambient air and core body temperature represented by the aural temperature are measured continuously. Heart rate monitoring is performed using a Polar H7 ECG (Polar, Kempele, Finland) strap that is placed under the chest, with a sampling frequency of 128 Hz. The metabolic rate, as metabolic equivalent tasks (METs) of each test subject, is calculated based on indirect calorimetry using a MetaMAX 3B (CORTEX-Medical, Leipzig, Germany) spiroergometer sensor. The average skin temperature is calculated based on measurements from three body-placed sensors, namely, scapula, chest and arm (Figure 2). The skin temperature measurements are performed using one Shimmer (Shimmer-Sensing, Dublin, Ireland) temperature sensor and two gSKIN® bodyTEMP patches (greenTEG, Zurich, Switzerland). Two heat flux gSKIN® patches are placed on both the chest and the left arm (Figure 2). The skin temperatures and heat flux measurements are acquired at a sampling frequency of 1 Hz. All the measured from the wearable sensors were received and saved on a smart phone. Core body temperature is estimated based on aural temperature measure measurements, which is performed using in-ear wireless (Bluetooth) temperature sensor (Cosinuss One, Düsseldorf, Germany) with a sampling rate if 1 Hz. At the end of each applied temperature level during the course of both experimental phases, a thermal sensation questionnaire, based on ASHRAE seven-point thermal scale, is performed for each test subject.

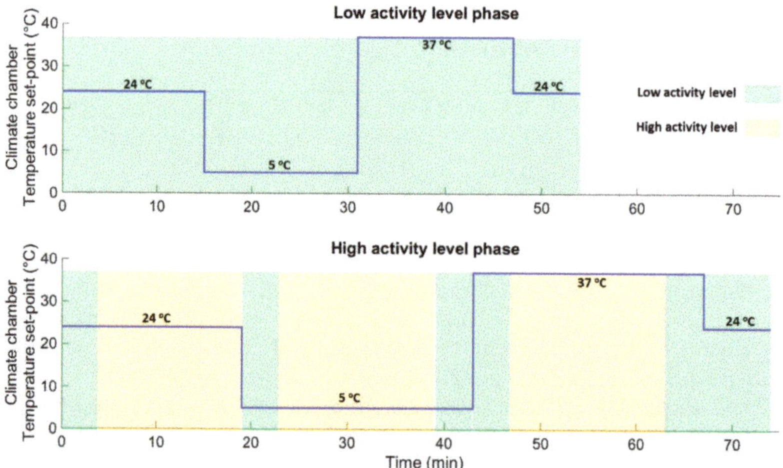

Figure 1. Plots showing the climate chambers' set-point temperatures programed during the 55 min low activity phase (upper graph) and the 75 min high activity phase (lower graph).

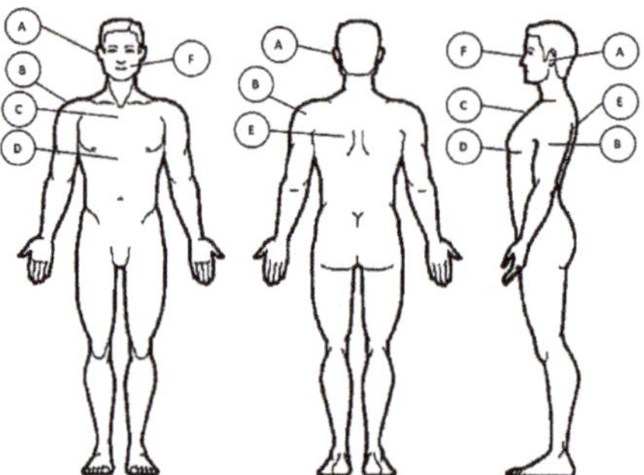

Figure 2. Sensor placement. (**A**) Ear channel for aural temperature measurement via the Cosinuss One; (**B**) upper arm where the skin temperature and heat flux are measured with the gSKIN patch; (**C**) middle upper chest where the skin temperature and heat flux are measured with the gSKIN patch; (**D**) lower chest where the heart rate is measured with the Polar H7; (**E**) scapula where skin temperature is measured with the Shimmer sensor; (**F**) mouth and nose where metabolic rate is measured via a MetaMAX-3B spiroergometer sensor.

2.2. Modelling and Classification

For the sake of present study, the measured variables are divided into wearables, which are easily measured variables using wearable sensors and gold standards (reference) variables, which are not suitable for wearable technologies. The wearables include heart rate H_R, aural temperature T_{er}, average skin temperature $\overline{T}_{sk}$, skin heat flux q_{sk} and ambient air temperature T_∞. On the other hand, the gold standards consist of the core temperature T_c, which is driven from the aural temperature $[T_c = f(T_{er})]$, metabolic rate M_r and personal thermal sensation votes TS. The ultimate goal of this work is to develop an adaptive classification model to predict the individual thermal sensation depending, solely, on the wearables or estimated variables. Hence, both of the metabolic rate and core body temperature are estimated using an online dynamic modelling approach (Figure 3). Then, the individual thermal sensation is predicted using a classification model (classifier) whose inputs are the wearables and estimated the metabolic rate and core body temperature (Figure 3).

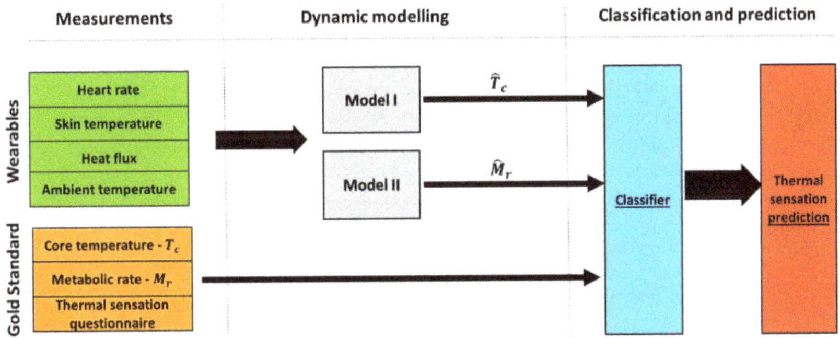

Figure 3. Overview of the main steps to predict the individual thermal sensation.

2.2.1. Dynamic Modelling

Although the system under study (occupant's thermoregulation) is inherently a non-linear system, the essential perturbation behaviour can often be approximated well by simple linearized transfer function (TF) models [17–19]. For the purposes of the present paper, therefore, the following linear, multi-input, single-output (MISO) discrete time systems are considered to estimate metabolic rate and core body temperature [18]:

$$y(k) = \sum_{r=1}^{r=R} \frac{B_r(z^{-1})}{A_r(z^{-1})} u_r(k - \delta_r) + \xi(k), \tag{1}$$

where k denotes the value of the associated variable at the kth sampling instant; $y(k)$ is the output variable; $u_r(k)$, $r = 1, 2, \ldots, R$ are input variables, while $A(z^{-1})$ and $B(z^{-1})$ are appropriately defined polynomials in the backshift operator z^{-1}, i.e., $z^{-i}y(k) = y(k - i)$ and $\xi(k)$ is additive noise, a serially uncorrelated sequence of random variables with variance σ^2 that accounts for measurement noise. The simplified refined instrumental variable (SRIV) algorithm was utilised in the identification and estimation of the models (model parameters and model structure) [20]. Two main statistical measures were employed to determine the most appropriate model structure. Namely, the coefficient of determination R_2^T, based on the response error; and YIC (Young's information criterion), which provides a combined measure of model fit and parametric efficiency, with large negative values indicating a model which explains the output data well and yet avoids over-parameterisation [21]. Additionally, the estimation performance of the selected models is evaluated used the mean absolute error (MAE) value.

2.2.2. Classification Model

To predict the individual thermal sensation, a classification model (classifier) is developed and trained based on the wearables and estimated variables (metabolic rate and core body temperature), together with the thermal sensation votes (gold standard). A modified support vector machine (SVM) technique, namely, the least squares support vector machine (LS-SVM), is used to develop and train the thermal sensation classifier [22,23]. SVMs are originally presented as binary classifiers [22] that assign each data instance $X \in \mathbb{R}^d$ to one of two classes described by a class label $y \in \{-1, 1\}$ based on the decision boundary that maximises the margin $2/\|\mathbf{w}\|_2$ between the two classes. Generally, a feature map $\varphi : \mathbb{R}^d \Rightarrow \mathbb{R}^p$ is used to transform the geometric boundary between the two classes to a linear boundary $L : \mathbf{w}^T \varphi(x) + b = 0$ in feature space, for some weight vector $\mathbf{w} \in \mathbb{R}^{p \times 1}$ *and* $b \in \mathbb{R}$. The class of each instance can then be found by $y = sign\big(\mathbf{w}^T \varphi(x) + b\big)$, where *sign* refers to the sign function. Due to some computational complexities of standard SVM because of the quadratic programming problem, the least squares support vector machine (LS-SVM) is presented to overcome such problem. LS-SVM, in contrast with standard SVM, relies on a least squares cost function as follows:

$$\min_{w,\,b;\,e} \frac{1}{2}\,\mathbf{w}^T\mathbf{w} + \gamma \sum_{i=1}^{N} e_i^2, \tag{2}$$

such that $y_i\big(\mathbf{w}^T \varphi(x_i) + b\big) \geq 1 - e_i$ *and* $e_i \geq 0$, $i = 1,\,2,\,\ldots,\,N$, where e_i errors such that $1 - e_i$ is proportional to the signed distance of x_i from the decision boundary, and γ represents the regularisation constant. In LS-SVM, instead of solving the quadratic programming problem, a set of linear equations to be solved is sufficient to find the optimal solution of the classifier. The LS-SVMlab (Least Squares Support Vector Machine lab) Matlab-based toolbox is used to implement the LS-SVM classification algorithm [22].

The performance of the classification model is determined based on accuracy, sensitivity, precision and F1-score as follows:

$$Accuracy = \frac{TP + TN}{TP + TN + FP + FN},\ Sensitivity = \frac{TP}{TP + FN}$$

$$Precision = \frac{TP}{TP + FP},\ F1Score = \frac{2 * Precision * Sensitivity}{Precision + Sensitivty}$$

where TP, TN, FP and FN are the true positive, true negative, false positive and false negative, respectively.

3. Results and Discussion

3.1. Dynamic Modelling and Estimation of an Individual's Metabolic Rate

The average metabolic rate obtained from the 25 participants at the temperature levels (24, 5 and 37 °C) during low and high activity phases are presented in Table 2.

Table 2. Average (±standard deviation) of the measured metabolic rate obtained from the 25 test subjects during low and high activity phases.

	Measured Metabolic Rate (MET *)	
Temperature	Low Activity Phase	High Activity Phase
24 °C	1.19 ± 0.19	5.11 ± 1.18
5 °C	1.18 ± 0.23	5.32 ± 1.24
37 °C	1.22 ± 0.17	5.48 ± 1.20

* 1MET = 1.163 W·kg^{-1}.

Different combinations of input variables (wearables) are tested for the best estimation of an individual's metabolic rate. The SRIV algorithm, combined with YIC and R_2^T selection criteria, suggested that a second-order MISO discrete-time TF with heart rate (H_R), average skin heat flux ($\bar{q}_{sk}$) and aural temperature (T_{er}) as input variables is the best (with average $R_2^T = 0.89 \pm 0.04$ and $YIC = -13.62 \pm 2.33$) to describe and estimate the dynamic behaviour of the individual's metabolic rate. More specifically, the SRIV algorithm identified the following general MISO discrete-time TF model structure:

$$\hat{M}_r(k) = \left[\frac{B_1(z^{-1})}{A(z^{-1})} \frac{B_2(z^{-1})}{A(z^{-1})} \frac{B_3(z^{-1})}{A(z^{-1})} \right] \cdot \left[\begin{array}{c} T_{er}(k - \delta_1) \\ H_R(k - \delta_2) \\ \bar{q}_{sk}(k - \delta_3) \end{array} \right] + \xi(k) \tag{3}$$

where $\hat{M}_r(k)$ is the estimated metabolic rate and the numerator polynomials B_1, B_2 and B_3 are of the following orders (number of zeros) 2, 3 and 2, respectively. The system delays δ_1, δ_2 and δ_3 are varied from person to another (inter-personal) with average values of 1.4, 0.20 and 0.21 min, respectively (Table 3). A simulation example of the developed estimation model (Equation (3)) for one test subject during the low activity experimental phase at normal temperature (24 °C) is depicted in Figure 4.

Table 3. Average R_2^T, *YIC*, model delays and MAPE for the selected MISO-DTF model to estimate the individual's metabolic rate obtained from the 25 test subjects during low and high activity phases.

	Average $R_2^T \pm$ std	Average *YIC* $\pm$ std	Model Delays Average $[\delta_1\ \delta_2\ \delta_3]$	Average MAPE $\pm$ std
Low activity phase	0.85 ± 0.02	-12.32 ± 3.4	[1.5 0.3 0.25] min	$10 \pm 2.2\%$
High activity phase	0.94 ± 0.03	-14.43 ± 2.8	[1.2 0.18 0.20] min	$7.6 \pm 2.6\%$

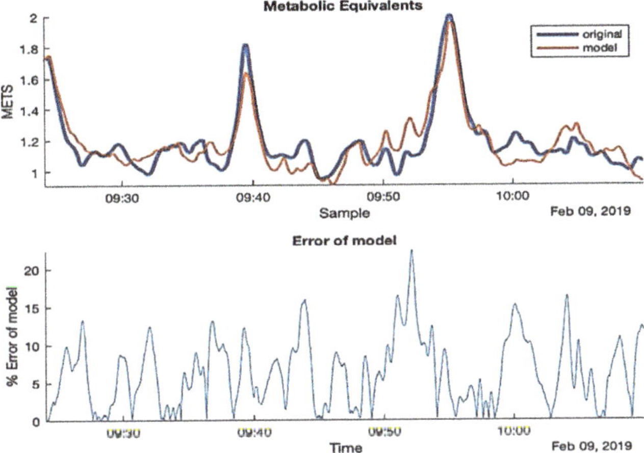

Figure 4. A simulation example of the developed MISO-DTF model (Equation (3)) to estimate the metabolic rate during the low activity experimental phase at normal temperature (24 °C).

The estimation performance of the selected general MISO-DTF (Equation (3)) is evaluated based on the mean absolute percentage error ($MAPE = \frac{100\%}{N} \sum_{k=1}^{N} \left| \frac{\hat{M}_r(k) - M_r(k)}{M_r(k)} \right|$) value.

The results have shown that the developed general model (Table 3), for all test subjects, a higher average MAPE value ($10 \pm 2.2\%$) during the low activity phases than the average MAPE value ($7.6 \pm 2.6\%$) resulted during the high activity phases. The METs (metabolic equivalent tasks) are a measure which accounts for a normalized form of energy expenditure per kilogram of mass. There is a consensus that the measurement of the metabolic rate might vary among individuals (interpersonal) up to 75% [24], even within the same day from morning to afternoon for the same subject (intrapersonal) up to 6%, though measurements on different days might be comparable on fasted subjects [25]. Hence, a general estimation model of individual metabolic rate will not be efficient in this case. However, the general estimation performance of the suggested general MISO model can be enhanced by using the online adaptive form of the SRIV algorithm [26]. The online adaptive (closed-loop) SRIV algorithm provides the possibility to personalise the developed general model by retuning the model parameters and model delays based on the streaming data acquired from the wearable sensors.

3.2. Classification Model and Prediction of Individual's Thermal Sensation

In order to give an idea about the interaction relationship between considered variables, the correlation between all measured variables are calculated and represented in a colour-map Pearson correlation coefficient (r) matrix, as shown in Figure 5. High positive or negative correlation coefficient values, such as that between heat flux and skin temperature, are reflected as a strong interaction between these variables, which can affect the feature selection of the classification model.

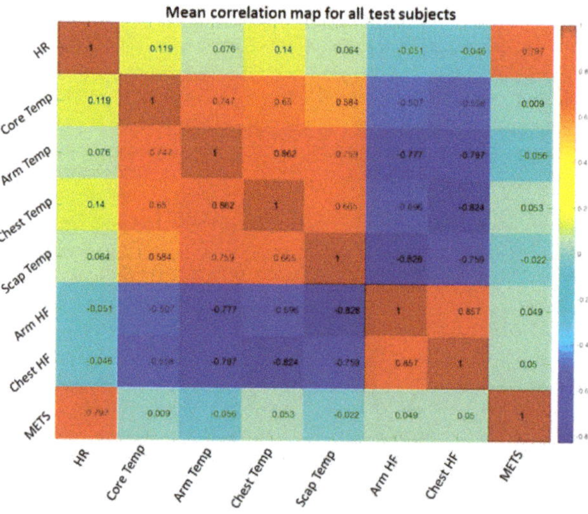

Figure 5. Colour-map of the correlation matrix representing the correlation levels ($r \in [-1, 1]$) between the mean values of all measured variables, namely, heart rate (HR), aural (core) temperature, arm temperature, chest temperature, scapula temperature, heat flux from the arm skin (Arm HF), heat flux from the chest skin (Chest HF) and metabolic rate (METs).

The classification model for predicting the individual's thermal sensation is developed, based on the LS-SVM approach, by training the classifier on 80% of the data points, while the rest of the data (20%) is used for testing. The model accuracy, sensitivity F1-score and overall confusion matrix are computed to evaluate the performance of the developed classifier. The feature space includes all the measured and estimated input variables, namely, T_{er}, H_R, $\overline{q}_{sk}$, $\overline{T}_{sk}$, $\Delta\overline{T} = T_{er} - \overline{T}_{sk}$ and $\hat{M}_r$. Additionally, other features are extracted by computing the variance, min, max, root mean squares (RMS) and first derivative ($\frac{dx}{dt}$, where x is the variable) of the aforementioned measured and estimated variables. The age and gender of the test subjects are also included in the feature spaces.

Figure 6 is showing the distribution of the participants' thermal sensation votes at the three environment temperatures (24, 5 and 37 °C). The aforementioned figure shows the 'confusion space', or the area in which the reported thermal sensation votes at the three environmental temperatures are overlapping. Such observation shows that the thermal perception may overlap even with large differences in the surrounding environmental temperatures. Such a confusion space is a great challenge for any predictive model of thermal sensation, especially for static models such as PMV.

For the sake of the main objective of the present work, the computational cost of the developed algorithm should be low enough to be compatible with wearable technology and online modelling. Hence, a feature selection procedure is employed to obtain the most reduced-dimension model yet with the best error performance. Feature selection here is based on evaluating all possible feature combinations and selecting the combination with best error performance. The feature selection step results in a feature space including 25 features, as shown in Table 4.

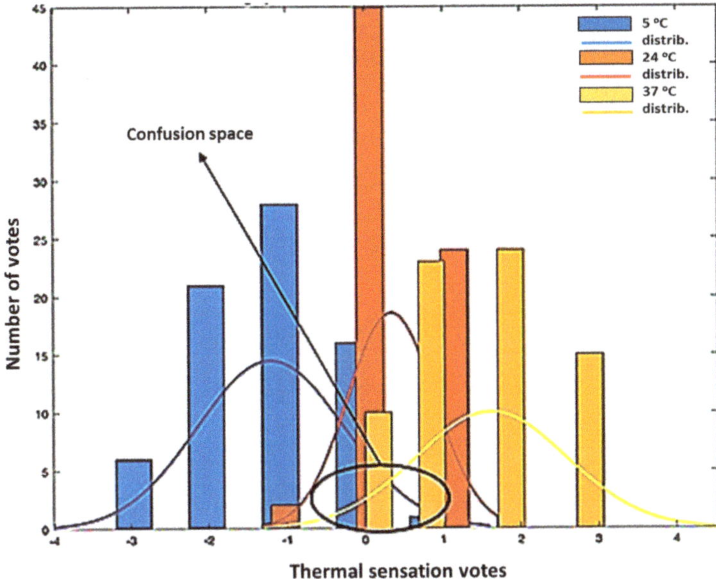

Figure 6. Distribution (distrib.) of the participants' (25) thermal sensation votes at the three environment temperatures (24, 5 and 37 °C) showing the confusion space.

Table 4. An overview of the selected feature space including the measured and estimated variables (six variables) and some operations on these variables (× = selected).

	Variance	Min	Max	RMS	$\frac{d}{dt}$
T_{er}	×	×	×	×	-
H_R	×	×	×	×	-
$\bar{q}_{sk}$	×	×	×	×	×
$\bar{T}_{sk}$	×	-	-	×	-
$\Delta\bar{T}$	×	×	×	×	-
$\hat{M}_r$	-	-	-	-	-

A classification model is developed based on the selected 25 input features and trained using the LS-SVM approach. The resulting confusion matrix from the developed classification model based on the selected feature space is shown in Figure 7.

The overall error performance results of the developed classification model are presented in Table 5. For a seven-class classification problem, the developed classifier have shown an overall accuracy of 76% to predict the individual's thermal sensation. The developed classifier has shown a high (84%) F1-score, which reflects low false-positives and -negatives.

Table 5. Overall error performance (accuracy, sensitivity, precision and F1-score) of the developed general LS-SVM classification model.

Measure	Value
Accuracy	0.76
Sensitivity	0.82
Precision	0.87
F1 score	0.84

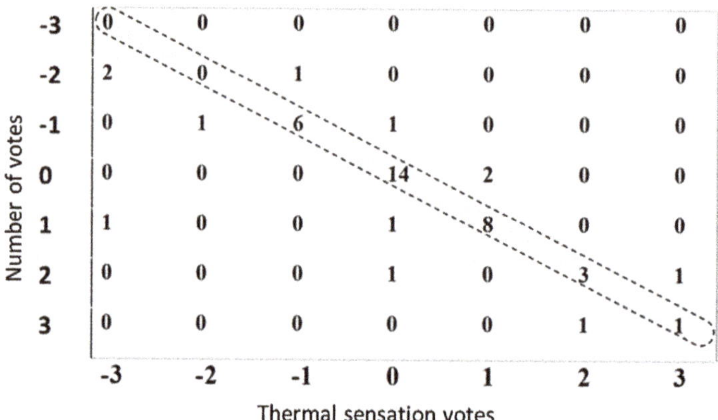

Figure 7. The resulted confusion matrix from testing the developed LS-SVM classifier. The diagonal represents the correctly-classified data points.

The error performance results of the developed general classification for each class separately are shown in Table 6. The results showed that the error performances of classes 1, 2, 6 and 7 are very low (see Table 6), which can be attributed to the low number (0, 2, 4 and 2, respectively) of obtained votes for these classes, or, in other words, due to the uneven class distribution. Therefore, the overall F1-score is a more reliable and efficient measure of performance than the accuracy in this case.

Table 6. The error performance (precision, sensitivity and F1-score) of the developed general LS-SVM model for the seven-class classification problem.

	Class 1	Class 2	Class 3	Class 4	Class 5	Class 6	Class 7
Sensitivty	-	0	0.75	0.88	0.80	0.60	0.50
Precision	0	0	0.86	0.82	0.80	0.75	0.50
F1-score	0	0	0.80	0.85	0.80	0.67	0.50

SVM is used in recent studies to assess the occupant's thermal demands [12] and to predict thermal comfort/sensation [11]. In these studies, the results have shown that SVM is able to predict thermal comfort/sensation with accuracy and F1 scores of 76.7% and 84%, respectively. However, these results is only obtained by reducing the seven-class classification problem to a three-class problem. Hence, we believe that reducing the number of classes will improve our suggested general model performance. Moreover, based on streaming data obtained from wearable sensor technologies, a personalised adaptive classification model, based on the same extracted features, will enhance the model performance to predict the individual's thermal sensation. Different related works investigated the problem of thermal sensation and comfort prediction via machine learning algorithms. Ghahramani et al. [27] applied the hidden Markov model (HMM) technique to the thermal comfort prediction problem with three levels of thermal comfort. The main issue in the used dataset in this study is the class imbalance, which is not tackled by the proposed methodology. A recent study by Lu et al. [7] proposed a personalised model, however, the study strictly investigated two subjects and developed a dedicated model for each subject.

Testing of the trained model for each test subject was implemented using the Matlab platform on a computer with Intel® 8 Core i7 (2.7 GHz) processor and 16 GB of RAM. The average computational time for one test run on this computer was 100 ms. The developed model should be trained using data from different populations (e.g., different ages, ethnicities and physical conditions) and different environmental conditions (e.g., broader ranges of temperature and humidity). In future research, the mental status of the participants should be taken into the account to investigate the capability of the developed model to comply with different mental conditions (e.g., stresses). Moreover, a sensitivity analysis should be performed considering different accuracy and sensitivity levels of the wearable sensors. On the other hand, the LS-SVM approach used in this paper is suitable for online adaptation with a flexibility to receive new data (streaming data) variables. Hence, the developed model can be used for adaptive real-time mode-based monitoring of individual thermal sensation. Additionally, as such, the developed mode is suitable for different applications such as the simulation of the human thermal state under different environmental conditions and for building design and control. In this paper, we present the possibility of using this model for adaptive HVAC control systems.

3.3. Adaptive Occupant-Based HVAC Predictive Controller

In modern buildings, it is very common that HVAC control systems are designed in such way to ensure parsimonious energy use and cost-effective building operation. This often happens by tuning HVAC control parameters (e.g., set points) to exploit the inherent trade-off between energy consumption and thermal comfort, with the latter acting as a constraint defining a theoretical and practical upper bound on potential energy savings [28–30]. In this paper, we suggested a model-predictive control (MPC) strategy, which is based on continuous feedback of occupant's thermal state (sensation/comfort) with main control objective to achieve occupant's thermal comfort. Then the energy use can be employed in the controller's cost-function as constraints.

In this section, we introduced to an adaptive occupant-based HVAC predictive controller using the developed LS-SVM predictive classification model. The general framework of the proposed HVAC predictive controller approach is depicted in Figure 8. In this paper, we only describe the main methodology to use the LS-SVM predictive classification model for the occupant's thermal state in the generalized predictive control (GPC) scheme, which will be studied and investigated further in future work. The GPC is one of the most popular model predictive controlling (MPC) methods in broad number of fields. The basic principle of GPC is to calculate a sequence (control horizon, N_c) of future control signals that minimizes a cost function defined over a prediction horizon (N_p) [31]. In general, the GPC algorithm consists of two main subsystems, namely, a prediction model and an optimizer. As shown in Figure 8, two main components, namely, the adaptive algorithm for LS-SVM predictive model and the GPC algorithm, are suggested.

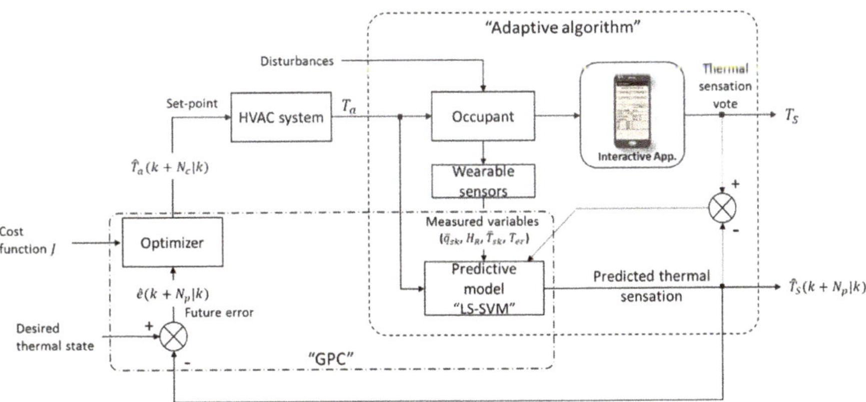

Figure 8. Block diagram of the proposed adaptive HVAC LS-SVM-based model predictive controller.

3.3.1. Adaptive LS-SVM-Based Algorithm for Predicting the Occupant's Thermal Sensation

The availability of the real-time sensors data, from the wearable technologies, has given the possibility of streaming data, which are processed via the proposed online streaming algorithm to adapt the classifier model. This adaptive algorithm is needed to handle the newly arrived data (streaming data) in the training set. Different approaches are available in the literature to handle these challenges, such as incremental learning methods [32], which work on adapting and retuning the parameters of the general model based on the newly collected data. Another approach is the localized learning, which is based on developing a local model for each test point or subset of the test set [33]. The streaming data includes:

i.   Wearable sensor data, which consists of the continuously (easily) measured variables, namely, occupant's heart rate, skin heat flux, skin temperature, ambient temperature and aural temperature.

ii.   Data obtained from the interactive mobile application, which consists of the occupant's data, namely, age and gender. Additionally, the occupant's thermal sensation vote (T_S) is to be obtained via mobile application-based questioner (interactive application).

3.3.2. The GPC Algorithm

In general, the goal of any controller is to calculate the input (control signal) to the controlled system (plant) such that the output follows the desired reference. However, in case of the predictive controller, the GPC algorithm aims to find the best-predicted output sequence (using the prediction model) that is the closest to a predefined reference trajectory (desired thermal state in our case). The *prediction model* in our case is the adaptive LS-SVM classification model that predicts the occupant's thermal sensation (T_S). The algorithm simulates multiple future scenarios (predicted output sequence) in a systematic way using the *optimizer*, then the predicted output $\hat{T}_S(k + N_p|k)$ is used to calculate the optimal future input (ambient temperature, $\hat{T}_a(k + N_c|k)$). The optimizer solves an online optimization problem based on a defined *cost function* (Figure 8), which minimizes the predicted error $\hat{e}(k + N_p|k)$ between the reference $R_S(k)$ and the predicted output $\hat{T}_S(k + N_p|k)$. The cost-function is given as follows [34]:

$$J\left(N_1, N_p, N_c\right) = \sum_{j=N_1}^{N_p} \alpha_j \left[\hat{T}_S(k + j \mid k) - R_S(k + j)\right]^2 + \sum_{j=1}^{N_c} \lambda_j [\Delta T_a(k + j - 1)]^2 \tag{4}$$

where $\Delta T_a(k)$ is the change in the control signal (ambient temperature), $\hat{T}_S(k + j \mid k)$ is the predicted output (thermal sensation) sequence, $R_S(k)$ is the reference (desired level of thermal sensation), N_1 is the minimum of the prediction horizon and α and λ are the weighting factors. The suggested control signal ($\hat{T}_a$) can be incorporated into the HVAC system by feeding it as a set-point to the HVAC controller. The sequence of predicted thermal sensation is crucial in the optimisation (cost function) of the control (manipulated) variables [35,36]. In the presented approach, we have considered the air temperature as the only manipulated variable; however, more HVAC-related variables can be added to the optimisation step (e.g., ventilation rate and energy consumption).

The proposed approach (Figure 8) depends on the extracted features from easily measured variables (T_{er}, H_R, $\bar{q}_{sk}$ and $\overline{T}_{sk}$,) that can be collected from already available (off-the-shelf) wearable sensors (e.g., smart watches and on-body smart tags). As such, this proposed approach has the advantage over other models (e.g., [37]), which depend on difficult and/or invasive measurements (core body temperature and metabolic rate) and, consequently, not convenient for long-term monitoring. Moreover, the used LS-SVM used in this approach is suitable for online prediction of an individual's thermal state with minimum computational cost (100 ms for the prediction of the thermal sensation of each individual).

4. Conclusions

In this present paper, 25 participants are subjected to three different environmental temperatures, namely 5 °C (cold), 20 °C (moderate) and 37 °C (hot), at two different activity levels, namely, at low level (rest) and high level (cycling at 80 W power). Metabolic rate, heart rate, average skin temperature (from three different body locations), heat flux and aural temperature are measured continuously during the course of the experiments. The thermal sensation votes are collected from each test subject based on the ASHRAE seven-point questionnaire. The results have shown that a reduced-ordered (second-order) MISO-DTF including three input variables (wearables), namely, aural temperature, heart rate and average heat flux, is best to estimate the individual's metabolic rate (non-wearable) with an average MAPE of 8.7%. A general classification model based on the LS-SVM technique is developed to predict the individual's thermal sensation. For a seven-class classification problem, the results have shown that the overall model accuracy and F1-score of the developed classifier are 76% and 84%, respectively. It is suggested in this paper that the overall performance of the model can be enhanced by using a personalised adaptive classification algorithm based on streaming data from wearable sensors. The developed LS-SVM classification model for the prediction of the occupant's thermal sensation can be integrated in the HVAC system to provide an occupant thermal state-based climate controller. In this paper, we introduced an adaptive occupant-based HVAC predictive controller using the developed LS-SVM predictive classification model.

Author Contributions: Conceptualisation: A.Y. and J.-M.A.; methodology: A.Y. and N.C.; software: A.Y.; validation: A.Y.; formal analysis: A.Y. and N.C.; investigation: A.Y. and N.C.; resources: J.-M.A.; data curation: A.Y. and N.C.; writing—original draft preparation: A.Y. and N.C.; writing—review and editing: A.Y. and J.-M.A.; visualisation: A.Y.; supervision: J.-M.A. and A.Y.; project administration: J.-M.A.; funding acquisition: J.-M.A.

Funding: This research received no external funding.

Conflicts of Interest: The authors declare no conflict of interest.

References

1. ASHRAE. *Thermal Environmental Conditions for Human Occupancy*; American Society of Heating, Refrigeration and Air Conditioning Engineers, Inc.: Atlanta, GA, USA, 2004.
2. ISO-10551. *Ergonomics of the Thermal Environment—Assessment of the Influence of the Thermal Environment Using Subjective Judgement Scales*; ISO: Brussels, Belgium, 1995.
3. Koelblen, B.; Psikuta, A.; Bogdan, A.; Annaheim, S.; Rossi, R.M. Thermal sensation models: A systematic comparison. *Indoor Air* **2017**, *27*, 680–689. [CrossRef] [PubMed]
4. Enescu, D. Models and Indicators to Assess Thermal Sensation Under Steady-State and Transient Conditions. *Energies* **2019**, *12*, 841. [CrossRef]
5. Kenneth, C. *Parsons, Human Thermal Environments: The Effects of Hot, Moderate, and Cold Environments on Human Health, Comfort, and Performance*, 3rd ed.; CRC Press: Boca Raton, FL, USA, 2014.
6. Nikolopoulou, M.; Steemers, K. Thermal comfort and psychological adaptation as a guide for designing urban spaces. *Energy Build.* **2003**, *35*, 95–101. [CrossRef]
7. Lu, S.; Wang, W.; Wang, S.; Hameen, E.C. Thermal Comfort-Based Personalized Models with Non-Intrusive Sensing Technique in Office Buildings. *Appl. Sci.* **2019**, *9*, 1768. [CrossRef]
8. De Dear, R.; Brager, G.S. Developing an adaptive model of thermal comfort and preference. *ASHRAE Trans.* **1998**, *104*, 145–167.
9. Kim, J.; Zhou, Y.; Schiavon, S.; Raftery, P.; Brager, G. Personal comfort models: Predicting individuals' thermal preference using occupant heating and cooling behavior and machine learning. *Build. Environ.* **2018**, *129*, 96–106. [CrossRef]
10. Chaudhuri, T.; Soh, Y.C.; Li, H.; Xie, L. Machine learning based prediction of thermal comfort in buildings of equatorial Singapore. In Proceedings of the 2017 IEEE International Conference on Smart Grid and Smart Cities (ICSGSC), Singapore, 23–26 July 2017; pp. 72–77.
11. Farhan, A.A.; Pattipati, K.; Wang, B.; Luh, P. Predicting individual thermal comfort using machine learning algorithms. In Proceedings of the 2015 IEEE International Conference on Automation Science and Engineering (CASE), Gothenburg, Sweden, 24–28 August 2015; pp. 708–713.
12. Dai, C.; Zhang, H.; Arens, E.; Lian, Z. Machine learning approaches to predict thermal demands using skin temperatures: Steady-state conditions. *Build. Environ.* **2017**, *114*, 1–10. [CrossRef]

13. Huang, C.-C.J.; Yang, R.; Newman, M.W. The potential and challenges of inferring thermal comfort at home using commodity sensors. In Proceedings of the 2015 ACM International Joint Conference on Pervasive and Ubiquitous Computing-UbiComp'15, Osaka, Japan, 7–11 September 2015; pp. 1089–1100.

14. Hussain, S.; Kang, B.H.; Lee, S. *A Wearable Device-Based Personalized Big Data Analysis Model*; Information Technology in Bio- and Medical Informatics; Springer: Cham, Switzerland, 2014; Volume 8867, pp. 236–242.

15. ASHRAE. *ASHRAE Standard 55. Atlanta GA: American Society of Heating*; Refrigerating and Air-Conditioning Engineers, Inc.: Atlanta, GA, USA, 2017.

16. Dewhirst, M.W.; Viglianti, B.L.; Lora-Michiels, M.; Hanson, M.; Hoopes, P.J. Basic principles of thermal dosimetry and thermal thresholds for tissue damage from hyperthermia. *Int. J. Hyperth.* **2003**, *19*, 267–294. [CrossRef]

17. Young, P.C. *Control and Dynamic Systems: Advances in Theory and Applications*; Academic Press Inc., Elsevier Science: New York, NY, USA, 1989.

18. Young, P.C. *Concise Encyclopedia of Environmental Systems*, 1st ed.; Pergamon-Elsevier Science Ltd.: Oxford, UK, 1993.

19. Youssef, A.; D'Haene, M.; Vleugels, J.; De Bruyne, G.; Aerts, J.-M. Localised Model-Based Active Controlling of Blood Flow During Chemotherapy to Prevent Nail Toxicity and Onycholysis. *J. Med. Boil. Eng.* **2018**, *39*, 139–150. [CrossRef]

20. Young, P.; Jakeman, A. Refined instrumental variable methods of recursive time-series analysis Part III. Extensions. *Int. J. Control* **1980**, *31*, 741–764. [CrossRef]

21. Young, P.C.; Chotai, A.; Tych, W. *Identification, Estimation and Control of Continuous-Time Systems Described by Delta Operator Models*; Kluwer Academic Publishers: Dordrecht, The Netherlands, 1991.

22. Suykens, J.A.K.; Van Gestel, T.; De Brabanter, J.; De Moor, B.; Vandewalle, J. *Least Squares Support Vector Machines*; World Scientific Pub Co Pte Ltd.: Singapore, 2002.

23. Suykens, J.; Vandewalle, J. Least Squares Support Vector Machine Classifiers. *Neural Process. Lett.* **1999**, *9*, 293–300. [CrossRef]

24. Byrne, N.M.; Hills, A.P.; Hunter, G.R.; Weinsier, R.L.; Schutz, Y. Metabolic equivalent: One size does not fit all. *J. Appl. Physiol.* **2005**, *99*, 1112–1119. [CrossRef] [PubMed]

25. Haugen, H.A.; Melanson, E.L.; Tran, Z.V.; Kearney, J.T.; Hill, J.O. Variability of measured resting metabolic rate. *Am. J. Clin. Nutr.* **2003**, *78*, 1141–1144. [CrossRef] [PubMed]

26. Garnier, H.; Young, P.C.; Gilson, M. Simple Refined IV Methods of Closed-Loop System Identification. *IFAC Proc. Vol.* **2009**, *42*, 1151–1156. [CrossRef]

27. Ghahramani, A.; Castro, G.; Karvigh, S.A.; Becerik-Gerber, B. Towards unsupervised learning of thermal comfort using infrared thermography. *Appl. Energy* **2018**, *211*, 41–49. [CrossRef]

28. Kontes, G.D.; Giannakis, G.I.; Horn, P.; Steiger, S.; Rovas, D.V. Using Thermostats for Indoor Climate Control in Office Buildings: The Effect on Thermal Comfort. *Energies* **2017**, *10*, 1368. [CrossRef]

29. Ghahramani, A.; Zhang, K.; Dutta, K.; Yang, Z.; Becerik-Gerber, B. Energy savings from temperature setpoints and deadband: Quantifying the influence of building and system properties on savings. *Appl. Energy* **2016**, *165*, 930–942. [CrossRef]

30. Zavala, V.M. Real-Time Optimization Strategies for Building Systems. *Ind. Eng. Chem. Res.* **2013**, *52*, 3137–3150. [CrossRef]

31. Camacho, E.F.; Bordons, C. Model predictive control. *Int. J. Robust Nonlinear Control* **1999**, *13*, 280.

32. Losing, V.; Hammer, B.; Wersing, H. Incremental on-line learning: A review and comparison of state of the art algorithms. *Neurocomputing* **2018**, *275*, 1261–1274. [CrossRef]

33. Léon, B.; Vapnik, V. Local learning algorithms. *Neural Comput.* **1992**, *4*, 888–900.

34. Camacho, E.F.; Bordons, C. *Model Predictive Control*; Springer: London, UK, 2007.

35. Hosseini, S.M.; Carli, R.; Dotoli, M. Model Predictive Control for Real-Time Residential Energy Scheduling under Uncertainties. In Proceedings of the 2018 IEEE International Conference on Systems, Man, and Cybernetics (SMC), Miyazaki, Japan, 7–10 October 2018; pp. 1386–1391.

36. Afram, A.; Janabi-Sharifi, F. Theory and applications of HVAC control systems—A review of model predictive control (MPC). *Build. Environ.* **2014**, *72*, 343–355. [CrossRef]
37. Fiala, D.; Lomas, K.J.; Stohrer, M. Computer prediction of human thermoregulatory and temperature responses to a wide range of environmental conditions. *Int. J. Biometeorol.* **2001**, *45*, 143–159. [CrossRef] [PubMed]

Review

Well-Defined Conjugated Macromolecules Based on Oligo(Arylene Ethynylene)s in Sensing [†]

Agata Krywko-Cendrowska [1], Dawid Szweda [2] and Roza Szweda [2,*]

[1] Department of Chemistry, University of Basel, Mattenstrasse 24a, BPR 1096, 4058 Basel, Switzerland; a.krywko@unibas.ch
[2] Łukasiewicz Research Network–PORT Polish Center for Technology Development, Stabłowicka 147, 54-066 Wrocław, Poland; dawidszweda@gmail.com
* Correspondence: roza.szweda@port.org.pl
† Special Issue on "Bioprocess Monitoring and Control".

Received: 8 April 2020; Accepted: 29 April 2020; Published: 3 May 2020

Abstract: Macromolecules with well-defined structures in terms of molar mass and monomer sequence became interesting building blocks for modern materials. The precision of the macromolecular structure makes fine-tuning of the properties of resulting materials possible. Conjugated macromolecules exhibit excellent optoelectronic properties that make them exceptional candidates for sensor construction. The importance of chain length and monomer sequence is particularly important in conjugated systems. The oligomer length, monomer sequence, and structural modification often influence the energy bang gap between the highest occupied molecular orbital (HOMO) and the lowest unoccupied molecular orbital (LUMO) of the molecules that reflect in their properties. Moreover, the supramolecular aggregation that is often observed in oligo-conjugated systems is usually strongly affected by even minor structural changes that are used for sensor designs. This review discusses the examples of well-defined conjugated macromolecules based on oligo(arylene ethynylene) skeleton used for sensor applications. Here, exclusively examples of uniform macromolecules are summarized. The sensing mechanisms and importance of uniformity of structure are deliberated.

Keywords: well-defined macromolecules; sequence-defined macromolecules; sequence-defined polymers; conjugated oligomers; oligo(arylene ethynylene)s; biosensors; sensors; process monitoring

---

1. Introduction

Nowadays, facing the development of precise polymer chemistry, in particular new synthetic methods that allow for monomer sequence control we are looking for new areas of application of macromolecules where the sequence matters. To design new applications of macromolecules the sequence–property relationship has to be well understood. Sensing and process monitoring are expanding areas where the structure of macromolecules and the sequence of monomers became a crucial parameter to achieve specificity and selectivity of the detection and bioprocess monitoring.

The monitoring of bioprocesses in an organism is performed by cascade communication between the network of biomolecules [1]. Biological components often react very sensitively to environmental changes (e.g., pH, temperature, nutrients), which may result in adverse effects on the activity of the cells or the reproducibility of the process. Uniform, sequence-defined macromolecules such as proteins and DNA are key features in the regulation of biological processes. The well-defined and sequence-controlled structures of those biomolecules enable precise recognition of specific molecular patterns or environmental changes (e.g., temperature, pH) to regulate the cascade events in the living organisms. The monomer sequence, for instance, amino acids in proteins or nucleotides in DNA, determines functions and is responsible for regulation of thousands of cascade events in our body. The whole mammalian immune system relies on a large array of nucleic acid sensors [2,3].

The precision of the primary structure is also very important in man-made (bio)sensor systems based on conjugated macromolecules [4]. Over the last years, the synthetic strategies leading to well/sequence-defined macromolecules, that enable better control of the properties of resulting materials, were developed [5]. The monomer composition [6–19], sequence [4,6,12,19], and the oligomer length [6,9] are parameters that influence the energy levels of HOMO and LUMO orbitals in conjugated molecules. The structure indicates the energy gap between HOMO and LUMO, absorption, and emission properties in conjugated systems [6,19]. Even a slight modification of oligomer structure, e.g., in the side [20] or end groups [21,22], can affect its optoelectronic properties and influence sensing.

Conjugated macromolecules, due to their excellent optoelectronic properties, found great use in the construction of different types of sensors [23–25], e.g., monitoring of enzymatic activity [26], chirality sensors [27], protein sensing [28], material self-healing [29], diagnosis and drug discovery [25], biosensing and therapeutics [30]. The π-conjugated structure, allowing communication between monomers in molecules backbone, is responsible for excellent optoelectronic properties susceptible to environmental changes. The signal can be revealed in one or more dimensions [24] that induce selectivity of the read-out, e.g., chemical nose approach [25,31]. Multidimensional sensor response delivered by multiple sensor elements can selectively interact with the sample and produce a distinct pattern of response enabling specific identification of target components.

Among conjugated macromolecules, oligo(arylene ethynylene)s (OArEs) have gained considerable attention due to their excellent optoelectronic properties and emerging applications [32–35]. The current synthesis methods provide access to uniform OArEs of precise length and full sequence control [6,12,36]. OArEs are an important class of sensory materials [26,30,37–40]. OArEs sensors can be used in organic solvents as well as in the aqueous environment or as solid films. Their sensing mechanism usually takes advantage of their fluorescence properties, but not exclusively.

This review aims to summarize the examples of linear oligo(arylene ethynylene)s applied in sensing and to discuss their relevance and perspectives. Our study highlights the importance of adjustment and manipulation of their sequence and length to improve the performance of oligomer-based fluorescent sensors. The OArEs sensors in solution and solid films are discussed in the context of their applications in detection and process monitoring. The examples of oligomers that change properties upon the presence of a particular analyte or environmental changes are also included. Here, we focused on well-defined, uniform oligomers built from at least three arylene units connected via ethynylene linkage.

2. Synthesis of Well-Defined Conjugated Macromolecules

The synthetic strategies leading to uniform macromolecules that enable control over the monomer sequence have been introduced to polymer science during the last decades [41–43]. The nature-inspired need of defined macromolecular structures was a driving force to develop new synthetic methodologies, that combine current achievements of polymer chemistry, organic synthesis, and biochemistry to develop methods yielding uniform, sequence-defined macromolecules (Figure 1) [41,42,44,45].

In general, discrete macromolecules can be accessed by iterative synthesis. In principle, the synthesis relay on the stepwise attachment of protected monomers followed by deprotection. These two steps are repeated cyclically until the desired molecule is obtained. When the monomers are equipped with orthogonal functional groups there is no need for use of protecting groups [46,47]. Three main approaches can be distinguished: classical solution synthesis (Figure 1b), synthesis using a soluble polymer as a support (Figure 1c), or solid-phase synthesis (Figure 1d) [48]. The syntheses performed by classical organic chemistry methods are associated with cumbersome purification after each step. The use of polymeric supports in a soluble [6] or a solid phase [49] significantly simplifies the purification process.

In iterative synthesis it is very important to achieve high stepwise yields. The physical limitation of iterative synthesis is stepwise yield that determines possible length of macromolecules. The total yield equals the product of actual step yield multiplying and, thus its value decreases dramatically with the number of steps, according to the formula:

$$Y_{total}[\%] = \prod_{i=1}^{n}\left(\frac{Y_n}{Y_{nth}} \times 100\right)$$

(1)

where: *n*—number of steps, Y_n—actual yield of step *n* [g], Y_{nth}—theoretical yield of step *n* expressed in mass units.

The Y_{total} drops dramatically with the number of steps. For example, if the stepwise yield Y_n will be 95%, after 50 steps total yield will be only 7%.

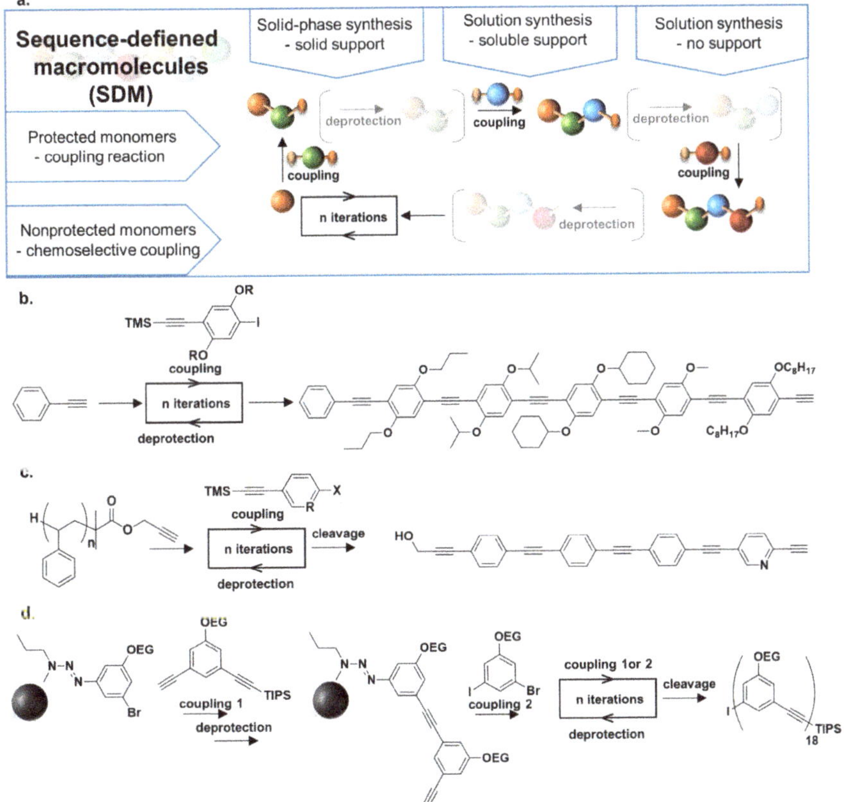

Figure 1. (**a**) The main approaches for the synthesis of sequence-defined conjugated macromolecules. Sequence-defined macromolecules can be obtained by multistep-growth synthesis using three main approaches: solid-phase synthesis, synthesis on soluble support, or solution synthesis. The monomers are used in a protected form that demands performance of the deprotection step after each coupling or by chemoselective reactions where monomers are equipped with orthogonal functional groups. The examples are of (**b**) solution synthesis [42], (**c**) synthesis on soluble support [6], and (**d**) solid-phase synthesis [11].

To obtain discrete conjugated oligomers (COs) based on oligo(arylene ethynylene)s skeleton, several stepwise synthetic strategies have been developed [42,50]. Usually, the oligomers are obtained by iterative solution synthesis that involves protected monomers. Due to poor solubility of arylene ethynylene oligomers they are usually synthesized from monomers functionalized by solubilizing substituents. Oligomers are produced by successive coupling and deprotection steps that are repeated in the cycle until the preferred macromolecule is obtained [8,9,11,51–53]. An interesting alternative to classical solution-phase synthesis is the use of soluble or solid supports [7,10], in particular for longer oligomers synthesis. Those approaches significantly simplify purification after each step. The synthesis can be facilitated by divergent/convergent approaches [54] of bidirectional growth [55].

For example, the group of M.A.R. Meier established a solution synthesis protocol for sequence-ordered, uniform pentamers built from five different monomers [12]. The oligomers were synthesized by Sonogashira cross-coupling reaction followed by deprotection (Figure 1b). The photophysical properties of the monodisperse oligomers differed only slightly, but the sequence had an impact on their thermal properties and the hydrodynamic volume.

Oligo(arylene ethynylene)s without solubilizing substituents can be obtained by a soluble-support approach. An interesting example of the polystyrene-tethered synthesis of uniform and sequence-defined oligomers without solubilizing substituents was reported by R. Szweda et al. [6]. The use of an ATRP-made, tailored polystyrene support enabled the synthesis of OArEs containing sequence ordered pyridine and benzene units. For oligomer synthesis, Sonogashira cross-coupling reaction was used (Figure 1c). The use of the soluble support approach gave access to the unsubstituted oligomers that are inaccessible using other methods due to their limited solubility. The UV and fluorescent properties were changing with the oligomer length and composition.

The solid-phase synthesis of oligo(phenylene ethylene)s (OPEs) was reported by the group of J. M. Tour [9]. The synthesis of 16-units oligo(phenylene ethylene)s was demonstrated [9]. The authors applied an iterative divergent/convergent approach on Merrifield resin leading to oligo(2-alkyl-1,4-phenylene ethynylene)s. At each stage of the iteration, the length of the oligomer doubled. Another example of solid-phase synthesis leading to oligomers of 18 repeating units was developed by J. Moore [11]. In this approach, the monoprotected bisethynylarene and a 3-bromo-5-iodo arene monomers bearing orthogonal reactivity were used.

Alternatively, the discrete conjugated macromolecules can be obtained by purification of oligomers mixtures e.g., using reverse-phase chromatography [56] or automated flash chromatography [19] as demonstrated by the group of C. J. Hawker. The automated flash chromatography is an efficient separation method and can be used for the separation of thiophene oligomers of 2–14 units in grams scale [19].

Other methods often employed for the synthesis of COs as step-growth polymerization lead to mixtures of products of dispersity often higher than 1.2 [57]. Dispersity polymers consist of a mixture of structures which has a significant impact on the properties and optical behavior that may influence their sensing performance [58–60]. Moreover, the polymerization technique is not easily reproducible and in the context of sensor application that might be a crucial factor to obtain the same properties of materials. The resulting polymer even in the best-controlled conditions does not consist of uniform macromolecules. It is unlikely to obtain exactly the same mixture of macromolecules in two different polymerization processes.

Although lots of effort has been made, all of the existing methods show many limitations. The usual problems are low yields, limited molar mass (chain length), small synthesis scale, high synthesis cost, time, and labor consumption, etc. In the context of applications, those problems pose new challenges to the chemists to optimize the synthetic strategies, that will overcome existing limitations and enable further development of the field.

3. Macromolecular Conjugated Sensor Probes

Sequence-defined, uniform macromolecules based on linear oligo(arylene ethynylene)s backbone have been used in different types of sensors that can be classified into two main approaches: (i) solution probes, where oligomers are dissolved in a medium (Section 3.1) and (ii) solid-phase probes (Section 3.2), where oligomers are used as films or they are immobilized on a solid support. In the solution phase, we can distinguish two main sensor categories: classical OArEs (Section 3.1.1) and oligo(arylene ethynylene) electrolytes (Section 3.1.2), that possess ionic side-chains or end groups and exhibit water solubility. In this review, the compounds were divided according to the applied approach.

The oligo(arylene ethynylene)-based sensors can be also categorized according to the type of signal used in sensing, e.g., fluorescence, electrochemical, UV-vis, circularly polarized light (CD). Among the typical detection methods, the most popular is based on fluorescence that takes advantage of excellent optoelectronic properties of OArEs.

The sensing properties of OArEs strongly depends on their structure and chosen monomers. It was demonstrated that certain oligomers have specific structures and can selectively respond to the presence of particular species, e.g., selective detection of fibrillar and prefibrillar amyloid protein aggregates [61]. Sensing can be influenced even by little structural changes like oligomers length and monomer composition. In the following sections, the examples presenting the influence of structural factors on detection efficiency are discussed.

The examples of linear, discrete oligo(arylene ethynylene)s applied in sensing for detection and process monitoring are listed in Table 1. The oligomers that change properties upon the presence of a particular analyte or environmental changes are also included.

Table 1. Oligo(arylene ethynylene)s applied in sensing.

No.	OArEs Structures	Sensed Species (Potential Application)	Solvent/Medium (Concentration Range/LOD)	Sensing Mechanism	Detection Method	Ref.
		Oligo(phenylene ethynylene) in solutions				
1	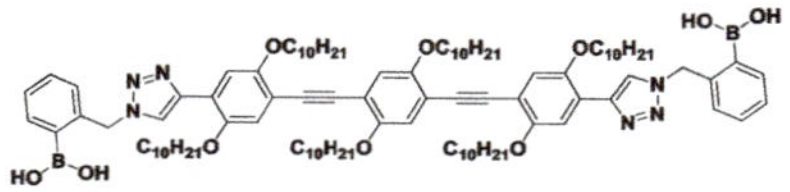	D-fructose (fluorescent saccharide sensors)	Aqueous buffer pH 8.21 (D-fructose - 85-390 mM)	Reversible and pH-dependent complexation of phenylboronic acid with saccharides	FL	[38]
2	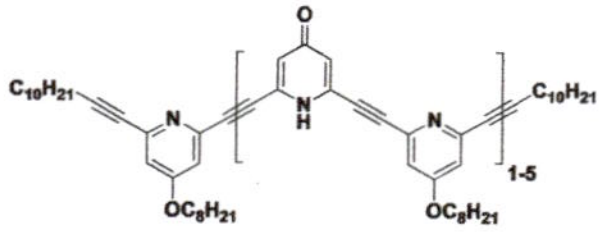	Octyl β-D-glucopyranoside	Dichloromethane and chloroform	Dynamic transformation between CD-silent dimer and CD-active helical oligomer-saccharide	CD, ^{1}H NMR	[62]
3	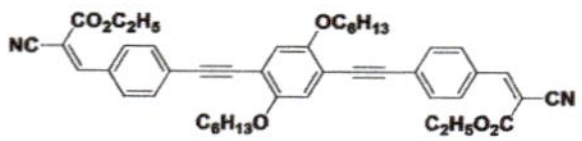	Cysteine (fluorescent chemical sensor for cysteine)	THF, THF/water (1:1, v/v) (1.0–10.0 μM)	Fluorescence quenching by blocking the photo-induced electron PET mechanism by free electron pair at N atom in Cys	FL	[63]
4		Enantioselective sensing of carboxylic acids, e.g., non-racemic tartaric acid (chiroptical sensing of carboxylic acids)	Chloroform (0.2 μM or 0.8 mM)	Chiroptical signals induced upon hydrogen bonding between urea protons and carboxylate groups in analytes	CD	[64]
5	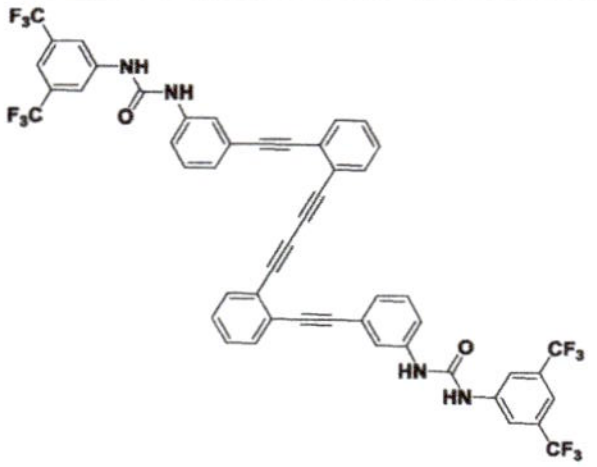	Enantioselective sensing of carboxylic acids, e.g., non-racemic tartaric acid (chiroptical sensing of carboxylic acids)	Chloroform (0.2 μM or 0.8 mM)	Chiroptical signals induced upon hydrogen bonding between urea protons and carboxylate groups in analytes	CD	[64]

Table 1. *Cont.*

No.	OArEs Structures	Sensed Species (Potential Application)	Solvent/Medium (Concentration Range/LOD)	Sensing Mechanism	Detection Method	Ref.
6		Chiral organic ions, i.e., camphorsulfonates, adenosine 3′,5′-cyclic monophosphate (cyclic monophosphate chirality sensors)	CH_2Cl_2, MeOH/CH_2Cl_2 (1%, *v/v*) (1 mg/mL)	CD signal controlled/switchable through acid-base chemistry	CD	[65]
7		Silver (I) compounds: $AgBF_4$	CH_2Cl_2 or 95:5 mixture of CH_2Cl_2/acetone (0-2.5×10^{-4} M)	Organic monomolecular emitter which behaves as a circularly polarized luminescence (CPL)-based ratiometric probe	CD, FL	[66]
8		Silver (I) compounds: $AgBF_4$	CH_2Cl_2 or 95:5 mixture of CH_2Cl_2/acetone (0-2.5×10^{-4} M)	Organic monomolecular emitter which behaves as a circularly polarized luminescence (CPL)-based ratiometric probe	CD, FL	[66]
9		Temperature (fluorescent thermosensors)	$CHCl_3$; 2-methyltetrahydrofuran (80 to 320 K)	Change of the twist angles in the π-conjugated backbones depending on the temperature	FL	[67]
10		Temperature (fluorescent thermosensors)	$CHCl_3$; 2-methyltetrahydrofuran (80 to 320 K)	Change of the twist angles in the π-conjugated backbones depending on the temperature	FL	[67]

Table 1. *Cont.*

No.	OArEs Structures	Sensed Species (Potential Application)	Solvent/Medium (Concentration Range/LOD)	Sensing Mechanism	Detection Method	Ref.
11		Scratch healing efficiency in mussel-inspired polymer (monitoring of thermally triggered self-healing systems)	Air; temperature treatment at 60 °C	Reversible Zn−histidine interactions	FL, CLSM	[29]
		Oligo(phenylene ethynylene)s electrolytes				
12		Methyl viologen (MV^{2+}), 3,3′-diethyloxacarbocyanine iodide (DOC) (optical materials in chemo- and biodetection)	Water pH 8 (MV^{2+} 0.01–0.3 µM; DOC 0.02–5 mM)	Amplified fluorescence quenching due to ion pairing between the oligomer and the quencher cation	FL, UV-vis	[68]
13		Ca^{2+}	Water pH 8 (0.02–10 mM)	Fluorescence shift	FL, UV-vis	[68]
14		bacteria: *Amycolatopsis azurea*, *Amycolatopsis orientalis* subsp. *lurida*, *Bacillus lichenformis*, *Bacillus subtilis*, *Escherichia coli* (BL21(DE3)), *Escherichia coli* (*E. coli* (DH5α)), *Escherichia coli* (*E. coli* (XL1 Blue)), *Lactococcus lactis*, *Lactococcus plantarum*, *Pseudomonas putida*, *Streptomyces coelicolor*, and *Streptomyces griseus* (efficient identification of bacteria)	5 mM phosphate buffer (pH 7.4); (OD_{600} = 1.0)	Fluorescence recovery by replacement of conjugated polymers coupled on gold nanoparticles with bacteria	FL	[69]
15		4-nitrophenyl phosphate, bis(cyclohexylammonium) salt hydrate (NPP), disodium hydrogen phosphate (DHP), 9,10-anthraquinone-2,6-disulfonic acid disodium salt (AQS) (optoelectronic sensor devices)	Water, pH 7.0 (DHP < 0.25 mM, AQS 3.2 µM, NPP 10 µM)	Quenching effect due to formation of non-emissive aggregates	FL, UV-vis	[70]

Table 1. *Cont.*

No.	OArEs Structures	Sensed Species (Potential Application)	Solvent/Medium (Concentration Range/LOD)	Sensing Mechanism	Detection Method	Ref.
16	R = OCH₂CH₂N⁺(CH₃)₃, OCH₂CH₂CH₂N⁺(CH₃)₃ or OCH₂CH₂CH₂SO₃⁻	Light-activated biocides against *Escherichia coli*, *Staphylococcus epidermidis*, and *Staphylococcus aureus*	0.85% NaCl in water	Under UV the conjugate photosensitize the generation of singlet oxygen which triggers the cytotoxicity	FL	[35]
17	R = OCH₂CH₂CH₂N⁺(CH₃)₃	Light-activated biocides against *Escherichia coli*, *Staphylococcus epidermidis*, and *Staphylococcus aureus*	0.85% NaCl in water	Under UV the conjugate photosensitize the generation of singlet oxygen which triggers the cytotoxicity	FL	[35]
18		Oxygen sensor for photosplitting of water	Water, argon-degassed water	Photoaddition of water across triple bond of ethynyl group in absence of oxygen, the addition of singlet-oxygen across a triple bond in presence of oxygen; formation of phenols via cleavage of alkoxy side chains in both cases	UV-vis, MS	[71]
19		Sodium dodecyl sulfate (SDS), carboxymethyl amylose (CMA), and carboxymethyl cellulose (CMeC)	Water, deuterium oxide (50 µM)	Fluorescence quenching by water; fluorescence enhancement due to formation of oligomer-surfactant complex	FL, UV-vis	[72]
20		4-nitrophenyl phosphate, bis(cyclohexylammonium) salt hydrate (NPP), 9,10-anthraquinone-2,6-disulfonic acid disodium salt (AQS) (optoelectronic sensor devices)	Water, pH 7.0 (AQS 6.1 µM, NPP 23 µM)	Strong fluorescence quenching in the presence of electron deficient species	FL, UV-vis	[70]

Table 1. *Cont.*

No.	OArEs Structures	Sensed Species (Potential Application)	Solvent/Medium (Concentration Range/LOD)	Sensing Mechanism	Detection Method	Ref.
21	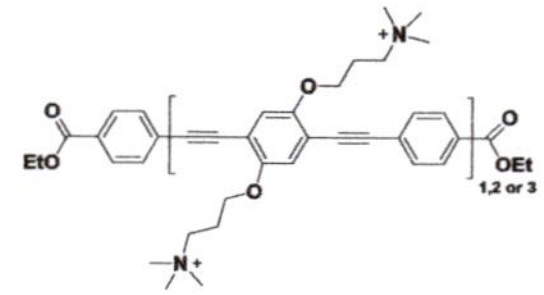	Fibril formation from native hen egg white lysozyme (HEWL)	10 mM citrate buffer in water (pH 3)	Significant fluorescence enhancement in solution with HEWL amyloids	FL, CD	[73]
22	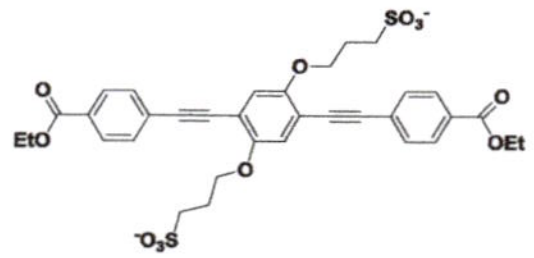	Amyloid protein aggregates Aβ40	pH 8.0 Tris (0.52 µM) pH 7.4 PB (0.48 µM)	Aggregate-specific binding inducing fluorescence turn on	FL	[61]
23	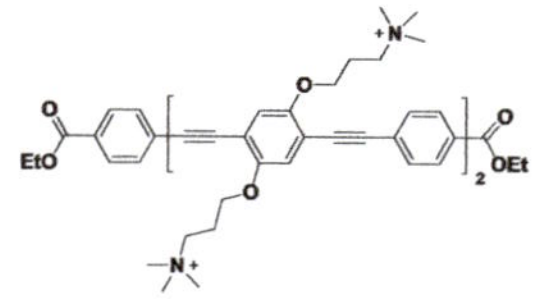	Amyloid protein aggregates Aβ40	pH 8.0 Tris (0.45 µM) pH 7.4 PB (0.24 µM)	Aggregate-specific binding inducing fluorescence turn on	FL	[61]
24	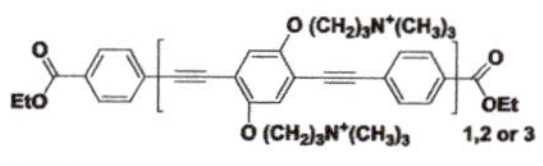	Activity of phospholipases (type A2, A1, C) and acetylcholinesterase (sensors of enzymes as biomarkers for pollution or disease)	Water, pH 7.5	Formation of molecular aggregates or conformational changes leading to change of photochemical properties	FL, UV-vis	[74]
25	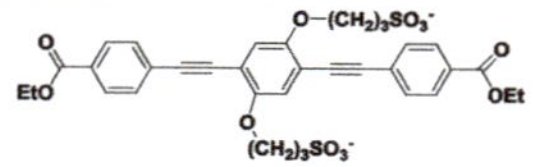	Activity of phospholipases (type A2, A1, C) and acetylcholinesterase (sensors of enzymes as biomarkers for pollution or disease)	Water, pH 7.5	Formation of molecular aggregates or conformational changes leading to change of photochemical properties	FL, UV-vis	[74]

Table 1. *Cont.*

No.	OArEs Structures	Sensed Species (Potential Application)	Solvent/Medium (Concentration Range/LOD)	Sensing Mechanism	Detection Method	Ref.
26	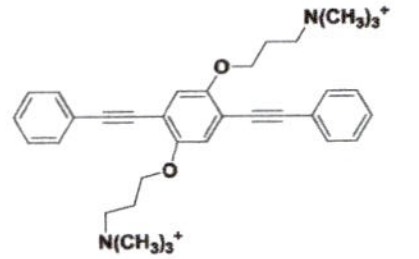	Presented on the example of surfactants SDS and TTAB	Water	Formation of aggregates with strongly red-shifted absorbance	FL, UV-vis	[75]
27	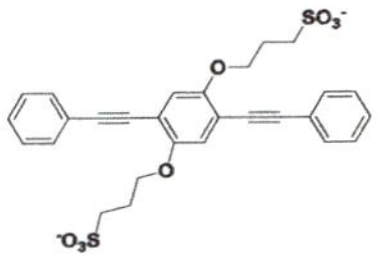	Presented on the example of surfactants SDS and TTAB	Water	Formation of aggregates with strongly red-shifted absorbance	FL, UV-vis	[75]
28	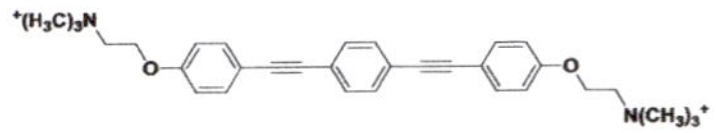	Presented on the example of surfactants SDS and TTAB	Water (TTAB 0–200 μM, SDS 0–30 μM)	Fluorescence loss resulting from rapid internal conversion between singlet states	FL, UV-vis	[75]
29	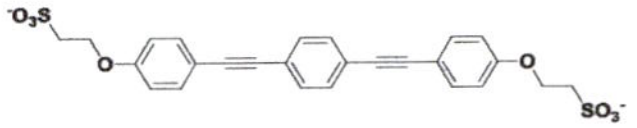	Presented on the example of surfactants SDS and TTAB	Water (TTAB 0–200 μM, SDS 0–30 μM)	Fluorescence loss resulting from rapid internal conversion between singlet states	FL, UV-vis	[75]
30	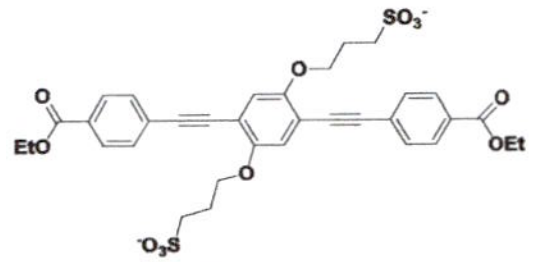	Cationic surfactant cetyl trimethylammonium bromide (CTAB) (theranostic agent e.g., amyloid diseases)	Water (0–1500 μM)	Under UV the conjugate photosensitize the generation of singlet oxygen which triggers the cytotoxicity	FL, UV-vis	[76]
31	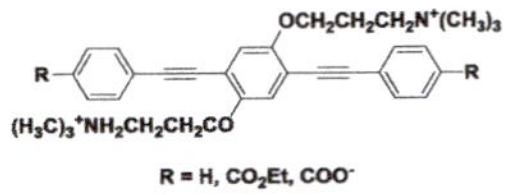	Anionic biomacromolecules; anionic biopolymer carboxymethylcellulose (CMC)	Water (0–42 μM)	strong spectral red shifts in both absorption and fluorescence coupled with the increase in fluorescence efficiency upon complexation with CMC	FL, UV-vis	[77]

Table 1. *Cont.*

No.	OArEs Structures	Sensed Species (Potential Application)	Solvent/Medium (Concentration Range/LOD)	Sensing Mechanism	Detection Method	Ref.
32	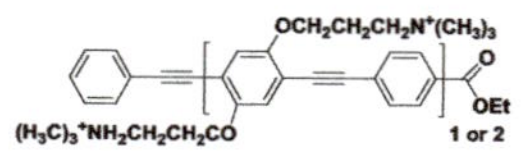	Anionic biomacromolecules; presented the example of carboxymethylcellulose (CMC), carboxymethylamylose (CMA) and synthetic Laponite clay	Water (CMC 0–50 μM, CMA 0–0.28 μM, Laponite 0–31 μg)	Strong spectral red shifts caused by effective increase in the conjugation length upon template-induced formation of linear J-dimers or possibly because of planarization	CD, FL, UV-vis	[78]
33	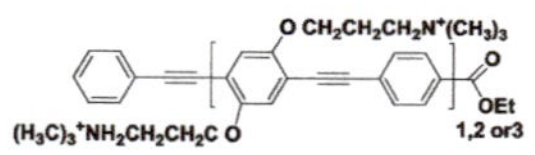	Amyloid protein aggregates; detergents (i.e., carboxymethyl amylose (CMA) and carboxymethyl cellulose (CMeC)); activity of phospholipases (PL A2, A1, C) and acetylcholinesterase; phospahatase/kinase)	Organic solvents (e.g., methanol); water	Red/blue shift in the emission spectrum on the interface between the oligo-electrolyte and the analyte; reversible fluorescence turn-on (phospahatase–kinase; LOD of 0.05 units/mL)	FL	[79]
34	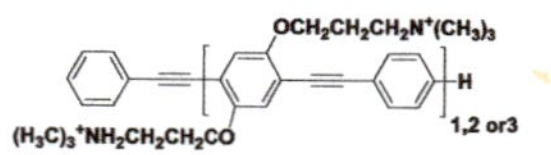	Amyloid protein aggregates; detergents (i.e., carboxymethyl amylose (CMA) and carboxymethyl cellulose (CMeC)); activity of phospholipases (PL A2, A1, C) and acetylcholinesterase; phospahatase/kinase)	Organic solvents (e.g., methanol); water	Red/blue shift in the emission spectrum on the interface between the oligo-electrolyte and the analyte; reversible fluorescence turn-on (phospahatase–kinase; LOD of 0.05 units/mL)	FL	[79]
35	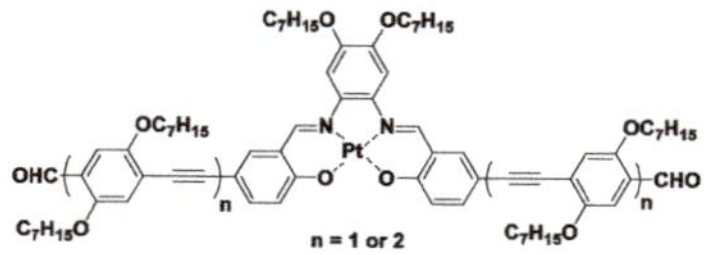	Signal dependent on polarity of solvents (tunable luminescent sensory materials)	CH_2Cl_2, THF, 2-methyltetrahydrofuran	Controllable integration of functionalized phosphorescent signaling units into well-defined conjugated materials	FL, UV-vis, TAS	[80]
Oligo(phenylene ethynylene)s films						
36	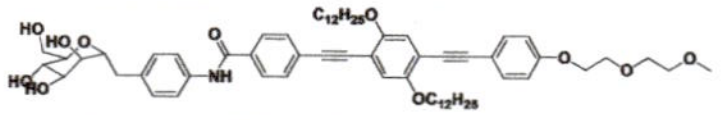	*Escherichia coli* bacteria (new family of sensitive and selective biochips to detect *E. coli*)	Sample in a culture medium, (LOD = 10^4 CFU/mL)	Staining of bacteria by conjugated polymers	LSCM, SPR	[81]
37	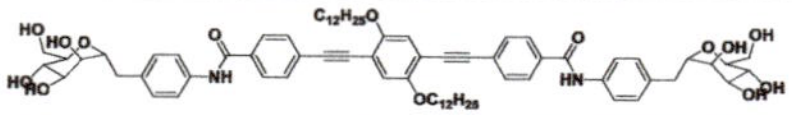	*Escherichia coli* bacteria (new family of sensitive and selective biochips to detect *E. coli*)	Sample in a culture medium, (LOD = 10^4 CFU/mL)	Staining of bacteria by conjugated polymers	LSCM, SPR	[81]

Table 1. *Cont.*

No.	OArEs Structures	Sensed Species (Potential Application)	Solvent/Medium (Concentration Range/LOD)	Sensing Mechanism	Detection Method	Ref.
38		Gram-positive *Bacillus subtilis* and Gram-negative *Escherichia coli*	Sample in miliQ water	Staining of bacteria by conjugated polymers	FL, LSCM, mRS,	[82]
39	R = 5,11,17,23-tetrakis(1,1-dimethylethyl)-25-(oxymethyl)-26,27,28-tripropyloxycalix[4]arene	Detection of high explosive materials and explosive markers: 2,4,6-trinitrotoluene (TNT) and 2,4-dinitrotoluene (DNT), nitromethane (NM), 2,3-dimethyl-2,3-dinitrobutane (DMNB)	Vapors in air	Amplified fluorescence detection response thanks to the establishment of a tridimensional network of strong π–π and CH–π interactions with electron-deficient guests developed near the transduction centers	FL	[83]
40		Detection of nitroaromatic explosives: trinitotoluene (TNT), 2,4-dinitrotoluene (DNT),	Vapors in air, detection threshold—0.75 ppbv for TNT, 9 ppbv for DNT	In the presence of nitroaromatic explosives, conjugated fluorescent materials exhibit excellent fluorescence extinction properties due to a charge transfer mechanism	FL	[84]
41		Inorganic acids: HCl, H_2SO_4, HNO_3, H_3PO_4 (monomolecular layer chemistry-based fluorescent sensing films)	Acetone (5–40 µM)	Fluorescence quenching due to protonation of the imino group next to the conjugated segment	FL	[85]
42	R = cholic acid	Inorganic acids: HCl, H_2SO_4, HNO_3, H_3PO_4 (monomolecular layer chemistry-based fluorescent sensing films)	Acetone (5–40 µM)	Fluorescence quenching due to protonation of the imino group next to the conjugated segment	FL	[85]
43		pH (thin-film ratiometric chemosensors)	Water (pH 2–10)	Shift in the emission spectrum of fluorescein (formation of anionic isomers)	FL	[86]

Table 1. *Cont.*

No.	OArEs Structures	Sensed Species (Potential Application)	Solvent/Medium (Concentration Range/LOD)	Sensing Mechanism	Detection Method	Ref.
44		2,4,6-trinitrophenol (PA), 2,4,6-trinitrotoluene (TNT), 2,4-dinitrotoluene (DNT), nitrobenzene (NB) (sensor for nitro compounds)	Water (PA, 0.2–5 μM; TNT, DNT, NB, 50 μM)	Reversible fluorescence quenching upon oligomer-analyte complex formation	FL	[87]
45		2,4,6-trinitrophenol (PA), 2,4,6-trinitrotoluene (TNT), 2,4-dinitrotoluene (DNT), nitrobenzene (NB) (sensor for nitro compounds)	Water (PA, 0.2–5 μM; TNT, DNT, NB, 50 μM)	Reversible fluorescence quenching upon oligomer-analyte complex formation	FL	[87]
46		H_2 (gas sensor for environmental monitoring)	-	Theoretical DFT-based model: adsorbed H_2 molecule significantly changes the characteristics of the current–voltage curve of the OPE molecular junction	CAL (DFT)	[88]
47		Data encoding	-	Electrically controllable local heating mechanism for the forward reaction and catalyzed by a single charge transferprocess for the reverse switching	CV	[89]
48		Cysteine (fluorescent chemical sensor for cysteine)	THF (1.0–10.0 μM)	Fluorescent "turn-on" upon electrostatic attraction between carboxylic groups of oligomer and ammonium groups of Cys	FL	[63]
49		Dopamine (electrochemical sensors)	Human serum in aqueous PBS (0.01–60 μM, LOD 5 nM)	Oxidation/reduction of dopamine on the film-modified electrode	CV	[90]
50		Perfluorocyclopent-1-ene-1,2 -diyl)bis(5-methylthiophene-2- carbaldehyde (PBMC)	Solid state, xerogel, liquid, i.e., ethanol, water, methanol–water mixture (1:1)	Photoswitch based on pcFRET (on/off)	FL	[91]

Abbreviations: ^{1}H NMR—proton nuclear magnetic resonance spectroscopy, FL—fluorescence spectroscopy, CAL—calculations, CD—circular dichroism spectroscopy, CLSM—confocal laser scanning microscopy, CV—cyclic voltammetry, LOD—lower limit of detection, mRS—microRaman spectroscopy, MS—mass spectrometry, SPR—surface plasmon resonance, TAS—transient-absorption spectroscopy, THF—tetrahydrofuran, UV-vis—ultraviolet-visible spectroscopy.

3.1. Oligo(Arylene Ethynylene)s as Sensors Probes in Solution

3.1.1. Oligo(Arylene Ethynylene) Sensors

Oligo(arylene ethynylene)s have been used as sensors to detect different species e.g., chemicals [65], saccharides [38,62], amino acids [63], explosives [84], ions [66], and physical changes e.g., temperature [67] or solvent polarity [67]. OArEs were also used for process monitoring to track self-healing of polymers [29].

For example, the oligo(phenylene ethynylene)s foldamers with urea end-groups (Table 1, no. 4) were used for the detection of chiral carboxylic acids e.g., tartaric acid. [64] The stereodynamic oligomer-carboxylic acid complexes formed chiral structures easily detectable by CD measurements. It was demonstrated that the chiroptical signal could be used for quantitative analyses providing a fast and simple method for chirality sensing assays (Figure 2).

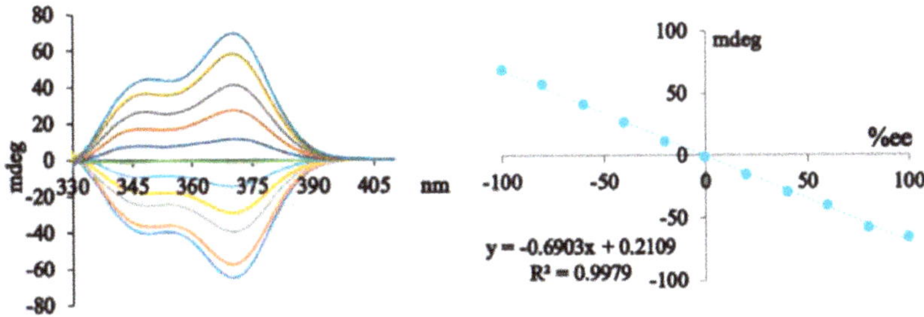

Figure 2. Circular dichroism (CD) spectra of the mixture obtained with oligo(phenylene ethynylene)s (Table 1, no. 4), Et₃N, and samples of tartaric acid and linear relationship between the CD amplitude at 370 nm and the sample enantiomeric excess. Reprinted from [64] with permission from Elsevier.

For instance, conjugated oligomers functionalized by boronic acid have been used as sensors to detect different saccharides: D-fructose, D-galactose, D-ribose, and D-glucose, in potassium phosphate buffer/DMSO (99/1, *v/v*) [38]. By the addition of saccharides, significant fluorescence enhancement was observed, and the response was different depending on saccharide. However, it was shown that the fluorescence response can be observed only for well-designed oligomer structures. The oligo(phenylene ethynylene)s with $OC_{10}H_{21}$ side chains (Table 1, no. 1) and boronic acid groups attached via triazole linker were sensitive exclusively to fructose presence. It was found that the fluorescence response depends on supramolecular interactions between sensors and analyte molecules which are very structure dependent. This study highlighted the need for a specific design of oligomer fluorophore in the development of effective saccharide sensors.

Ortho-oligo(phenylene ethynylene)s (Table 1, no. 7, 8) were used as a circularly polarized luminescence probe for the detection of silver ions [66]. The enantiopure helical core has been prepared by a new macrocyclization reaction. The combination of such o-OPE helical skeleton and pyrene reporter units lead to two characteristic circularly polarized emission features. The intensity of the bands linearly corresponds with silver(I) concentration.

Interestingly, the temperature change that is a physical process can be followed by oligo(arylene ethynylene)s [67]. The acetylene-bridged pentiptycene (n = 2, 3, and 4) (Table 1, no. 9) and phenylene–pentiptycene–phenylene three-ring system (Table 1, no. 10) were evaluated as fluorescent temperature sensors. The trimer and tetramer showed a unique response to temperature and solvent polarity driven by intramolecular interactions (Figure 3). It was found that the twisted region of their rotational potentials occurs at the local energy minimum, and the distribution of rotational conformers is sensitive to the temperature and solvent polarity. Twisting of the π-conjugated backbones reflected in 40 nm blue-shifted fluorescence spectra. It was demonstrated that upon temperature change in the range between 80 and 320 K, the fluorescence emission of acetylene-bridged pentiptycene tetramer shifted significantly. This property can be used for the development of low-temperature sensors.

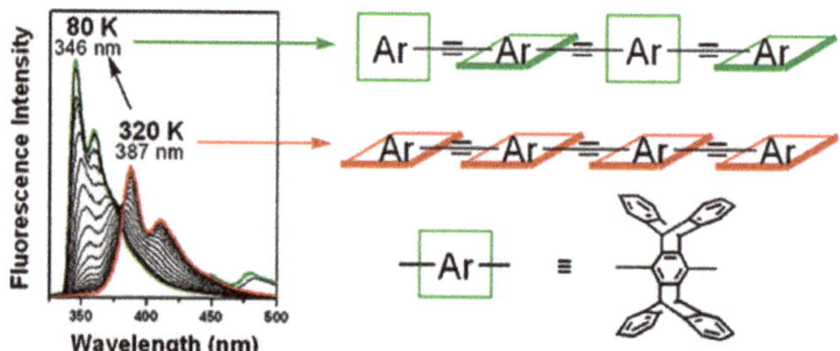

Figure 3. The change of conformation caused by temperature difference influence the fluorescence properties. Temperature dependence of fluorescence spectra (λ ex = 303 nm) of tetramer in methylotetrafuran at 20- and 10-K intervals, respectively, between 80 and 320 K. Reprinted with permission from [67]. Copyright 2006 American Chemical Society.

Recently it was demonstrated that oligo(arylene ethynylene)s (Table 1, no. 11) can be used as a fluorescent probe for monitoring of the self-healing process [29]. The OPE incorporated into a mussel-inspired scratch-healing polymer network helped to determine detailed depth- and time-dependent self-healing efficiency using confocal laser scanning microscopy (Figure 4). The damage of the network resulted in decreased fluorescence emission of polymer within the scratch. The mobility of the fluorescence marker is connected with the plasticity of the polymeric material, thus during scratch refilling, no independent migration of dye within the polymeric material was detected.

3.1.2. Oligo(Arylene Ethynylene) Electrolytes

Oligo(arylene ethynylene) electrolytes are very attractive macromolecules for application in sensing. As sensor probes, they combine the excellent fluorescence properties of a conjugated aryl-alkyne system with electrolyte advantages especially water solubility [35]. Due to the presence of ionic groups, these oligomers are very sensitive to the environment changes, e.g., ionic strength, pH, presence of ions, presence of electrolytes. The charged pendant groups can induce electrostatic interactions with oppositely charged (macro)molecules that reflect in fluorescence properties variation [37]. Moreover, the charges distributed along the oligomer molecules affect their aggregation thus they exhibit high fluorescence response to alterations of aggregates structure and conformational changes. Those changes caused by the presence of individual charged molecules may reveal a unique response in the photophysical properties of the conjugated chromophore. The resulting fluorescence quenching or enhanced emission can indicate presence of ions [70], oxygen [71], surfactants [70,72,75,79,92], detergents [76,79], MV^{2+} ions [68], solvent polarity [80], anionic biopolymer carboxymethylcellulose [77,78], biomolecules [93], and bacteria [69]. The oligomers were used for processes monitoring of e.g., amyloid formation [61,73], enzymatic activity [74], and photochemical reaction processes [71].

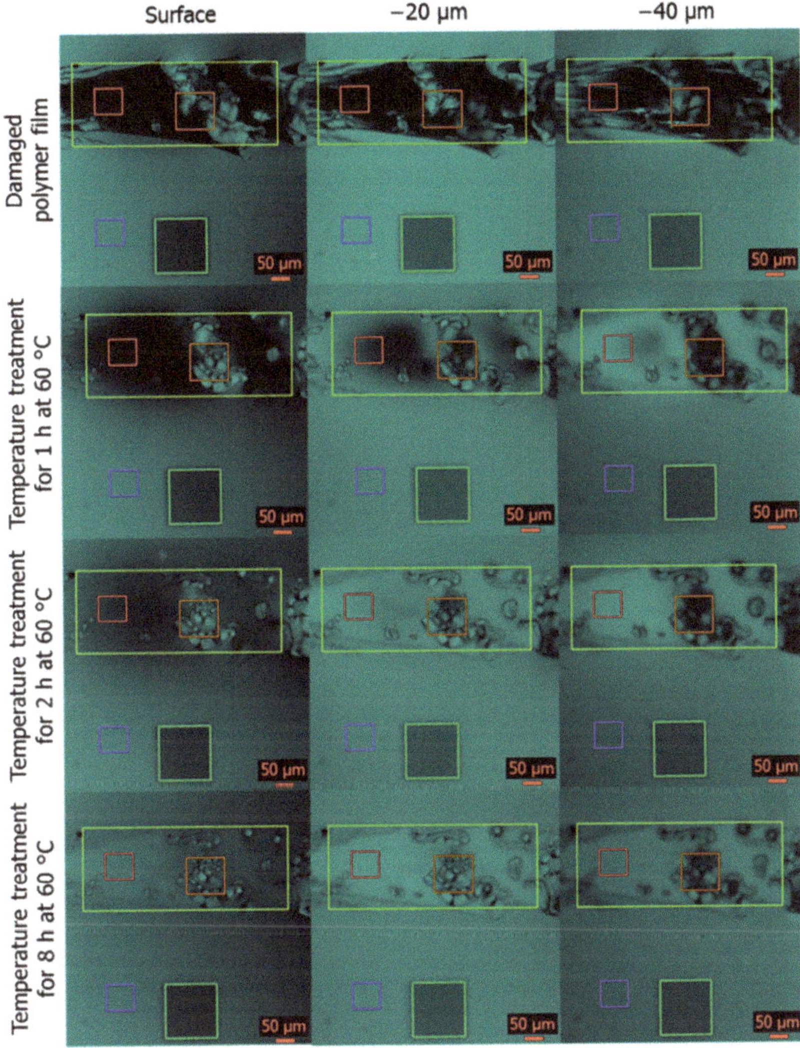

Figure 4. Time- and depth-dependent confocal laser scanning microscopy (CLSM) measurements in fluorescence mode (λ_{ex} = 405 nm) with the fluorescence channel (λ_{em} = 406–510 nm) monitoring thermally triggered self-healing procedure, in particular the virgin damaged cross-linked copolymer film, after 1, 2, and 8 h of thermal treatment at 60 °C: (red) homogeneous area within the scratch, (yellow) heterogeneous area covering the majority of the analyzed defect, (orange) specific area with residual removed film material, (blue) intact and undamaged reference area for each measurement, and (green) photo-bleached marker area. Reprinted with permission from [29]. Copyright 2018 American Chemical Society.

For instance, the oligo(p-phenylene ethynylene) electrolytes (OPE) were successfully applied to track amyloid formation [61,73]. Oligomers with ester terminal moieties and positively charged $-(CH_2)_3N(CH_3)_3^+$ pendant groups of different length OPEn (n = 1, 2, and 3) (Table 1, no. 21) and OPE1 negatively charged with pendant $-(CH_2)_3SO_3^-$ groups (Table 1, no. 22) were evaluated as probes for monitoring of the fibrillation process (Figure 5) [61,73]. The carboxyester terminal groups of OPEs cause high fluorescence quenching due to the strong interactions with the solvent, on the other hand, the oligomers show strong fluorescence emission when in a water-poor environment. These environment-dependent fluorescence properties were used for the sensor design. It was demonstrated that positively charged OPEs used in 10:1 (protein:OPE) molar ratio are effective molecular taggants for selective sensing of the amyloid fibril of the model protein HEWL. Upon fibril formation, OPEs form clusters with the fibrils, where the carboxyester terminal groups are isolated from water. In a non-water environment, they form superluminescent chiral J aggregates [94] and significant fluorescence enhancement is observed (Figure 5). It was found that due to the energy transfer the excitation at 280 nm characteristic for HEWL results in the emission of OPE only in solutions containing OPEs and HEWL amyloids that indicate amyloid formation.

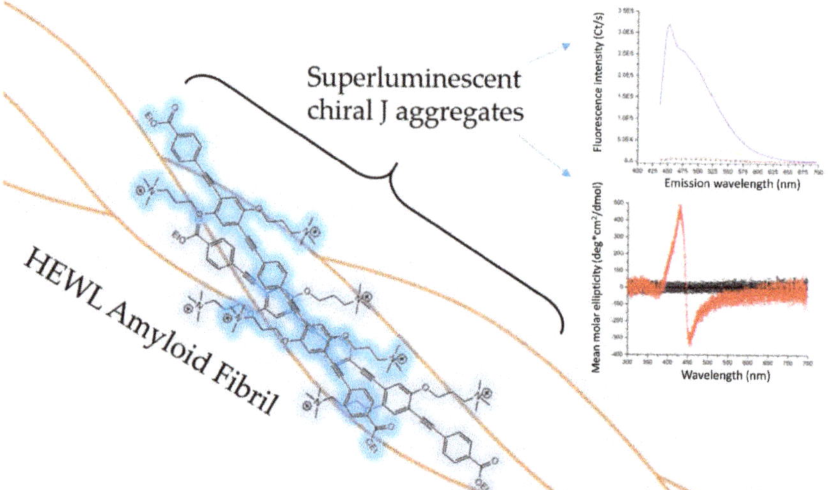

Figure 5. Oligo(p-phenylene ethynylene) (OPEs) electrolytes are forming specific chiral constructs together with amyloid fibrils. The construct exhibits enhanced fluorescence quenching and a unique CD signal. Circular dichroism spectra of OPE n = 3 (10 µM) in phosphate buffer with hen egg white lysozyme (HEWL) monomer (black trace) and with HEWL (10 µM) amyloid (red trace). Emission spectra of OPE n = 3 in phosphate buffer (PB, pH 7.4, 10 mM) alone (black long dashed line) with HEWL monomers (red short dashed line) and with HEWL amyloids (blue solid line), concentration: 500 nM, protein concentration: 5 µM monomer basis/0.25 mg/mL. Reprinted with permission from [73]. Copyright 2015 American Chemical Society.

p-Phenylene ethynylene oligomers can be also used for monitoring of enzymatic processes. Complexes of oligomers (Table, no. 24, 25) with enzyme substrates were successfully used to follow activity and inhibition of two biomarkers, phospholipase indicating heart and circulatory disease, and acetylcholinesterase for Alzheimer's diagnosis (Figure 6) [74]. In a buffer solution, oligomers form complexes with positively charged substrates e.g., 1,2-dilauroyl-sn-glycero-3-phosphoglycerol (DLPG) and lauroyl choline (LaCh). The DLPG-oligomer (Table, no. 24) complex upon phospholipases undergoes transformation due to the cleavage of DLPG phosphate bond that resulted in a swift of the fluorescence quenching. The aggregates of an anionic oligomer (Table, no. 25) with lauroyl choline were used as a sensor to detect the activity and inhibition of acetylcholinesterase.

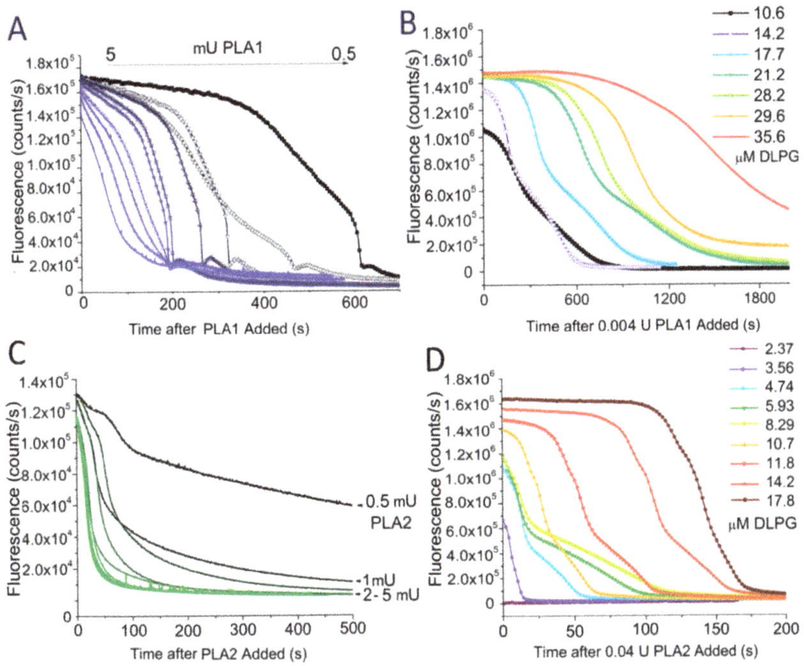

Figure 6. (**A**) Fluorescence of the oligomer/1,2-dilauroyl-sn-glycero-3-phosphoglycerol (DLPG) (Table 1, no. 24) aggregates over the course of Phospholipase A1 activity with 1.4 µMOPE and a DLPG concentration of 7.27 µM, with enzyme added ranging from 0.5 to 5 mU of Phospholipase A1. (**B**) A concentration of 1.4 µM of +2C with DLPG at a series of concentrations from 10.6 to 35.6 µM (7.5–25.4 DLPG:OPE ratio), followed by the addition of 4 mU of Phospholipase A1. (**C**) Fluorescence of the oligomer/DLPG aggregates over the course of Phospholipase A2 activity with 1.4 µM oligomer and a DLPG concentration of 7.27 µM, with enzyme added ranging from 0.5 to 5 mU of Phospholipase A2. (**D**) A concentration of 1.4 µM of +2C with DLPG at a series of concentrations from 2.37 to 17.8 µM (1.7–12.7 DLPG:oligomer ratio), followed by addition of 40 mU of Phospholipase A2. t = −1 s is the time of enzyme addition. Wavelength of excitation is 375 nm, emission is 440 nm. Reprinted with permission from [74]. Copyright 2015 American Chemical Society.

OArEs (Table 1, no. 18) can be used to monitor chemical processes, e.g., photolysis [71]. For example, the photo-induced degradation process of oligomer (Table 1, no. 18) occurred by three main routes: the photoprotonation of the triple bond followed by the addition of water, the addition of singlet oxygen across the triple-bond, and the cleavage of the quaternary ammonium side-chains. The degradation led to the formation of different products depending on the reaction atmosphere (argon or oxygen). All those structural changes reflected in fluorescence properties indicating the rate and mechanism of the degradation. Whenever the process was performed in the presence or absence of oxygen, different products of different fluorescence properties were formed. The dependence of fluorescence properties on the reaction atmosphere led to developing an oxygen-sensing methodology based on fluorescence read-out of OArE photo-degradation.

Constructs of oligo(phenylene ethynylene)s electrolyte and gold nanoparticles can be used for selective bacteria identification using the "chemical nose" sensing concept (Table 1, no. 14) [69]. In a solution, positively charged gold nanoparticles form complexes with negatively charged oligo(phenylene ethynylene)s, and oligomer fluorescence is quenched. In the presence of bacteria, some OPEs are released to the solution and fluorescence is restored as a consequence of the presence of free oligomers (Figure 7). The applied oligomer with branched oligo(ethylene glycol) side chain reduces the non-specific interaction of oligomer and bacteria. Depending on bacteria the oligomer replacement is different which results in selective fluorescence response.

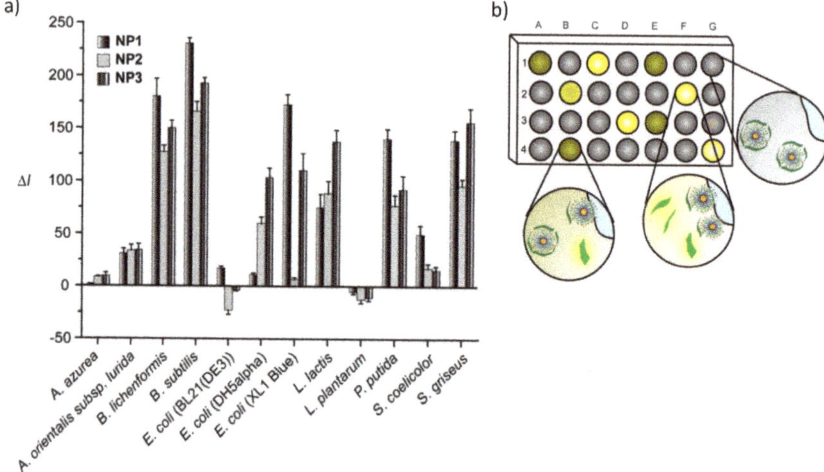

Figure 7. (**a**) Fluorescence intensity patterns of nanoparticle–oligomer (Table 1, no. 14) constructs in the presence of various bacteria strains. (**b**) The schematic presentation of sensor design. Bacteria interact with gold nanoparticle-oligomer construct and as oligomers macromolecules are released to the solution, fluorescence enhancement is observed. For each bacteria, interactions with nanoparticles are unique. In the figure, columns represent bacteria of different types, and rows represent the oligomer–nanoparticle constructs. Reprinted with permission from [69]. Copyright 2008 John Wiley and Sons.

The π-conjugated oligo(phenylene ethynylene) backbones with two negatively charged $-CH_2COO^-$ groups on each repeating unit and lengths of n = 5, 7, and 9 (Table 1, no. 13) were used to detect Ca^{2+} ions and quenching ionic agents [68]. In the presence of bivalent calcium ions, the oligomers aggregated causing fluorescence shift. The shift depended on oligomer length and for the shorter oligomers (n = 5, 7), the effects are less pronounced than for longer ones n = 9. This shift can be explained by the planarization of the phenylene ethynylene backbone and formation of "excimer-like" excited states, that are not observed in the absence of Cu^{2+} ions. The ligomers were also evaluated for fluorescence quenching in the presence of methyl viologen and 3,3′-diethyloxacarbocyanine–well-known fluorescence quenching agents. It was found that the quenching efficiency depends on oligomer length. Taken together, the elongation of oligomer increased the ionic charge of macromolecules that in presence of counter ions favor the formation of ordered and backbone-overlapped aggregates.

For example, while the fluorescence of cationic OPEs with amine end groups is quenched in water, the addition of a small amount of oppositely charged detergent, sodium dodecyl sulfate (SDS), causes a significant increase in the OPE fluorescence due to the formation of a complex (Table 1, no. 16, 17) [35]. These OPE-detergent complexes exhibited antimicrobial properties [95], which, in addition to the fluorescence emission during their formation, can be utilized for the development of multifunctional biosensors.

3.2. Oligo(Arylene Ethynylene) Sensor Films

Oligo(arylene ethynylene) films consist of packed macromolecules with π-conjugated backbone thus exhibit high fluorescence emission which can be altered upon binding of an analyte molecule. OArEs films are an excellent materials for detection of amino acids [63], bacteria [81,82], explosives [35,83,84,87], pH [86], inorganic acids [85], gas [88], digital information [89], or chemicals [90,91]. Usually, the detection of an analyte is based on fluorescence changes, its enhancement or quenching upon binding of the sensed molecule. The OArEs films can be obtained by covalent immobilization e.g., reaction between an aldehyde and amine-functionalized surface [85,87], triethoxysilane group and glass [86], electrostatic binding [96], or drop-casting [90].

For instance, oligo(phenylene ethynylene)s bearing 4-aminophenyl-D-mannopyranoside groups (Table 1, no. 36, 37) in combination with laser scanning confocal microscopy have been used for the detection of *E. coli* bacteria [81]. Oligomer probes with two mannose groups enable discrimination between uropathogenic and the non-uropathogenic *E. coli* mutant. Moreover, the films of oligomer on aluminum support together with SPR allowed for quantitative biosensing of uropathogenic *E. coli* achieving a LOD of 10^4 CFU/mL. Those findings showed the direction towards robust biochips to detect bacteria.

For example, oligo(p-phenylene ethynylene) (Table 1, no. 44, 45) films have been examined in sensing of common explosive nitroaromatic compounds (NACs) i.e., 2,4,6-trinitrophenol (picric acid, PA), 2,4,6-trinitrotoluene (TNT), 2,4-dinitrotoluene (DNT), and nitrobenzene (NB) [87]. Interestingly, the film with cholesterol side groups (Table 1, no. 45) exhibited sensitivity to changes of water/THF solvents ratio (Figure 8a). In water, the film adapted a compacted structure causing a decrease in fluorescence intensity whereas in THF the chains attained extended conformation. In the presence of NACs molecules, complete fluorescence quenching was observed as the effect of the formation of nonfluorescent OPE-NACs complexes. This effect was not interfered by the presence of other compounds, including methanol, THF, toluene, dichloromethane, ammonia, HCl, NaOH, NaCl, copper salts, or seawater (Figure 8b). The experiments revealed that the cholesterol chains incorporated in the oligomer structure induced the sensitivity of the films towards the detected molecules by at least one order of magnitude. Thus, the films of oligo(p-phenylene ethynylene) with cholesterol groups can be used as an effective sensor for explosives.

Surface-immobilized monolayers of defined in length, short oligo(p-phenylene ethynylene) oligomers end-capped by fluorescein (Table 1, no. 43) have been used as narrow-range threshold fluorescent pH indicators (Figure 9a) [86]. At low pH, fluorescein was in its lactone form and the observed emission mostly originated from the oligomer. Upon pH increase fluorescein form change to anionic that favors electron delocalization with a respective decrease in HOMO-LUMO gap. A smaller energy gap facilitates the exciton migration that results in fluorescence enhancement. Moreover, an increase of pH causes a bathochromic shift of oligomer emission due to energy transfer from the oligomer backbone to fluorescein (Figure 9b). This unique pH-dependent response was observed only for oligomer-fluorescein dyad structures immobilized on the surface. The dyad structure was crucial for sensor selectivity. Experiments performed for immobilized fluorescein did not reveal such a selective sensor response. For comparison the same experiment was performed for dyad oligomers in solution, however, the fluorescence signal was much weaker in intensity and the pH validation range was significantly narrowed (pH 8 to 10).

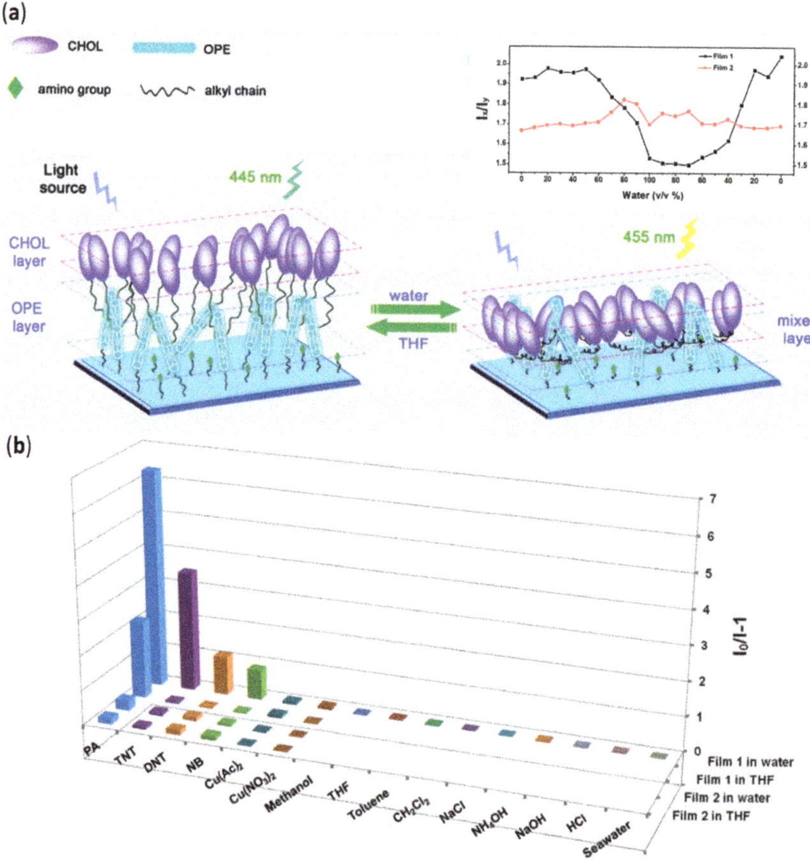

Figure 8. (**a**) The two states adopted by the oligomers containing cholesterol side chains (Table 1, no. 45) in water and tetrahydrofuran (THF), respectively. THF is a good solvent for oligomer and its cholesterol side chains and macromolecules in the film attain extended conformation. In contrast, water is a poor solvent for both the oligomer backbone and the side chains, thus the oligomer film is collapsed. Plots of the ratios of Ix/Iy of a given fluorescent film (Film 1-oligo(p-phenyleneethynylene) with cholesterol moieties and Film 2-pristin oligo(p-phenyleneethynylene)) against the compositions of the mixture solvents in which the fluorescence measurements were conducted (for Film 1 (Table 1, no. 45), λ_{em} = 445 nm, y λ_{ex} 500 nm; for Film 2 (Table 1, no. 44), λ_{em} 374 nm, λ_{ex} 420 nm). (**b**) Quenching efficiencies of various common explosive nitroaromatic compounds (NACs) on the fluorescence emission of Film 1 and Film 2 in water and THF, respectively, and the interferences of commonly found interferents in the sensing of Film 1 (concentration of NACs and interferents are 50 mM). Adapted from [87] with permission from The Royal Society of Chemistry.

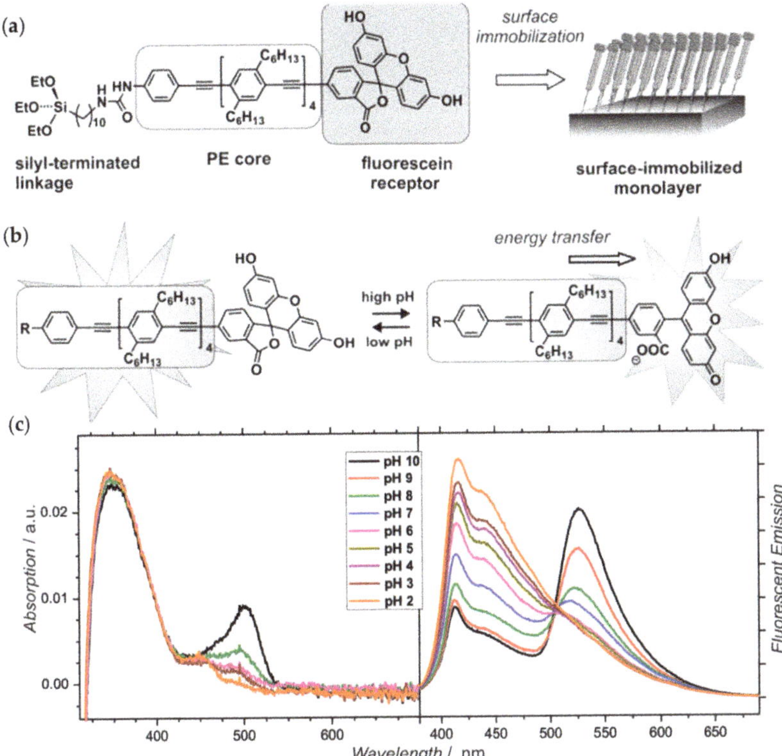

Figure 9. (**a**) Structure of the sensor and its assembly into a surface-immobilized monolayer. (**b**) The general principle of generating pH-dependent fluorescent response. (**c**) pH-dependent absorption (left) and fluorescence (right) spectra of monolayer fluorescein-oligomer film. Reprinted with permission from [86]. Copyright 2013 John Wiley and Sons.

Additionally, immobilized oligo(p-phenylene ethynylene) can act as chemosensors for the detection of polar species in an aprotic solvent. For example, a self-assembled monolayer of oligo(p-phenylene ethynylene) with cholic acid moieties (Table 1, no. 42) immobilized onto a glass slide, has been used as a sensor for trace amounts of inorganic acids, such as HCl, H_2SO_4, HNO_3, and H_3PO_4, in acetone medium [85]. The presence of a cholic acid unit induced the formation of hydrophobic pockets in the upper part of the layer (Figure 10a). This pocket containing imino group was able to trap ions that interacted with the imino groups. Basing on the comparative studies performed for different acids, it was revealed that for the anaerobic acids, the quenching efficiency depended on the size of the molecule and hydrogen bonds between the anions (Figure 10b). In other words, to observe efficient quenching the acid ions had to fit the cavity of the hydrophilic pocket. When chloride anion was trapped in the pocket the fluorescence quenching originated from the protonation of the imino group next to the phenylene ethynylene segment was observed.

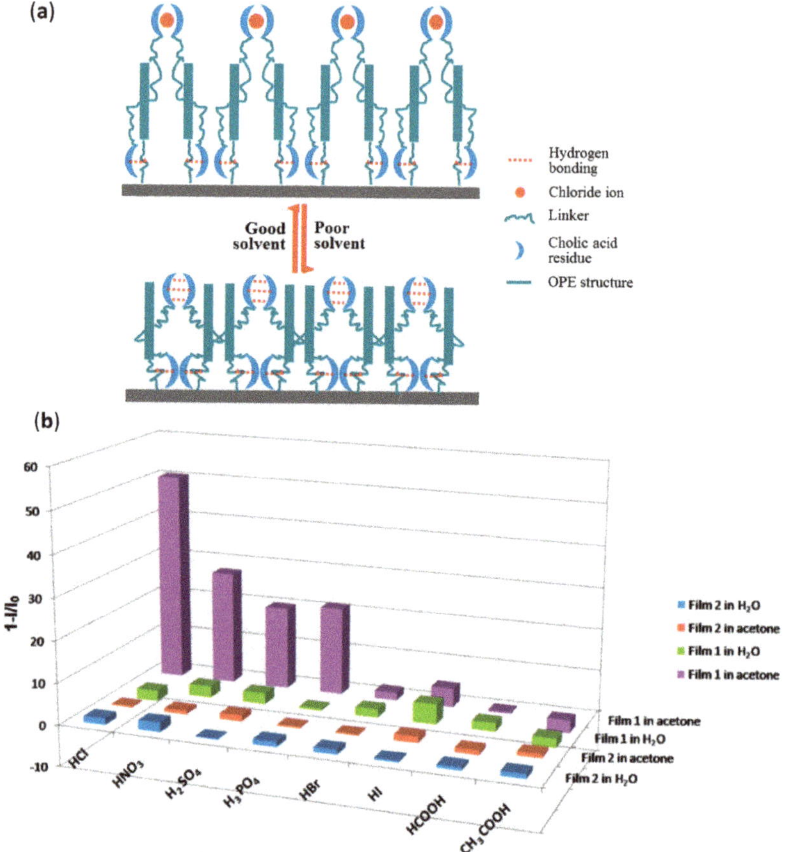

Figure 10. (**a**) Illustration of immobilized oligomer (Table 1, no. 42) conformations in different medium representing good (acetone as an example) or poor solvent (water, for example). In a good solvent, the hydrophilic pocket is formed as an upper layer of the film. (**b**) Quenching efficiencies of various acids to the fluorescence emission of Film 1-oligo(p-phenylene ethynylene) with cholic acid side chains (Table 1, no. 42) and Film 2-oligo(p-phenylene ethynylene) (Table 1, no. 41) in water and acetone, respectively (concentration of acids are 20 μM). Reprinted with permission from [85]. Copyright 2012 American Chemical Society.

Very sensitive sensor response can be achieved using electrochemical sensing methods. An electrochemical sensor based on an oligo(phenylene ethynylene) (Table 1, no. 49) and chemically reduced graphene oxide (rGO) nanocomposite was used for the quantification of dopamine (DA) [90]. This nanocomposite was synthesized by a simple ultrasonication method and then drop-casted onto a polished glassy carbon electrode and followed by casting of a Nafion ethanol solution (0.25 wt%). The formation of the oligomer nanocomposite was attributed to the π–π stacking interaction between the conjugated structure of oligo(phenylene ethynylene) and rGO as well as the electrostatic force between the amino group of oligomer and the carboxyl group on rGO. Anchoring of the oligomer changed the configuration of the multiple bonds so that a conjugated system represented a characteristic feature of conducting polymers. The developed sensor exhibited significantly enhanced electrocatalytic activity toward the oxidation of DA in a human serum PBS solution in the concentration range of 0.01–60 μM with LOD of 5 nM, a significantly lower value than those reported for the other DA sensors [97].

A chemical sensor based on GO-oligo(phenylene ethynylene) (Table 1, no. 48) nanocomposites was developed for amino acid detection [63]. The oligo(phenylene ethynylene) with cyanoacrylate groups in presence of cysteine residue change fluorescence properties. As a result of the interaction between oligomer and cysteine blue-shifted and decreased fluorescence emission was observed. For oligomer-GO nanocomposite the behavior was opposite and fluorescence enhancement occurred. The strong response of oligomer to cysteine can be used as a highly sensitive sensor.

Oligo(phenylene ethynylene)-based temperature sensors have been used to encode digital information [89]. The oligomers (Table 1, no. 47) were used as junctions between two Au electrodes (Figure 11a). Interestingly, during local temperature changes, the oligomers were able to change their structure between norbornadiene (NB)-state and quadricyclane (QC)-state (Figure 11d). The molecule states exhibited different conductance values that can be assigned to "1" and "0" digital symbols. The temperature-dependent conducting properties of oligomers could be used for local temperature monitoring. This system due to the clear response, translated into two states can be further exploited as a new approach for encoding digital information.

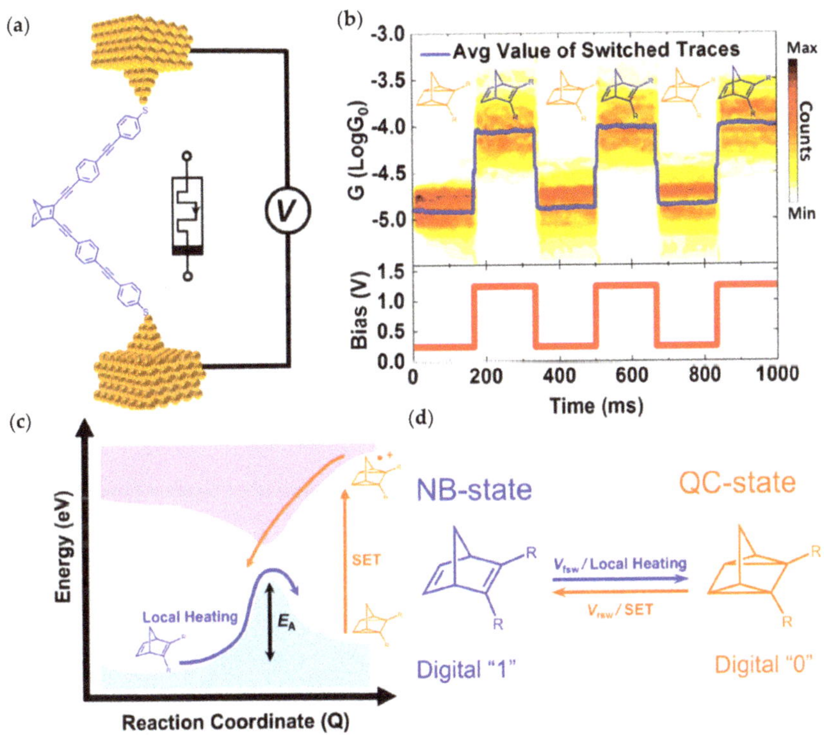

Figure 11. (**a**) Schematic of the molecular device with a modulating bias. (**b**) Reversible switching behavior of single-molecule devices and the applied waveform. (**c**) Energy landscape of isomerization processes. Blue and orange arrows indicate the electrically controlled forward and reverse switching processes, respectively. (**d**) Schematic describing the processes for controlling the norbornadiene (NB)-quadricyclane (QC) switching within a molecular junction (blue and orange arrows). The switching processes within a molecular junction are controlled in the forward direction (NB to QC) by electrically controlling the local temperature and in the reverse process (QC to NB) by catalyzing the reaction through a single electron transfer (SET) process. These two states possess different conductance values and can be used to encode digital information. Reprinted with permission from [89]. Copyright 2020 John Wiley and Sons.

4. Conclusions

Uniform, π-conjugated oligomers based on an arylene ethynylene core are attractive sensory materials. They can respond to the environment changes (polarity, temperature), presence of chemicals (amino acids, saccharides, ions), macromolecules (proteins, polymers), bacteria, and process monitoring. The successfully designed oligo(arylene ethynylene)-based sensors can be used as selective probes to detect particular analytes in the mixture and they can be used for selective process monitoring. However, their huge potential has not been explored, yet.

Well-defined conjugated arylene ethynylene can be accessed by iterative chemistry protocols that permit for full structure precision and sequence definition. The solubility issues occurring for oligo(arylene ethynylene)s can be overcome by the synthesis approach that uses soluble support. Nevertheless, the high synthesis scale and yields remain a challenge.

The sensing parameters (sensitivity, selectivity, specificity) are strongly connected with the oligomer structure. Even a small difference in structure, e.g., one unit length difference may result in loss of sensor selectivity and sensitivity. Although a variety of examples were described, it has been still difficult to rationally design the arylene ethynylene oligomers with high selectivity and affinity, though more systematic studies in the field are needed.

In the near future sensing and process monitoring can become an interesting and emerging application for sequence-defined polymers built from π-conjugated segments. As it was shown by many examples in this review sensing is one of the applications where monomer sequence, composition, and length matter. Systematic studies on the sequence–property relationship can open an avenue for more specific and selective sensors relevant to biological samples.

Author Contributions: R.S. developed the concept of the paper and wrote the manuscript. A.K.-C., D.S. contributed to the preparation of the manuscript, commented on the content, and helped with editing. All authors have read and agreed to the published version of the manuscript.

Funding: This research received funding from the project No 2018/31/D/ST5/01365 of Polish National Science Centre.

Acknowledgments: R. Szweda acknowledges Polish National Science Centre project No. 2018/31/D/ST5/01365 for received funding.

Conflicts of Interest: The authors declare no conflict of interest.

References

1. Biechele, P.; Busse, C.; Solle, D.; Scheper, T.; Reardon, K. Sensor systems for bioprocess monitoring. *Eng. Life Sci.* **2015**, *15*, 469–488. [CrossRef]

2. Vabret, N.; Bhardwaj, N.; Greenbaum, B.D. Sequence-Specific Sensing of Nucleic Acids. *Trends Immunol.* **2017**, *38*, 53–65. [CrossRef]

3. Lee, M.J.; Yaffe, M.B. Protein Regulation in Signal Transduction. *CSH Perspect. Biol.* **2016**, *8*, a005918. [CrossRef]

4. Scheer, P.V.D.; Laar, T.V.D.; Sprakel, J. Chain length-dependent luminescence in acceptor-doped conjugated polymers. *Sci. Rep.* **2019**, *9*, 11217. [CrossRef]

5. Lutz, J.-F.; Lehn, J.-M.; Meijer, E.W.; Matyjaszewski, K. From precision polymers to complex materials and systems. *Nat. Rev. Mater.* **2016**, *1*, 16024. [CrossRef]

6. Szweda, R.; Chendo, C.; Charles, L.; Baxter, P.N.W.; Lutz, J.-F. Synthesis of oligoarylacetylenes with defined conjugated sequences using tailor-made soluble polymer supports. *Chem. Commun.* **2017**, *53*, 8312–8315. [CrossRef]

7. Young, J.K.; Nelson, J.C.; Moore, J.S. Synthesis of Sequence Specific Phenylacetylene Oligomers on an Insoluble Solid Support. *J. Am. Chem. Soc.* **1994**, *116*, 10841–10842. [CrossRef]

8. Lavastre, O.; Ollivier, L.; Dixneuf, P.H.; Sibandhit, S. Sequential catalytic synthesis of rod-like conjugated poly-ynes. *Tetrahedron* **1996**, *52*, 5495–5504. [CrossRef]

9. Jones, L.; Schumm, J.S.; Tour, J.M. Rapid Solution and Solid Phase Syntheses of Oligo(1,4-phenylene ethynylene)s with Thioester Termini: Molecular Scale Wires with Alligator Clips. Derivation of Iterative Reaction Efficiencies on a Polymer Support. *J. Org. Chem.* **1997**, *62*, 1388–1410. [CrossRef]

10. Anderson, S. Phenylene Ethynylene Pentamers for Organic Electroluminescence. *Chem. Eur. J.* **2001**, *7*, 4706–4714. [CrossRef]

11. Elliott, E.L.; Ray, C.R.; Kraft, S.; Atkins, J.R.; Moore, J.S. Solid-Phase Synthesis of m-Phenylene Ethynylene Heterosequence Oligomers. *J. Org. Chem.* **2006**, *71*, 5282–5290. [CrossRef]

12. Schneider, R.V.; Waibel, K.A.; Arndt, A.P.; Lang, M.; Seim, R.; Busko, D.; Bräse, S.; Lemmer, U.; Meier, M.A.R. Sequence-definition in stiff conjugated oligomers. *Sci. Rep.* **2018**, *8*, 17483. [CrossRef]

13. Norris, B.N.; Zhang, S.; Campbell, C.M.; Auletta, J.T.; Calvo-Marzal, P.; Hutchison, G.R.; Meyer, T.Y. Sequence Matters: Modulating Electronic and Optical Properties of Conjugated Oligomers via Tailored Sequence. *Macromolecules* **2013**, *46*, 1384–1392. [CrossRef]

14. Zhang, S.; Bauer, N.E.; Kanal, I.Y.; You, W.; Hutchison, G.R.; Meyer, T.Y. Sequence Effects in Donor–Acceptor Oligomeric Semiconductors Comprising Benzothiadiazole and Phenylenevinylene Monomers. *Macromolecules* **2017**, *50*, 151–161. [CrossRef]

15. Zhang, S.; Hutchison, G.R.; Meyer, T.Y. Sequence Effects in Conjugated Donor–Acceptor Trimers and Polymers. *Macromol. Rapid Commun.* **2016**, *37*, 882–887. [CrossRef]

16. Nishide, Y.; Osuga, H.; Saito, M.; Aiba, T.; Inagaki, Y.; Doge, Y.; Tanaka, K. Synthesis and Properties of a Series of Well-Defined and Polydisperse Benzo[1,2-b:4,3-b']dithiophene Oligomers. *J. Org. Chem.* **2007**, *72*, 9141–9151. [CrossRef]

17. Wang, S.; Liu, B.; Gaylord, B.S.; Bazan, G.C. Size-Specific Interactions Between Single- and Double-Stranded Oligonucleotides and Cationic Water-Soluble Oligofluorenes. *Adv. Funct. Mater.* **2003**, *13*, 463–467. [CrossRef]

18. Liu, B.; Gaylord, B.S.; Wang, S.; Bazan, G.C. Effect of Chromophore-Charge Distance on the Energy Transfer Properties of Water-Soluble Conjugated Oligomers. *J. Am. Chem. Soc.* **2003**, *125*, 6705–6714. [CrossRef]

19. Lawrence, J.; Goto, E.; Ren, J.M.; McDearmon, B.; Kim, D.S.; Ochiai, Y.; Clark, P.G.; Laitar, D.; Higashihara, T.; Hawker, C.J. A Versatile and Efficient Strategy to Discrete Conjugated Oligomers. *J. Am. Chem. Soc.* **2017**, *139*, 13735–13739. [CrossRef]

20. Johansson, L.B.G.; Simon, R.; Bergström, G.; Eriksson, M.; Prokop, S.; Mandenius, C.-F.; Heppner, F.L.; Åslund, A.K.O.; Nilsson, K.P.R. An azide functionalized oligothiophene ligand—A versatile tool for multimodal detection of disease associated protein aggregates. *Biosens. Bioelectron.* **2015**, *63*, 204–211. [CrossRef] [PubMed]

21. Sun, T.; Niu, Q.; Li, Y.; Li, T.; Hu, T.; Wang, E.; Liu, H. A novel oligothiophene-based colorimetric and fluorescent "turn on" sensor for highly selective and sensitive detection of cyanide in aqueous media and its practical applications in water and food samples. *Sens. Actuators B Chem.* **2018**, *258*, 64–71. [CrossRef]

22. Lan, L.; Li, T.; Wei, T.; Pang, H.; Sun, T.; Wang, E.; Liu, H.; Niu, Q. Oligothiophene-based colorimetric and ratiometric fluorescence dual-channel cyanide chemosensor: Sensing ability, TD-DFT calculations and its application as an efficient solid state sensor. *Spectrochim. Acta A* **2018**, *193*, 289–296. [CrossRef] [PubMed]

23. Miao, R.; Peng, J.; Fang, Y. Recent advances in fluorescent film sensing from the perspective of both molecular design and film engineering. *Mol. Syst. Des. Eng.* **2016**, *1*, 242–257. [CrossRef]

24. Freudenberg, J.; Hinkel, F.; Jänsch, D.; Bunz, U.H.F. Chemical Tongues and Noses Based upon Conjugated Polymers. *Top. Curr. Chem.* **2017**, *375*, 67. [CrossRef] [PubMed]

25. Geng, Y.; Peveler, W.J.; Rotello, V.M. Array-based "Chemical Nose" Sensing in Diagnostics and Drug Discovery. *Angew. Chem. Int. Ed.* **2019**, *58*, 5190–5200. [CrossRef] [PubMed]

26. Feng, F.; Liu, L.; Yang, Q.; Wang, S. Water-Soluble Conjugated Polymers for Fluorescent-Enzyme Assays. *Macromol. Rapid Commun.* **2010**, *31*, 1405–1421. [CrossRef]

27. Chen, Z.; Wang, Q.; Wu, X.; Li, Z.; Jiang, Y.-B. Optical chirality sensing using macrocycles, synthetic and supramolecular oligomers/polymers, and nanoparticle based sensors. *Chem. Soc. Rev.* **2015**, *44*, 4249–4263. [CrossRef]

28. Kubota, R.; Hamachi, I. Protein recognition using synthetic small-molecular binders toward optical protein sensing in vitro and in live cells. *Chem. Soc. Rev.* **2015**, *44*, 4454–4471. [CrossRef]

29. Ahner, J.; Pretzel, D.; Enke, M.; Geitner, R.; Zechel, S.; Popp, J.; Schubert, U.S.; Hager, M.D. Conjugated Oligomers as Fluorescence Marker for the Determination of the Self-Healing Efficiency in Mussel-Inspired Polymers. *Chem. Mater.* **2018**, *30*, 2791–2799. [CrossRef]

30. Wang, B.; Queenan, B.N.; Wang, S.; Nilsson, K.P.R.; Bazan, G.C. Precisely Defined Conjugated Oligoelectrolytes for Biosensing and Therapeutics. *Adv. Mater.* **2019**, *31*, 1806701. [CrossRef]

31. Li, Z.; Askim, J.R.; Suslick, K.S. The Optoelectronic Nose: Colorimetric and Fluorometric Sensor Arrays. *Chem. Rev.* **2019**, *119*, 231–292. [CrossRef] [PubMed]

32. Bunz, U.H.F. Poly(aryleneethynylene)s: Syntheses, Properties, Structures, and Applications. *Chem. Rev.* **2000**, *100*, 1605–1644. [CrossRef] [PubMed]

33. Wang, H.; Cui, H.; Liu, X.; Li, L.; Cao, Y.; Liu, T.; Fang, Y. Alternative Copolymerization of a Conjugated Segment and a Flexible Segment and Fabrication of a Fluorescent Sensing Film for HCl in the Vapor Phase. *Chem. Asian J.* **2013**, *8*, 101–107. [CrossRef] [PubMed]

34. Zyryanov, G.V.; Kopchuk, D.S.; Kovalev, I.S.; Nosova, E.V.; Rusinov, V.L.; Chupakhin, O.N. Chemosensors for detection of nitroaromatic compounds (explosives). *Russ. Chem. Rev.* **2014**, *83*, 783–819. [CrossRef]

35. Whitten, D.G.; Tang, Y.; Zhou, Z.; Yang, J.; Wang, Y.; Hill, E.H.; Pappas, H.C.; Donabedian, P.L.; Chi, E.Y. A Retrospective: 10 Years of Oligo(phenylene-ethynylene) Electrolytes: Demystifying Nanomaterials. *Langmuir* **2019**, *35*, 307–325. [CrossRef]

36. Lehnherr, D.; Chen, C.; Pedramrazi, Z.; DeBlase, C.R.; Alzola, J.M.; Keresztes, I.; Lobkovsky, E.B.; Crommie, M.F.; Dichtel, W.R. Sequence-defined oligo(ortho-arylene) foldamers derived from the benzannulation of ortho(arylene ethynylene)s. *Chem. Sci.* **2016**, *7*, 6357–6364. [CrossRef]

37. Löffler, S.; Antypas, H.; Choong, F.X.; Nilsson, K.P.R.; Richter-Dahlfors, A. Conjugated Oligo- and Polymers for Bacterial Sensing. *Front. Chem.* **2019**, *7*. [CrossRef]

38. Mulla, K.; Dongare, P.; Zhou, N.; Chen, G.; Thompson, D.W.; Zhao, Y. Highly sensitive detection of saccharides under physiological conditions with click synthesized boronic acid-oligomer fluorophores. *Org. Biomol. Chem.* **2011**, *9*, 1332–1336. [CrossRef]

39. Bunz, U.H.F.; Seehafer, K.; Bender, M.; Porz, M. Poly(aryleneethynylene)s (PAE) as paradigmatic sensor cores. *Chem. Soc. Rev.* **2015**, *44*, 4322–4336. [CrossRef]

40. Mako, T.L.; Racicot, J.M.; Levine, M. Supramolecular Luminescent Sensors. *Chem. Rev.* **2019**, *119*, 322–477. [CrossRef]

41. Lutz, J.-F.; Ouchi, M.; Liu, D.R.; Sawamoto, M. Sequence-Controlled Polymers. *Science* **2013**, *341*, 1238149. [CrossRef] [PubMed]

42. Solleder, S.C.; Schneider, R.V.; Wetzel, K.S.; Boukis, A.C.; Meier, M.A.R. Recent Progress in the Design of Monodisperse, Sequence-Defined Macromolecules. *Macromol. Rapid Commun.* **2017**, *38*, 1600711. [CrossRef] [PubMed]

43. Lutz, J.-F. Defining the Field of Sequence-Controlled Polymers. *Macromol. Rapid Commun.* **2017**, *38*, 1700582. [CrossRef] [PubMed]

44. De Neve, J.; Haven, J.J.; Maes, L.; Junkers, T. Sequence-definition from controlled polymerization: The next generation of materials. *Polym. Chem.* **2018**, *9*, 4692–4705. [CrossRef]

45. Feng, H.-T.; Yuan, Y.-X.; Xiong, J.-B.; Zheng, Y.-S.; Tang, B.Z. Macrocycles and cages based on tetraphenylethylene with aggregation-induced emission effect. *Chem. Soc. Rev.* **2018**, *47*, 7452–7476. [CrossRef] [PubMed]

46. Wong, C.-H.; Zimmerman, S.C. Orthogonality in organic, polymer, and supramolecular chemistry: From Merrifield to click chemistry. *Chem. Commun.* **2013**, *49*, 1679–1695. [CrossRef] [PubMed]

47. Cavallo, G.; Al Ouahabi, A.; Oswald, L.; Charles, L.; Lutz, J.-F. Orthogonal Synthesis of "Easy-to-Read" Information-Containing Polymers Using Phosphoramidite and Radical Coupling Steps. *J. Am. Chem. Soc.* **2016**, *138*, 9417–9420. [CrossRef]

48. Hill, S.A.; Gerke, C.; Hartmann, L. Recent Developments in Solid-Phase Strategies towards Synthetic, Sequence-Defined Macromolecules. *Chem. Asian J.* **2018**, *13*, 3611–3622. [CrossRef]

49. Trinh, T.T.; Laure, C.; Lutz, J.-F. Synthesis of Monodisperse Sequence-Defined Polymers Using Protecting-Group-Free Iterative Strategies. *Macromol. Chem. Phys.* **2015**, *216*, 1498–1506. [CrossRef]

50. Tour, J.M. Conjugated Macromolecules of Precise Length and Constitution. Organic Synthesis for the Construction of Nanoarchitectures. *Chem. Rev.* **1996**, *96*, 537–554. [CrossRef]

51. Zhang, L.; Colella, N.S.; Cherniawski, B.P.; Mannsfeld, S.C.B.; Briseno, A.L. Oligothiophene Semiconductors: Synthesis, Characterization, and Applications for Organic Devices. *ACS Appl. Mater.* **2014**, *6*, 5327–5343. [CrossRef] [PubMed]

52. Koch, F.P.V.; Smith, P.; Heeney, M. "Fibonacci's Route" to Regioregular Oligo(3-hexylthiophene)s. *J. Am. Chem. Soc.* **2013**, *135*, 13695–13698. [CrossRef] [PubMed]

53. Ten Hoeve, W.; Wynberg, H.; Havinga, E.E.; Meijer, E.W. Substituted 2,2′:5′,2′′:5′′,2′′′:5′′′,2′′′′:5′′′′,2′′′′′: 5′′′′′,2′′′′′′:5′′′′′′,2′′′′′′′:5′′′′′′′,2′′′′′′′′:5′′′′′′′′,2′′′′′′′′′:5′′′′′′′′′,2′′′′′′′′′′-undecithiophenes, the longest characterized oligothiophenes. *J. Am. Chem. Soc.* **1991**, *113*, 5887–5889. [CrossRef]

54. Schumm, J.S.; Pearson, D.L.; Tour, J.M. Iterative Divergent/Convergent Approach to Linear Conjugated Oligomers by Successive Doubling of the Molecular Length: A Rapid Route to a 128Å-Long Potential Molecular Wire. *Angew. Chem. Int. Ed.* **1994**, *33*, 1360–1363. [CrossRef]

55. Huang, S.; Tour, J.M. Rapid bi-directional synthesis of oligo(1, 4-phenylene ethynylene)s. *Tetrahedron Lett.* **1999**, *40*, 3447–3450. [CrossRef]

56. Vonk, E.C.; Langeveld-Voss, B.M.W.; van Dongen, J.L.J.; Janssen, R.A.J.; Claessens, H.A.; Cramers, C.A. Separation and characterization of oligomers by reversed-phase high-performance liquid chromatography; a study on well-defined oligothiophenes. *J. Chromatogr. A* **2001**, *911*, 13–26. [CrossRef]

57. Advincula, R.C. Review of Conjugated Polymer Synthesis: Methods and Reactions. *J. Am. Chem. Soc.* **2011**, *133*, 5622. [CrossRef]

58. Kline, R.J.; McGehee, M.D.; Kadnikova, E.N.; Liu, J.; Fréchet, J.M.J. Controlling the Field-Effect Mobility of Regioregular Polythiophene by Changing the Molecular Weight. *Adv. Mater.* **2003**, *15*, 1519–1522. [CrossRef]

59. Pingel, P.; Zen, A.; Neher, D.; Lieberwirth, I.; Wegner, G.; Allard, S.; Scherf, U. Unexpectedly high field-effect mobility of a soluble, low molecular weight oligoquaterthiophene fraction with low polydispersity. *Appl. Phys. A* **2009**, *95*, 67–72. [CrossRef]

60. Lan, Y.-K.; Huang, C.-I. Charge Mobility and Transport Behavior in the Ordered and Disordered States of the Regioregular Poly(3-hexylthiophene). *J. Phys. Chem. B* **2009**, *113*, 14555–14564. [CrossRef]

61. Fanni, A.M.; Monge, F.A.; Lin, C.-Y.; Thapa, A.; Bhaskar, K.; Whitten, D.G.; Chi, E.Y. High Selectivity and Sensitivity of Oligomeric p-Phenylene Ethynylenes for Detecting Fibrillar and Prefibrillar Amyloid Protein Aggregates. *ACS Chem. Neurosci.* **2019**, *10*, 1813–1825. [CrossRef] [PubMed]

62. Abe, H.; Machiguchi, H.; Matsumoto, S.; Inouye, M. Saccharide Recognition-Induced Transformation of Pyridine−Pyridone Alternate Oligomers from Self-Dimer to Helical Complex. *J. Org. Chem.* **2008**, *73*, 4650–4661. [CrossRef]

63. Adachi, N.; Yoshinari, M.; Suzuki, E.; Okada, M. Oligo(p-phenylene ethynylene) with Cyanoacrylate Terminal Groups and Graphene Composite as Fluorescent Chemical Sensor for Cysteine. *J. Fluoresc.* **2017**, *27*, 1449–1456. [CrossRef]

64. De los Santos, Z.A.; Yusin, G.; Wolf, C. Enantioselective sensing of carboxylic acids with a bis(urea)oligo(phenylene)ethynylene foldamer. *Tetrahedron* **2019**, *75*, 1504–1509. [CrossRef]

65. Suk, J.-M.; Kim, D.A.; Jeong, K.-S. Helicity Control of an Indolocarbazole Foldamer by Chiral Organic Anions. *Org. Lett.* **2012**, *14*, 5018–5021. [CrossRef] [PubMed]

66. Reiné, P.; Justicia, J.; Morcillo, S.P.; Abbate, S.; Vaz, B.; Ribagorda, M.; Orte, Á.; Álvarez de Cienfuegos, L.; Longhi, G.; Campaña, A.G.; et al. Pyrene-Containing ortho-Oligo(phenylene)ethynylene Foldamer as a Ratiometric Probe Based on Circularly Polarized Luminescence. *J. Org. Chem.* **2018**, *83*, 4455–4463. [CrossRef] [PubMed]

67. Yang, J.-S.; Yan, J.-L.; Hwang, C.-Y.; Chiou, S.-Y.; Liau, K.-L.; Gavin Tsai, H.-H.; Lee, G.-H.; Peng, S.-M. Probing the Intrachain and Interchain Effects on the Fluorescence Behavior of Pentiptycene-Derived Oligo(p-phenyleneethynylene)s. *J. Am. Chem. Soc.* **2006**, *128*, 14109–14119. [CrossRef]

68. Feng, F.; Yang, J.; Xie, D.; McCarley, T.D.; Schanze, K.S. Remarkable Photophysics and Amplified Quenching of Conjugated Polyelectrolyte Oligomers. *J. Phys. Chem. Lett.* **2013**, *4*, 1410–1414. [CrossRef]

69. Phillips, R.L.; Miranda, O.R.; You, C.-C.; Rotello, V.M.; Bunz, U.H.F. Rapid and Efficient Identification of Bacteria Using Gold-Nanoparticle–Poly(para-phenyleneethynylene) Constructs. *Angew. Chem. Int. Ed.* **2008**, *47*, 2590–2594. [CrossRef]

70. George, W.N.; Giles, M.; McCulloch, I.; Steinke, J.H.G.; deMello, J.C. Efficient Quenching of a Guanidinium-Containing Fluorescence Sensor. *ChemPhysChem* **2011**, *12*, 765–768. [CrossRef]

71. Hill, E.H.; Goswami, S.; Evans, D.G.; Schanze, K.S.; Whitten, D.G. Photochemistry of a Model Cationic p-Phenylene Ethynylene in Water. *J. Phys. Chem. Lett.* **2012**, *3*, 1363–1368. [CrossRef] [PubMed]

72. Hill, E.H.; Evans, D.G.; Whitten, D.G. The influence of structured interfacial water on the photoluminescence of carboxyester-terminated oligo-p-phenylene ethynylenes. *J. Phys. Org. Chem.* **2014**, *27*, 252–257. [CrossRef]

73. Donabedian, P.L.; Pham, T.K.; Whitten, D.G.; Chi, E.Y. Oligo(p-phenylene ethynylene) Electrolytes: A Novel Molecular Scaffold for Optical Tracking of Amyloids. *ACS Chem. Neurosci.* **2015**, *6*, 1526–1535. [CrossRef] [PubMed]

74. Hill, E.H.; Zhang, Y.; Evans, D.G.; Whitten, D.G. Enzyme-Specific Sensors via Aggregation of Charged p-Phenylene Ethynylenes. *ACS Appl. Mater.* **2015**, *7*, 5550–5560. [CrossRef] [PubMed]

75. Hill, E.H.; Sanchez, D.; Evans, D.G.; Whitten, D.G. Structural Basis for Aggregation Mode of oligo-p-Phenylene Ethynylenes with Ionic Surfactants. *Langmuir* **2013**, *29*, 15732–15737. [CrossRef]

76. Donabedian, P.L.; Creyer, M.N.; Monge, F.A.; Schanze, K.S.; Chi, E.Y.; Whitten, D.G. Detergent-induced self-assembly and controllable photosensitizer activity of diester phenylene ethynylenes. *Proc. Natl. Acad. Sci. USA* **2017**, *114*, 7278–7282. [CrossRef] [PubMed]

77. Tang, Y.; Zhou, Z.; Ogawa, K.; Lopez, G.P.; Schanze, K.S.; Whitten, D.G. Photophysics and self-assembly of symmetrical and unsymmetrical cationic oligophenylene ethynylenes. *J. Photochem. Photobiol. A* **2009**, *207*, 4–6. [CrossRef]

78. Tang, Y.; Zhou, Z.; Ogawa, K.; Lopez, G.P.; Schanze, K.S.; Whitten, D.G. Synthesis, Self-Assembly, and Photophysical Behavior of Oligo Phenylene Ethynylenes: From Molecular to Supramolecular Properties. *Langmuir* **2009**, *25*, 21–25. [CrossRef]

79. Tang, Y.; Corbitt, T.S.; Parthasarathy, A.; Zhou, Z.; Schanze, K.S.; Whitten, D.G. Light-Induced Antibacterial Activity of Symmetrical and Asymmetrical Oligophenylene Ethynylenes. *Langmuir* **2011**, *27*, 4956–4962. [CrossRef]

80. Tong, W.-L.; Lai, L.-M.; Chan, M.C.W. Platinum(ii) Schiff base as versatile phosphorescent core component in conjugated oligo(phenylene–ethynylene)s. *Dalton Trans.* **2008**, *1*, 1412–1414. [CrossRef]

81. Arias, E.; Méndez, M.T.; Arias, E.; Moggio, I.; Ledezma, A.; Romero, J.; Margheri, G.; Giorgetti, E. Supramolecular Recognition of Escherichia coli Bacteria by Fluorescent Oligo(Phenyleneethynylene)s with Mannopyranoside Termini Groups. *Sensors* **2017**, *17*, 1025. [CrossRef] [PubMed]

82. Concepción García, M.; Turlakov, G.; Moggio, I.; Arias, E.; Valenzuela, J.H.; Hernández, M.; Rodríguez, G.; Ziolo, R.F. Synthesis and photophysical properties of conjugated (dodecyl)benzoateethynylene macromolecules: Staining of Bacillus subtilis and Escherichia coli rhizobacteria. *New J. Chem.* **2019**, *43*, 3332–3340. [CrossRef]

83. Barata, P.D.; Prata, J.V. New entities for sensory chemistry based on calix[4]arene-carbazole conjugates: From synthesis to applications. *Supramol. Chem.* **2013**, *25*, 782–797. [CrossRef]

84. Caron, T.; Guillemot, M.; Montméat, P.; Veignal, F.; Perraut, F.; Prené, P.; Serein-Spirau, F. Ultra trace detection of explosives in air: Development of a portable fluorescent detector. *Talanta* **2010**, *81*, 543–548. [CrossRef]

85. Cui, H.; He, G.; Wang, H.; Sun, X.; Liu, T.; Ding, L.; Fang, Y. Fabrication of a Novel Cholic Acid Modified OPE-Based Fluorescent Film and Its Sensing Performances to Inorganic Acids in Acetone. *ACS Appl. Mater.* **2012**, *4*, 6935–6941. [CrossRef]

86. Imsick, B.G.; Acharya, J.R.; Nesterov, E.E. Surface-Immobilized Monolayers of Conjugated Oligomers as a Platform for Fluorescent Sensors Design: The Effect of Exciton Delocalization on Chemosensing Performance. *Adv. Mater.* **2013**, *25*, 120–124. [CrossRef]

87. Wang, H.; He, G.; Chen, X.; Liu, T.; Ding, L.; Fang, Y. Cholesterol modified OPE functionalized film: Fabrication, fluorescence behavior and sensing performance. *J. Mater. Chem.* **2012**, *22*, 7529–7536. [CrossRef]

88. Xin, J.-G.; Yang, C.-L.; Wang, M.-S.; Ma, X.-G. OPE molecular junction as a hydrogen gas sensor. *Curr. Appl. Phys.* **2018**, *18*, 273–279. [CrossRef]

89. Li, H.B.; Tebikachew, B.E.; Wiberg, C.; Moth-Poulsen, K.; Hihath, J. A Memristive Element based on Electrically Controlled Single-Molecule Reaction. *Angew. Chem. Int. Ed.* **2020**. [CrossRef]

90. Deng, J.; Liu, M.; Lin, F.; Zhang, Y.; Liu, Y.; Yao, S. Self-assembled oligo(phenylene ethynylene)s/graphene nanocomposite with improved electrochemical performances for dopamine determination. *Anal. Chim. Acta* **2013**, *767*, 59–65. [CrossRef]

91. Samanta, D.; Singh, A.; Verma, P.; Bhattacharyya, S.; Roy, S.; Maji, T.K. Photoswitchable J-Aggregated Processable Organogel by Integrating a Photochromic Acceptor. *J. Org. Chem.* **2019**, *84*, 10946–10952. [CrossRef] [PubMed]

92. Pappas, H.C.; Donabedian, P.L.; Schanze, K.S.; Whitten, D.G. Intended and Unintended Consequences and Applications of Unnatural Interfaces: Oligo p-Phenylene Ethynylene Electrolytes, Biological Cells and Biomacromolecules. *J. Braz. Chem. Soc.* **2016**, *27*, 256–266. [CrossRef]

93. Zhao, X.; Liu, Y.; Schanze, K.S. A conjugated polyelectrolyte-based fluorescence sensor for pyrophosphate. *Chem. Commun.* **2007**, *1*, 2914–2916. [CrossRef] [PubMed]

94. Ávila-Rovelo, N.R.; Ruiz-Carretero, A. Recent Progress in Hydrogen-Bonded π-Conjugated Systems Displaying J-Type Aggregates. *Org. Mater.* **2020**, *2*, 47–63. [CrossRef]

95. Dascier, D.; Ji, E.; Parthasarathy, A.; Schanze, K.S.; Whitten, D.G. Efficacy of End-Only-Functionalized Oligo(arylene-ethynylene)s in Killing Bacterial Biofilms. *Langmuir* **2012**, *28*, 11286–11290. [CrossRef] [PubMed]

96. Robinson, H.D.; Montazami, R.; Daengngam, C.; Zuo, Z.; Dong, W.; Metzman, J.; Heflin, R. Optoelectronic Materials and Devices Incorporating Polyelectrolyte Multilayers. In *Multilayer Thin Films*; Decher, G., Schlenoff, J.B., Eds.; Wiley-VCH Verlag GmbH & Co. KGaA: Weinheim, Germany, 2012; pp. 511–537.

97. Shrivastava, S.; Jadon, N.; Jain, R. Next-generation polymer nanocomposite-based electrochemical sensors and biosensors: A review. *TrAC Trends Anal. Chem.* **2016**, *82*, 55–67. [CrossRef]

Review

Modeling and Exploiting Microbial Temperature Response

Philipp Noll, Lars Lilge, Rudolf Hausmann and Marius Henkel *

Institute of Food Science and Biotechnology, Department of Bioprocess Engineering (150k), University of Hohenheim, Fruwirthstr. 12, 70599 Stuttgart, Germany; philipp.noll@uni-hohenheim.de (P.N.); lars.lilge@uni-hohenheim.de (L.L.); rudolf.hausmann@uni-hohenheim.de (R.H.)

* Correspondence: marius.henkel@uni-hohenheim.de

Received: 15 December 2019; Accepted: 14 January 2020; Published: 17 January 2020

Abstract: Temperature is an important parameter in bioprocesses, influencing the structure and functionality of almost every biomolecule, as well as affecting metabolic reaction rates. In industrial biotechnology, the temperature is usually tightly controlled at an optimum value. Smart variation of the temperature to optimize the performance of a bioprocess brings about multiple complex and interconnected metabolic changes and is so far only rarely applied. Mathematical descriptions and models facilitate a reduction in complexity, as well as an understanding, of these interconnections. Starting in the 19th century with the "primal" temperature model of Svante Arrhenius, a variety of models have evolved over time to describe growth and enzymatic reaction rates as functions of temperature. Data-driven empirical approaches, as well as complex mechanistic models based on thermodynamic knowledge of biomolecular behavior at different temperatures, have been developed. Even though underlying biological mechanisms and mathematical models have been well-described, temperature as a control variable is only scarcely applied in bioprocess engineering, and as a conclusion, an exploitation strategy merging both in context has not yet been established. In this review, the most important models for physiological, biochemical, and physical properties governed by temperature are presented and discussed, along with application perspectives. As such, this review provides a toolset for future exploitation perspectives of temperature in bioprocess engineering.

Keywords: thermal growth curve; temperature modeling; thermoregulation; monitoring and control; bioprocess engineering; calorimetry

---

1. Introduction

Predetermined by the applied system, bioprocesses are generally very sensitive to most changes in environmental conditions. It is for this reason that conditions such as temperature, pO_2, or pH are generally tightly controlled. In most cases, even small deviations from optimum values may lead to a significant reduction in the overall productivity and reproducibility of the process. Therefore, special consideration must be given to control tasks, which are typically defined by maintaining process variables within a narrow optimum [1]. In contrast to artificial laboratory conditions, microorganisms are usually exposed to a changing environment, with changes in pH, nutrient availability, competitors, and elevated or decreased temperature, etc. A crucial environmental factor for microorganisms is temperature. It affects the folding, structure, and stability of almost every biomolecule, as well as the metabolic reaction rate. Detection of temperature changes and subsequent adaptation of the metabolism are essential for microbial survival, such as by pathogens sensing intrusion into a host. Organisms can sense temperature shifts indirectly or by specialized sensing systems that evolved to detect changes in temperature in order to respond with adapted gene expression. This was extensively reviewed by Klinkert and Narberhaus [2]. Indirect temperature sensing is possible via the accumulation of aggregated proteins after a heat shock and via stalled ribosomes after a cold shock. Molecular thermosensors may consist of DNA, RNA, proteins, or lipids. DNA topology changes, e.g., supercoiling caused by heat stress, stable RNA structures preventing translation at sub-optimal temperatures, temperature-responsive regulatory proteins, and alterations in lipid membrane stability with respective to fluidity are just a few examples for direct temperature sensing. Temperature plays an essential role and has a crucial effect on biological processes. Targeted temperature adjustments for triggering a desired response may also be exploited for biotechnological applications. Noll et al. published an example for a thermosensitive structure to direct the carbon flow of a substrate into a product rather than into biomass, by exploiting an RNA thermometer to optimize the heterologous production of rhamnolipid biosurfactants [3].

Alterations in temperature lead to multiple, often complex, interconnected metabolic changes. Models describing a biological process as a function of temperature are therefore indispensable to reducing complexity and facilitating understanding of those interconnections. Mathematical descriptions of how (bio-)chemical reactions respond to high or low temperatures emerged as early as in the 1900s with the "primal" temperature model of Arrhenius [4]. He investigated the reaction kinetics for sugar cane inversion by acids, depending on temperature. Popularized by Arrhenius, a variety of temperature models evolved over time, as shown in Figure 1. These models range from data-driven empirical approaches to complex mechanistic models that are based on thermodynamic knowledge of biomolecular behavior at different temperatures. Models are readily available to be used as a tool for process control and design. An overview of the current state of thermo-modeling is crucial to reasonably selecting a suitable model for the bioprocess to be monitored or optimized. The aim of this review is to provide an overview of available temperature models to facilitate understanding of model intention and reasonable selection for application. So far, only a few applications for temperature in (industrial) process design, monitoring, and control have been described. Applied model approaches are usually based on fuzzy logic or artificial neural networks and do not harvest the full potential of deterministic approaches. There are a few examples for applied model-assisted temperature control strategies in industrial biotechnology. These include heat balancing for an estimation of metabolic activity to improve batch-to-batch reproducibility by applying a process control module which uses the difference between the culture temperature and temperature of a coolant to predict oxygen mass transfer rates and $k_L a$ values [5–7]. Furthermore, deterministic process models can be used to describe a biological process as a function of a physical condition, like the temperature. They may be used in the food sector to estimate product shelf life, determining critical control points of a process, or to maximize productivities and ensure safe distribution chains [8]. These examples highlight a potential for temperature and deterministic models in process design. An approach for experimental design to optimize processes depending on multiple parameters is the Response Surface Method (RSM): "RSM is a collection of statistical techniques used for studying the relationships between measured responses and independent input variables" [9]. This method may be used for optimization purposes in experimental design in the shape of a metamodel. It connects the response of an objective function to input variables and determines its relations, for example, by the means of first- or second-order polynomial equations. The Matlab software package from The Mathworks, Inc. (Natick, MA, USA) [10] or the Minitab statistic software (GMSL s.r.l., UK) [11] may be used to conduct an RSM analysis. In bioprocesses, usually very few process variables are available online that continuously display a process course. Furthermore, arguably the only control variable to direct a bioprocess towards a desired outcome is, in most cases, the addition of fresh media. Other variables like pH, temperature, or pO_2 are typically controlled at a constant value. Therefore, investigating and exploiting novel potentially available monitoring and control variables like temperature is a reasonable strategy to extend existing toolsets of bioprocess monitoring and control. Even though several systems inducible by temperature have been discovered and made available to biotechnologists in the last decades, only a few have been exploited for practical purposes, such as for process design or optimization [12]. For monitoring and control purposes, calorimetric approaches have been presented [5,6,13,14]. For control purposes, temperature may be used to directly address biological traits like RNA-thermometers to provoke a desired response, as previously evaluated [3]. Furthermore, indirect metabolic effects may be exploited, like elevated metabolic rates at high temperatures or the correct folding of proteins at low temperatures. Even though knowledge on microbial temperature responses and adaptation, along with descriptions developed by mathematical means, is available, its potential for applied industrial bioprocesses has not been sufficiently exploited. This review provides an overview on available thermo-models with the potential to develop model-assisted or model-derived process control strategies using temperature as a crucial parameter.

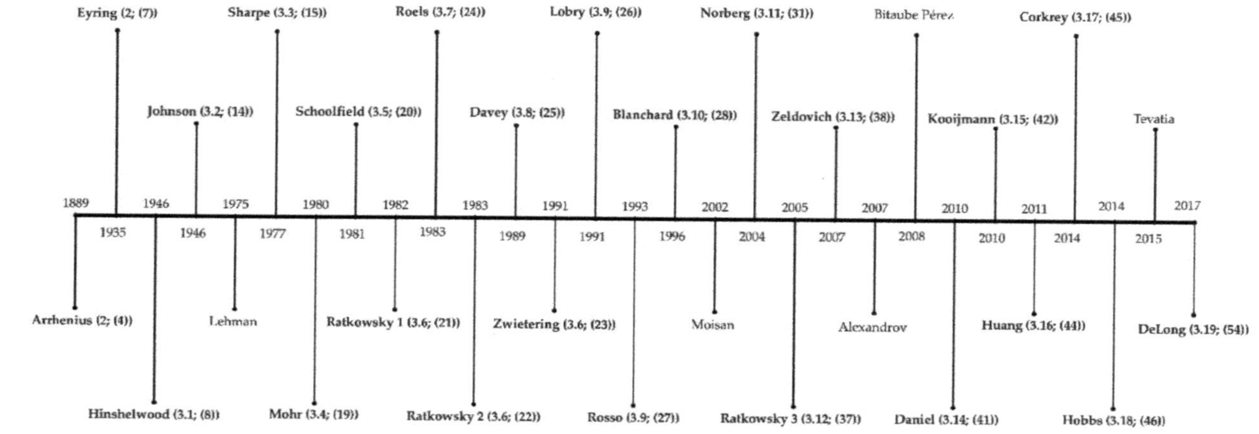

Figure 1. Timeline of temperature models described in paragraphs 2–3 starting in 1889 with the semi-empirical Arrhenius model. First author of the described model (paragraph number; equation number). Models marked with bold letters are described in detail and the remaining models in non-bold letters are summarized in Table 1.

2. History of Temperature Modeling—17th–20th Century

As early as the 17th century, there were theories on temperature being a form of particle movement. The kinetic theory of gases, with its origins in the 18th century, first specifically associated translational motions of molecules with heat and not with their vibrational or rotational motions [15]. Daniel Bernoulli was the pioneer of the kinetic theory of gases. He hypothesized that gases consist of a finite number of small spherical particles, which move through space along a straight line with high velocities. He assumed that heat increases the particle speed (v) and demonstrated that air pressure correlates with v^2. Air temperature can therefore be measured by this pressure at a constant density, making temperature proportional to v^2 [16,17]. The kinetic energies of the molecules are correlated with the ideal gas law of Equation (1), whose history began with the French engineer Émile Clapeyron in 1834 [18,19]. In the following, only SI units are used. Parameters with non-SI units, used by cited authors, were converted into SI units.

$$p \cdot V = n \cdot R \cdot T \tag{1}$$

where p is the pressure, Pa; V is the volume, m^3; n is the amount of substance, mol; R is the universal gas constant ~8.314, J mol^{-1} K^{-1}; and T is the absolute temperature, K. The Dutch chemist and first Nobel Prize laureate J. H. Van't Hoff observed that the chemical reaction rate doubles or triples when the temperature is increased by 10 K, which he expressed with the following equation:

$$Q_{10} = \left(\frac{k_2}{k_1}\right)^{\frac{10}{T_2 - T_1}} \tag{2}$$

This "rule of thumb" for chemical kinetics allows estimations for various phenomena in chemistry, biochemistry, and ecology. He furthermore described the change in the equilibrium constant K of a chemical reaction with respect to the change in temperature at constant pressure with the Van't Hoff Equation (3):

$$\frac{d}{dT} \ln K_{eq} = \frac{\Delta H}{R \cdot T^2} \tag{3}$$

where K is the dimensionless equilibrium constant; ΔH is the standard enthalpy change, J mol^{-1}; R is the universal gas constant ~8.314, J mol^{-1} K^{-1}; and T is the absolute temperature, K. Van't Hoff's student and the father of temperature models Svante Arrhenius continued the work of his teacher on the description of temperature-dependent specific reaction rate constants in chemical reactions with his essay "On the Reaction Velocity of the Inversion of Cane Sugars by Acids" [4]. Arrhenius observed that the reaction velocity of chemical reactions increased between 10% and 15% for each degree of rising temperature and postulated a semi-empirical model based on the Van't Hoff equation, which is shown in its integrated form in Equation (4).

$$k = A \cdot e^{\frac{-E_a}{R \cdot T}} \tag{4}$$

where k is the rate constant, s^{-1}, for a first-order rate constant; A is called a pre-exponential frequency or collision factor, s^{-1}, for a first-order rate constant; E_a is an empirical parameter, the (Arrhenius) activation energy, J mol^{-1}, characterizing the exponential temperature dependence of k; R is the universal gas constant ~8.314, J mol^{-1} K^{-1}; and T is the absolute temperature, K. In 1935, Henry Eyring formulated a statistical mechanistic equation following the transition state theory (former activated-complex theory) that assumed a transition state complex (‡TC) and a quasi-type equilibrium between educts (e_1; e_2), the transition state, and the product (P) [20,21]. The model has a similar way of describing the variance of the rate of a chemical reaction with temperature as the equation of Arrhenius. Therefore, it underlined Arrhenius' previous observations and assumptions with a mechanistic approach.

$$e_1 + e_2 \underset{k_2}{\overset{k_1}{\rightleftharpoons}} {}^{\ddagger}TC \overset{k_3}{\rightarrow} P \tag{5}$$

According to the transition state theory, the rate constant can be described as follows:

$$k(T) = \frac{k_B \cdot T}{h} \cdot K^{\ddagger} \tag{6}$$

where k_B is the Boltzmann constant $\sim 1.381 \cdot 10^{-23}$, J K^{-1}; T is the absolute temperature, K; h is the Planck's constant $\sim 6.626 \cdot 10^{-34}$, J s^{-1}; and $K^{\ddagger}$ is the dimensionless equilibrium constant. A different way to express the rate constant is by replacing the equilibrium constant with a term containing the standard molar changes of entropy and enthalpy:

$$k(T) = \frac{k_B \cdot T}{h} \cdot e^{\Delta^{\ddagger} S^{\circ}/R} \cdot e^{-\Delta^{\ddagger} H^{\circ}/(R \cdot T)} \tag{7}$$

where the entropy and enthalpy of activation are the standard molar change of entropy $\Delta^{\ddagger}S^{\circ}$, J K^{-1} mol^{-1}, when reactants form the transition state (activated) complex and standard molar change of enthalpy $\Delta^{\ddagger}H^{\circ}$, respectively, J mol^{-1}. R is the universal gas constant ~ 8.314, J mol^{-1} K^{-1}. The (Arrhenius) activation energy (E_a) and enthalpy of activation are not the same, but approximately equal, as they are convertible, depending on the molecularity [22].

2.1. Temperature in Biological Systems—The History Began with Arrhenius

Microbiologists have noticed a major effect of temperature on the growth rate of microbial populations and described this effect with the Arrhenius equation by simply replacing the rate constant k in Equation (4) with the growth rate (μ), meaning the reciprocal of the generation time. The so-called Arrhenius plot, where $\ln(\mu)$ is plotted against the reciprocal temperature, was used in the past and is still applied today to describe a relation between the temperature and growth of different bacteria and molds [23–26]. From this plot, Arrhenius parameters can easily be derived. Their plots show a good fit for lower temperatures. The Arrhenius model does not represent cell death, so a decrease of the growth curve at non-physiological temperatures. The lack of fit of the Arrhenius model for some temperature-dependent biological processes gave rise to the development of improved models describing growth as a function of temperature. Most of these models are based on Arrhenius' parental model and evolved over time.

2.2. Biological Mechanisms Involved in Temperature Responses

Microorganisms have developed molecular traits to respond to changing environmental temperatures. These traits have been extensively reviewed [2,27,28]. The principles of microbial thermo responses range from changing DNA topology, e.g., supercoiling caused by heat stress, stalled ribosomes, or stable RNA preventing translation during cold stress to proper folding of proteins, working optima of enzymes, or lipid membrane stability and fluidity. Biomolecules are generally thermos-sensitive. Therefore, various options for direct molecular thermosensing are possible. Molecular thermosensors may consist of DNA, RNA, proteins, or lipids [2]. The accessibility of DNA for the transcriptional machinery is crucial for transferring genetic information via RNA into a protein and is influenced by DNA topology [29]. DNA supercoiling, and thus accessibility, is altered in response to a shifting temperature, as has been reported for the plasmic DNA of meso- and thermophiles [30]. Mesophiles have negatively supercoiled plasmic DNA and hyperthermophilic archaea with a growth optimum ≥80 °C have relaxed or positively supercoiled plasmic DNA [30–32]. Proteins such as the histone-like structuring proteins (H-NS) work as temperature-dependent regulators, governing >200 temperature-regulated genes in *Salmonella* sp. and more than two third of *E. coli* K-12 temperature-regulated genes, respectively [33,34]. The inhibition of gene expression by H-NS is caused by trapping RNA polymerase and mediating DNA looping, thereby disturbing the progression of RNA polymerase [35–38]. Temperature-dependent gene expression is also influenced on the RNA level, where RNA can form inhibitory loop structures called RNA-thermometers (RNAT). Here, base pairing blocks the Shine-Dalgarno-sequence (SD) and AUG start codons, inhibiting ribosomal binding and translation initiation. By raising the temperature to a threshold (melting temperature), the hairpin structure opens and permits the access of ribosomes to the translation initiation site [2,39]. The secondary structure and thereby functionality of RNAT is characterized by canonical or non-canonical base paring, internal loops or mismatches, and the total number of loop structures. Based on these characteristics, RNA thermometers may be subdivided into three categories: (i) ROSE-like RNATs (repression of heat shock gene expression), (ii) FourU RNATs, and (iii) additional types of RNATs [40]. Most RNATs have been identified in the 5′-UTR of mRNA. The ROSE-like RNAT family is probably the most abundant temperature-sensing mRNA structure. ROSE-like RNATs usually control the repression of heat shock gene expression, but have also been reported to control expression of the rhamnosyl transferase, which is associated with *Pseudomonas aeruginosa* virulence [41]. ROSE-RNATs are located in the 5′UTR, are between 60 and 100 nt in length, and consist of 2–4 loop structures [41–43]. The majority of described RNATs of the second family, the FourU RNATs, govern the gene expression of virulence genes, and only two FourU RNAT's are known to control heat shock protein formation. FourU RNATs contain a sequence of four Uridines that occlude the SD sequence by canonical A-U and/or non-canonical G-U base pairing [40]. The virulence gene *lcrF* (*virF*) of *Yersina pestis* and the *agsA* small heat shock gene of *Salmonella enterica* were among the first genes described to be governed by a FourU RNAT [44]. Furthermore, attempts have been made to design synthetic RNATs with tailor-made characteristics to differ in up to 10-fold sensitivity- and around 3-fold threshold changes compared to a starting thermometer sequence [45]. On the protein level, global repressors, sensor kinases, methyl-accepting chemotaxis proteins (MCPs), M-like proteins, chaperones, and proteases are involved in microbial temperature responses [2]. The global transcriptional repressor CtsR has been termed a "protein thermosensor", and liberates DNA upon an up-shift in temperature connected to the expression of heat shock proteins. Due to a glycine-rich loop structure, CtsR exhibits intrinsic heat-sensing characteristics [46]. MCPs function as transmembrane receptors and consist of a periplasmic ligand-binding domain and a signaling domain in the cytosol that can interact with cytosolic sensor kinases [47]. The Tar MCP, for example, is convertible from a heat to a cold sensor in the presence of aspartate and consequent methylation at up to four sites [48]. The surface M and M-like proteins of the human pathogens group A streptococci bind to a variety of human plasma proteins in a temperature-dependent manner. The affinity of the M-like coild-coil protein Arp4 to IgA, is high at 10 and 20 °C, but low at 37 °C, due to a conformational change of Arp4 and consequent loss of the coild-coil conformation and binding activity [49]. The diverse class of small heat shock proteins (sHsp) can act as molecular chaperones upon heat shock. They are temperature sensors with different molecular mechanisms. For example, the sHsp Hsp26 of *Saccharomyces cerevisiae* consists of 24 subunits and changes its affinity state towards unfolded proteins at high temperatures by undergoing a conformational change [50,51].

2.3. Characteristic Graph for Growth as a Function of Temperature

Representative curvature of a model depicting the specific growth rate as a function of temperature, called the thermal growth curve, is shown in Figure 2. The simulated optimal specific growth rate (μ_{opt}) with the maximum turning point at the temperature optimum (T_{opt}) and growth rate at half of μ_{opt} ($\mu_{50\%opt}$) are marked in the model of a thermal growth curve. It has been pointed out that the term "optimal temperature" may need further specification to distinguish between the temperature for the optimal growth rate and the optimal temperature of the maximum biomass yield [52]. The minimum and maximum temperatures (T_{min} and T_{max}, respectively) for growth flank the asymmetric function and mark the thermal tolerance or thermal niche of an organism [53,54]. These three temperatures (T_{min}, T_{opt}, and T_{max}) are commonly referred to as cardinal temperatures. Bacteria can adapt to changing temperatures in the short run by producing cold- or heat-shock proteins. Furthermore, it was reported that the performance optimum of *E. coli* can be shifted when exposed to suboptimal temperatures for ~2000 generations. Conversely, the thermal niche breadth remained constant in that case [54]. The result is a reshaped thermal growth curve with the same upper and lower limits. The asymmetry of the thermal growth curve indicates that bacteria, which may be adapted to high temperatures, can survive in lower temperatures quite well. In contrary, fitness decreases sharply when temperatures exceed the optimum, resulting in thermal shock [55,56]. In one of the most recent approaches, the growth of psychrophiles, mesophiles, thermophiles, and hyperthermophiles was modeled, covering a temperature range of 124 °C, from −2 to +122 °C. The model was applied to 230 different strains of uni- and multicellular organisms with growth temperatures below freezing and the highest known temperature for biological growth so far [57,58].

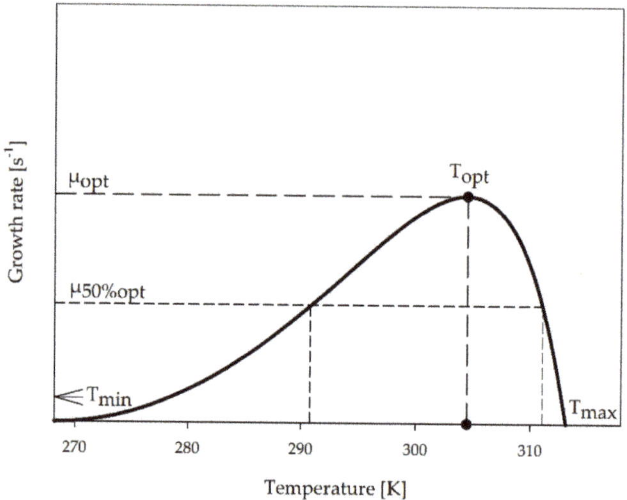

Figure 2. Scheme of the thermal growth curve where the temperature (K) is plotted against the growth rate (s^{-1}). Cardinal temperatures (T_{min}, T_{opt}, and T_{max}) with their corresponding growth rates (μ_{opt} and $\mu_{50\%opt}$) are indicated.

2.4. Mechanistic Versus Empirical Models

An often cited empirical approach for modeling the thermal growth curve of microorganisms is the approach of Ratkowsky et al. [59] (Scopus: cited by 615, 6 November 2019) and the semi-empirical model of Arrhenius. The development of mechanistic approaches for modeling the thermal growth curve of microorganisms started with the master reaction model of Johnson et al. in 1946 [60] (Scopus: cited by 80, 6 November 2019). The mechanistic models consider the description of single essential protein thermal stability (master reaction model) or the thermal stability of the whole proteome (the proteome model) as key for modeling the thermal growth curve. The transition of the native to an active and/or inactive state of the protein is considered. Grimaud et al. have extensively reviewed temperature growth models and concluded that empirical models display a better fit for balanced growth in non-limiting conditions than mechanistic models. Conversely, mechanistic models offer a complementary point of view for modeling thermal growth and can accurately represent temperature responses for growth under non-balanced conditions [61].

3. Temperature Modeling—From the 20th Century until Today

3.1. The Model of Hinshelwood (1946)

Hinshelwood expanded Arrhenius' model by adding a temperature-dependent term describing a "rate of degeneration" that becomes relevant at temperatures above T_{opt} [62]. Hinshelwood assumes a balanced growth for his model, saying that the total amount of compounds in a cell is constant. The model is based on the assumptions that just one enzymatic reaction is rate-controlling and the product of this reaction is a thermosensitive essential biomolecule which denatures irreversibly when temperatures are raised beyond the optimum. Temperature dependency and denaturation at high temperatures are of zero order and exhibit a temperature dependency similar to the Arrhenius model. The model represents the rate of synthesis in the minuend and degeneration in the subtrahend. At low temperatures, the subtrahend term is insignificantly small; in a small temperature region, both terms are almost equal, canceling each other out; and at high temperatures, the subtrahend term mostly accounts for the rapid decrease of the rate to zero.

$$u(T) = A_1 \cdot e^{-\frac{E_1}{R \cdot T}} - A_2 \cdot e^{-\frac{E_2}{R \cdot T}} \tag{8}$$

where A_1, and A_2 are referred to as pre/non-exponential, collision, or frequency actors, s^{-1}, related to entropy [62]; E_1 and E_2 are related to enthalpy [22], representing activation energies, J mol^{-1}, of the rate-determining enzyme reaction and high-temperature denaturation, respectively; R is the universal gas constant ~8.314, J mol^{-1} K^{-1}; and T is the temperature, K.

3.2. The Model of Johnson (1946)

In the same year, Johnson and Lewin [60] proposed another mechanistic model, which also assumes a simple case of a single reaction controlling growth, and called it their "master reaction model". In contrast to Hinshelwood, they assumed that a reversibly denaturable "master enzyme" E_0 controls an essential reaction for growth (assuming no substrate limitation). They reported their observation that *E. coli* stopped growing at non-viable 15 °C, but started growing exponentially again when transferred back to 37 °C. Increasing the exposition time of *E. coli* to non-viable temperatures led to lowered growth rates at 37 °C compared to the control. Hence, they assumed and described reversible protein denaturation damage as part of their model by integrating an equilibrium constant (K_1). The constant accounts for the equilibrium of reversibly denatured inactive (E_d) to native active enzymes (E_n).

$$K_1 = \frac{E_d}{E_n} \tag{9}$$

Hence, the amount of native active enzyme is given by Equation (10), with E_0 being the total amount of enzyme (native and denatured, mol).

$$E_n = \frac{E_0}{1 + K_1} = \frac{E_0}{1 + e^{\frac{-\Delta H}{R \cdot T}} \cdot e^{\frac{\Delta S}{R}}} \tag{10}$$

Johnson and Lewin then referred to Equation (7) proposed by Eyring, adding Equation (10) to account for the amount of active enzyme and substrate concentration and yielding Equation (11) for the temperature-dependent specific reaction rate (k).

$$k(T) = \frac{k_B \cdot T}{h} \cdot e^{\Delta^{\ddagger} S^{\circ}/R} \cdot e^{-\Delta^{\ddagger} H^{\circ}/(R \cdot T)} \cdot [S] \cdot [E_n] \tag{11}$$

By assuming that one single enzymatic reaction governs temperature-dependent growth at a constant substrate concentration and by substituting E_n in Equation (11) with rearranged Equation (10), temperature-dependent growth can be described as Equation (12):

$$u(T) = c \cdot T \cdot E_0 \cdot e^{\Delta^{\ddagger} S^{\circ}/R} \cdot e^{-\Delta^{\ddagger} H^{\circ}/(R \cdot T)} \cdot \frac{1}{1 + e^{-(\Delta H - T \cdot \Delta S)/(R \cdot T)}} \tag{12}$$

where c is a derived Boltzmann/Plancks constant, s K^{-1}, from the Eyring model of Equation (7); $\Delta^{\ddagger} H^{\circ}$ and $\Delta^{\ddagger} S^{\circ}$ are the standard molar change of enthalpy and entropy of activation, respectively (as described for Equation (7)); ΔH and ΔS are the enthalpy, J, and entropy change, J K^{-1}, respectively, between native and denatured enzymes; R is the universal gas constant ~8.314, J mol^{-1} K^{-1}; and T is the absolute temperature, K. The equation can then be shortened to the model in Equation (14) using the expression for Gibbs free energy change (between a catalytically active and reversibly denatured inactive state) at a constant temperature (Equation (13)):

$$\Delta G = \Delta H - T \cdot \Delta S \tag{13}$$

$$u(T) = C \cdot T \cdot e^{-\Delta^{\ddagger} H^{\circ}/(R \cdot T)} \cdot \frac{1}{1 + e^{-\Delta G/(R \cdot T)}} \tag{14}$$

where E_0 is assumed to be constant and $c \cdot e^{\Delta^{\ddagger} S^{\circ}/R} \cdot E_0$ is compressed to C. The fraction term containing the Gibbs free energy change can be assumed as the probability that the enzyme is in its native, and not its inactive, state. In the temperature region for a catalytically active enzyme, ΔG, J, has high positive values, yielding almost zero for the exponential term in the denominator of the probability term, and thus one for the probability for a catalytically active enzyme.

3.3. The Model of Sharpe (1977)

In 1977, Sharpe et al. [63,64] merged the models of Johnson and Lewin with the model of Hultin, which were both founded on Eyring's theory and modeled on the activated complex in chemical reactions [20,60,65,66]. The result was a unified rate model for the description of biological processes at physiological temperatures. Sharpe's model was originally developed for poikilotherms. Sharpe assumed balanced growth with a constant total amount of compounds per cell and just a single rate-controlling enzyme determining the development rate at all temperatures. Its reaction rate is of a zero order. The total concentration of enzyme (active + inactive) is assumed to be constant at all temperatures. Three enzyme states are considered and described: an inactivation state at low and high temperatures, as well as an active development state. Sharpe described transition between the states by his model, which depicts the thermal growth curve with the following equation:

$$u(T) = \frac{T \cdot e^{(\Phi - \Delta^{\ddagger} H^{\circ}/T)/R}}{1 + e^{(\Delta S_L - \Delta H_L/T)/R} + e^{(\Delta S_H - \Delta H_H/T)/R}} \tag{15}$$

where T is the temperature, K; is the universal gas constant ~8.314, J mol^{-1} K^{-1}; and the other parameters describe the rate-controlling enzyme reaction, where Φ is a dimensionless conversion factor; $\Delta^{\ddagger} H^{\circ}$ is the enthalpy of activation of the reaction catalyzed by the rate-controlling enzyme, J mol^{-1}; the subscript L accounts for low-temperature inactivation and the subscript H for high-temperature inactivation; and ΔS^{*}, J K^{-1} mol^{-1}, and ΔH^{*}, J mol^{-1}, mark the entropic and enthalpic change, respectively, upon high- or low-temperature inactivation specified by the subscript. In 1991, Zwietering et al. [8] rewrote the model of Sharpe exhibiting an Arrhenius-type of temperature dependency using activation energies rather than changes in enthalpy to describe growth. As described by the International Union of Pure and Applied Chemistry, IUPAC, the (Arrhenius) activation energy (E_a) and enthalpy of activation are not the same, but approximately equal, as they are convertible, depending on the molecularity [22].

$$u(T) = \frac{k_a \cdot e^{\frac{-E_a}{R \cdot T}}}{1 + k_l \cdot e^{-\frac{E_l}{R \cdot T}} + k_h \cdot e^{-\frac{E_h}{R \cdot T}}} \tag{16}$$

where k_a (s^{-1}), k_l (-), and k_h (-) are collision factors that are dimensionless in the denominator, as described by Zwietering et al. [8]; E, J mol^{-1}, represents the activation energy; subscript a accounts for the rate-determining enzyme reaction; subscripts h and l describe high- and low-temperature inactivation, respectively; R is the gas constant ~8.314, J mol^{-1} K^{-1}; and T is the temperature, K.

3.4. The Model of Mohr (1980)

Mohr and Krawiec [24] analyzed the thermal growth curves for 12 bacterial species. Among them were thermophiles, mesophiles, and psychrophiles. They used the Arrhenius plot for their data, where $\ln(\mu)$ is plotted against the reciprocal temperature in Kelvin. They reported two different slopes for the Arrhenius profiles for some mesophiles and thermophiles at suboptimal temperatures. The temperature at the interception point were both slopes meet is referred to as the "critical temperature" (T_{crit}) [24,67]. This point marks the turning point where the organization of an organism, and hence it growth behavior, changes. In order to describe the two different slopes, they proposed two equations, which they reduced to one with the assumption that "[…] a balance of organizations exists at any temperature":

$$u(T) = A_1 \cdot e^{-E_1/R \cdot T} - A_2 \cdot e^{-E_2/R \cdot T} \quad T_{crit} < T < T_{max} \tag{17}$$

$$u(T) = A'_1 \cdot e^{E'_1/R \cdot T} \quad T_{min} < T < T_{crit} \tag{18}$$

$$u(T) = \frac{1}{A_1^* \cdot e^{E'_1/R \cdot T} + A_1^{**} \cdot e^{E_1/R \cdot T}} - A_2 \cdot e^{-E_2/R \cdot T} \tag{19}$$

where A_1, A'_1, and A_2 are referred to as pre/non-exponential, collision, or frequency actors, s^{-1}; E_1, E'_1, and E_2 are referred to as temperature characteristics, J mol^{-1}; R is the gas constant ~8.314, J mol^{-1} K^{-1}; and T is the temperature, K. The parameters marked with a " ′ " are used to describe temperature-dependent growth for $T_{min} < T < T_{crit}$, whereas parameters without " ′ " are used to describe temperature-dependent growth for $T_{crit} < T < T_{max}$, $A_1^* = 1/A'_1$, and $A_1^{**} = 1/A_1$.

3.5. The Model of Schoolfield (1981)

Schoolfield developed a non-linear regression model [68] based on the model proposed by Sharpe. His group reformulated the model of Sharpe and eliminated the high correlations of Sharpe's parameters (e.g., 0.9996). Furthermore, Schoolfield et al. argued that there is no "readily apparent" initial guess for beginning iterations for parameter estimation. Hence, they also aimed to facilitate the regression process and parameter estimation.

$$u(T) = \frac{u_{25} \cdot \frac{T}{298} \cdot e^{[\frac{\Delta^{\ddagger}H^{\circ}}{R} \cdot (\frac{1}{298} - \frac{1}{T})]}}{1 + e^{[\frac{\Delta H_L}{R} \cdot (\frac{1}{T_{1/2L}} - \frac{1}{T})]} + e^{[\frac{\Delta H_H}{R} \cdot (\frac{1}{T_{1/2H}} - \frac{1}{T})]}} \tag{20}$$

Schoolfield et al. chose 25 °C (298 K) and the respective specific growth rate u_{25}, s^{-1}, as a reference because enzyme inactivation would be low or not present at that temperature in most biological systems. $\Delta^{\ddagger}H^{\circ}$ is the enthalpy of activation of the reaction catalyzed by the rate-controlling enzyme, J mol^{-1}; the subscript L accounts for low-temperature inactivation and the subscript H for high-temperature inactivation of the enzyme; and ΔH_* marks the enthalpic change upon high- or low-temperature inactivation specified by the subscript, J mol^{-1}. With an increasing or decreasing temperature, 50%

of the rate-controlling enzyme is inactivated by either a high $T_{1/2_H}$, K or low temperature $T_{1/2_L}$, K, as previously described by Hultin [66] and adopted by Schoolfield et al. [68].

3.6. The Models of Ratkowsky and Zwietering (1982–1991)

As extensively reviewed by Grimaud et al. [61], several models for temperature-dependent growth in biological systems have been developed. Most of these models were developed to describe food spoilage and medical applications. One often cited (802, Scopus, 8 August 2019) empirical model is the square root model proposed by Ratkowsky et al., as an alternative to the widely used Arrhenius model, to describe growth as a function of temperature [26]:

$$u(T) = [b_1 \cdot (T - T_{\min})]^2 \tag{21}$$

where b_1 is a Ratkowsky parameter, $K^{-1}\,s^{-0.5}$, and T_{min} is the minimum temperature of growth, K. This model was extended to the complete bio-kinetic range in 1983 by the same author [59]:

$$u(T) = \left(b_2 \cdot (T - T_{\min}) \cdot \left\{1 - e^{[c_2 \cdot (T - T_{\max})]}\right\}\right)^2 \tag{22}$$

where c is a Ratkowsky parameter, K^{-1}, and T_{max} is the maximum temperature, K, at which growth is observed. Zwietering et al. [8] argued that Ratkowsky's model could not be used for temperatures above T_{max} because the model predicted positive values for growth rates beyond the high-temperature end of the thermal niche. They therefore adapted the model accordingly and the result is shown in Equation (23).

$$u(T) = [b_3 \cdot (T - T_{\min})]^2 \cdot \left\{1 - e^{[c_3 \cdot (T - T_{\max})]}\right\} \tag{23}$$

3.7. The Model of Roels (1983)

In 1983, Roels et al. [69] developed a model to describe the growth rate as a function of temperature. The numerator has an Arrhenius-type appearance and the energy for activation was replaced by the Gibbs free energy change upon denaturation of a rate-controlling enzyme in the denumerator.

$$u(T) = \frac{A \cdot e^{\left(\frac{-E_G}{RT}\right)}}{1 + B \cdot e^{\left(\frac{-\Delta G_d}{RT}\right)}} \tag{24}$$

where A and B are pre-exponential factors, s^{-1}; R is the gas constant ~8.314, $J\,mol^{-1}\,K^{-1}$; and T is the temperature, K.

3.8. The Model of Davey (1989)

Davey proposed an empirical generalized predictive model based on a modified linear-Arrhenius equation, which combined the influence of temperature and water activity to describe microbial growth [70]. The activation energy parameter from the original Arrhenius model was replaced in Davey's model by two coefficients of inverse temperature. In his work, Davey provided and evaluated models for the influence of environmental factors like the a_w value or pH in combination with temperature, as well as temperature as the sole influencing factor, on growth [70–72]. The model describing temperature as the sole environmental factor is shown in Equation (25):

$$u(T) = e^{C_0 + C_1/T + C_2/T^2} \tag{25}$$

where C_0–C_2 are dimensionless Davey coefficients, - and the energy of activation given in the Arrhenius equation is replaced by two parameters of reciprocal temperature, K. Two years later, in 1991, Davey used the model for predicting the temperature-dependent lag time [72].

3.9. The Models of Lobry and Rosso (1991–1993)

In 1991, Lobry [73] developed an empirical model that includes the three cardinal temperatures (T_{min}, T_{opt}, and T_{max}) as parameters. The cardinal temperature model (CTM) estimates positive values for growth rates at temperatures between the low temperature and high temperature end (T_{min} and T_{max}), with the highest growth rate (u_{opt} (s^{-1})) at T_{opt}. Outside the thermal niche ($T < T_{min}$; $T > T_{max}$), negative values are predicted. Each microbial species exhibits these three characteristic cardinal temperatures, which permits direct biological interpretation of the model parameters and facilitates parameter estimation using experimental data for Lobry's CTM-model Equation (26). The authors emphasize the "absence of high structural correlations" between parameters of their model. In 1991, Rosso et al. [74] further elaborated Lobry's empirical model by including the point of inflection in the suboptimal range of temperatures, which was experimentally determined. Following this, the so-called cardinal temperature model with inflection (CTMI; Equation (27)) could be used to accurately predict growth in the suboptimal range of temperature. Rosso's group noted an "unexpected" high linear correlation between cardinal temperatures, especially between T_{max} and T_{opt}, with r = 0.991. They then argued that due to the correlations found, one instead of three cardinal temperatures could sufficiently describe the permissive temperature range for growth. They also mentioned an exception for the stated relationships between the cardinal temperatures for the growth behavior of *Vibrio* sp. In total, they analyzed 47 different data sets describing the growth of psychrophilic, mesophilic, and thermophilic strains.

$$u(T) = u_{opt} \cdot \left[1 - \frac{\left(T - T_{opt}\right)^2}{\left(T - T_{opt}\right)^2 + T \cdot (T_{max} + T_{min} - T) - T_{max} \cdot T_{min}} \right] \tag{26}$$

$$u(T) = u_{opt} \cdot \frac{(T - T_{max}) \cdot (T - T_{min})^2}{\left(T_{opt} - T_{min}\right) \cdot \left[\left(T_{opt} - T_{min}\right) \cdot \left(T - T_{opt}\right) - \left(T_{opt} - T_{max}\right) \cdot \left(T_{opt} + T_{min} - 2 \cdot T\right)\right]} \tag{27}$$

3.10. The Model of Blanchard (1996)

Blanchard et al. [75] originally developed their model to quantify the short-term temperature effect on natural assemblages of microphytobenthos' photosynthetic capacity. Blanchard described a progressive increase in photosynthetic capacity during a temperature increase up to an optimum temperature, with a rapid decrease when the temperature was raised beyond the optimum. For the model, cardinal temperatures (T_{min}, T_{opt}, and T_{max}) with a biological meaning are used to facilitate a reasonable initial guess for parameter estimation. Grimaud et al. [61] rewrote the Blanchard model to represent growth as a function of temperature instead of photosynthetic capacity, as shown in Equation (28):

$$u(T) = u_{opt} \left(\frac{T_{max} - T}{T_{max} - T_{opt}}\right)^{\beta} \cdot e^{-\beta \cdot (T_{opt} - T)/(T_{max} - T_{opt})} \tag{28}$$

where μ_{opt}, s^{-1}, is the maximal specific growth rate at optimal temperature T_{opt}, K; T_{max}, K, is the maximum temperature where growth is observed; and β is a dimensionless Blanchard parameter.

3.11. The Models of Eppley and Norberg (2004)

Eppley [76] proposed a simple function with a positive exponential correlation between temperature and maximum expected growth, as shown in Equation (29). He stated that this model may be used for a generalized estimate for μ_{max} in unicellular algae for temperatures <40 °C.

$$u(T) = 0.851 \cdot 1.066^T \tag{29}$$

The proposed Eppley curve or envelope function of Equation (29) shows the evolutionary interspecific upper limit for the maximum specific growth rate at any temperature up to 40 °C. The limit for the maximum growth rate increases exponentially until 40 °C in Eppley's function. For model assembly, Eppley used almost 200 data points of different species of unicellular algae [76]. Based on Eppley's findings, Jon Norberg developed a model for temperature-dependent growth in 2004 [77], shown in Equation (30).

$$u(T) = \left[1 - \left(\frac{T-Z}{w}\right)^2\right] \cdot 0.59 \cdot e^{0.0633 \cdot T} \tag{30}$$

An envelope function $0.59 \times e^{0.0633T}$ according to Eppley is contained in Norberg's model for temperature-dependent growth, where T, K, is the ambient temperature and Z, K, is the temperature with the maximum specific growth rate derived from the envelope function representing T_{opt} respectively. The width of the temperature response function is determined by the parameter w, K, meaning the width of the thermal niche. A generalized form of the Eppley–Norberg model Equation (31) would add a and b as dimensionless parameters, -, generalizing the Eppley envelope function.

$$u(T) = \left[1 - \left(\frac{T-Z}{w}\right)^2\right] \cdot a \cdot e^{b \cdot T} \tag{31}$$

3.12. The Modified Master Reaction Model (2005)

In 1967, Brandts recognized that the master reaction model proposed by Johnson and Lewin (see Equation (12)) failed to describe enzymatic reactions adequately when applied to the full bio kinetic temperature range [78,79]. Arguably, the limitations of the model arise from the assumed temperature independence of ΔG in the master reaction model upon protein denaturation. Therefore, Brandts et al. attributed the temperature dependency to ΔG by simply using an empirical polynomial expression relating ΔG to T. In 1974, Privalov et al. [80] reported a linear relation between enthalpy and entropy and temperature upon protein unfolding when assuming a specific constant heat capacity change for a specific protein. Almost 20 years later, it was reported that the change in enthalpy upon denaturation (ΔH_d) normalized to the number of amino acid residues (or molecular weight, respectively) at a specific temperature ($T_H^* \sim 373$ K) converged to a common value (ΔH^*). Likewise, the same convergence behavior to one common value (ΔS^*) at a specific temperature ($T_S^* \sim 385$ K) for entropy upon denaturation (ΔS_d) normalized to the number of amino acid residues was described for a number of homologous compounds [80–83]. The so-called convergence temperatures (T_H^* and T_S^*) were obtained as the temperatures where the apolar contributions (apolar hydrogen atoms CH) to the corresponding changes in entropy or enthalpy, respectively, upon denaturation approach zero [81,82,84]. Therefore, ΔH^* describes only polar and van der Waals interactions and ΔS^* primarily accounts for configurational entropy [83]. In 1990, Murphy et al. [81] analyzed the convergence behavior by plotting ΔH_d or ΔS_d normalized to mol amino acid residue against the normalized heat capacity change (ΔC_p) upon denaturation and obtained the following correlations:

$$\Delta S_d = \Delta S^* + \Delta C_p \cdot ln\left(\frac{T}{T_S^*}\right) \tag{32}$$

$$\Delta H_d = \Delta H^* + \Delta C_p \cdot \left(T - T_H^*\right) \tag{33}$$

The above-mentioned equations describe a temperature-dependent enthalpic and entropic change upon denaturation normalized to the number of amino acid residues in a protein (n). From Murphy's findings [81], the change in Gibbs free energy upon protein denaturation (~protein thermal stability) is given by

$$\Delta G_d(T) = n \cdot \left[\Delta H^* - T \cdot \Delta S^* + \Delta C_p \cdot \left[\left(T - T_H^*\right) - T \cdot ln\left(\frac{T}{T_S^*}\right)\right]\right] \tag{34}$$

where $\Delta C_p \cdot \left[\left(T - T_H^*\right) - T \cdot ln\left(\frac{T}{T_S^*}\right)\right]$ accounts for the hydrophobic contribution of the Gibbs free energy change upon denaturation of the rate-determining "master enzyme". Ross connected Murphy's findings with the rewritten master reaction model of Johnson and Lewis, where the change in enthalpy between the catalytically active and inactive state of the rate-limiting enzyme was replaced by the temperature-dependent Gibbs free energy change [79,85].

$$u(T) = \frac{c \cdot T \cdot e^{(-\Delta^{\ddagger}H^{\circ}/R \cdot T)}}{1 + e^{(-\Delta G_d/R \cdot T)}} \tag{35}$$

Replacing the description of the Gibbs free energy in the denominator of Equation (35) with the term in Equation (34) resulted in a modified master reaction model:

$$u(T) = \frac{c \cdot T \cdot e^{(-\Delta^{\ddagger}H^{\circ}/R \cdot T)}}{1 + e^{(-n \cdot \{\Delta H^* - T \cdot \Delta S^* + \Delta C_p \cdot [(T - T_H^*) - T \cdot ln(T/T_S^*)]\}/R \cdot T)}} \cdot \tag{36}$$

In the denominator of Equation (36), the thermodynamic parameters ΔH^*, ΔS^*, and ΔC_p are normalized to mol amino acid residue. In 2005, Ratkowsky et al. [79] reduced the eight-parameter model given in Equation (36) to a five-parameter model by simply applying the universal constants for globular proteins $T_H^* = 373.6$ K, $T_S^* = 385.2$ K, and $\Delta S^* = 18.1$ J K^{-1} found by Murphy et al. [81,82] to the model. Ratkowsky's group fitted the reduced five-parameter modified master reaction model to data from 35 bacterial strains. The universal constant ΔH^* suggested by Murphy's group with 5640 J mol^{-1} amino acid residue was found to be unsuitable for representing bacterial growth when applied to five data sets of Ross [86]. The reduced modified master reaction model with applied universal constants evaluated in the work of Ratkowsky et al. [79] is given in Equation (37).

$$u(T) = \frac{c \cdot T \cdot e^{(-\Delta^{\ddagger}H^{\circ}/8.314 \cdot T)}}{1 + e^{(-n \cdot \{\Delta H^* - T \cdot 18.1 + \Delta C_p \cdot [(T - 373.6) - T \cdot ln(T/385.2)]\}/8.314 \cdot T)}} \tag{37}$$

3.13. The Model of Zeldovich (2007–2016)

The group of Zeldovich [87] argued that the whole proteome has to be considered when describing the temperature response and sensitivity of an organism. Ghosh and Dill [88] continued the work of Zeldovich et al. and proposed a model that considers the folding stabilities across an organisms' proteome to describe temperature-dependent growth rates of bacteria. They assumed that the growth rate was a function of temperature composed of a product of two factors: First, Arrhenius-type low-temperature activation for one or more activated metabolic processes controlling the increase of growth rate at low temperatures, and second, a term accounting for the folded part of the proteome at any temperature, which also depicts the "denaturation catastrophe" when reaching high temperatures.

$$u(T) = u_0 \cdot e^{\left(\frac{-\Delta^{\ddagger}H^{\circ}}{k \cdot T}\right)} \prod_{i=1}^{\Gamma} \frac{1}{1 + e^{(-\Delta G_{un}(N_i, T)/R \cdot T)}}, \tag{38}$$

where μ_0 is a growth rate reference, s^{-1}. The parameter $\Delta^{\ddagger}H^{\circ}$, J mol^{-1}, describes the activation barrier for growth (e.g., an essential growth-limiting metabolic rate). The authors found that the activation barrier for growth (~68.2 kJ mol^{-1}) in *E. coli* approximately corresponds to the energy needed by ribosomes to form a peptide bond. Hence, the authors identified ribosomal action to grow protein chains as one of the key growth rate-limiting factors, along with protein motions necessary for enzymatic reactions. Γ describes the amount of essential proteins required for growth. The product term accounts for the probability that the ith essential protein composed of N_i amino acids is in its native state, which is expressed by ΔG_{un} in Equation (39):

$$\Delta G_{un} = -k_B \cdot T_0 \cdot n \cdot \left[\frac{g_0 + m_1 \cdot c}{k_B \cdot T_0} + \frac{\Delta C_p}{k_B \cdot T_0} \cdot \left(T - T_H^*\right) + \frac{T}{T_0} \cdot ln(z) - \frac{T}{T_0} \cdot \Delta C_p \cdot ln\left(\frac{T}{T_S^*}\right) + \frac{l_b}{2 \cdot n} \cdot \left(\frac{Q_n^2}{R_n \cdot (1 + \kappa \cdot R_n)} - \frac{Q_d^2}{R_d \cdot (1 + \kappa \cdot R_d)} \right) \right] \tag{39}$$

where k_B is the Boltzmann constant $\sim 1.381 \cdot 10^{-23}$, J K^{-1}; $T_0 = 300$ K; n is the number of amino acids with respective to the chain length of a protein; g_0 is the free energy upon amino acid desolvation and upon contact; c is the concentration of denaturant; ΔC_p is the heat capacity change upon denaturation, J K^{-1} mol-residue^{-1}; T is the absolute temperature, K; $T_H^* = 373$ K and $T_S^* = 385$ K are the enthalpic and entropic convergence temperatures, respectively; z is defined as the loss of average conformational freedom per backbone bond; l_b is the average Bjerrum length; Q_n and Q_d are the total net charge of native and denatured protein, respectively; R_n and R_d account for the radii of native and denatured protein, respectively; κ is the reciprocal of the Debye length (for further details, see [89,90]). To obtain the probability distribution for protein stabilities of a proteome p(ΔG), Equation (39) can be used. The equation accounts for the stability of an average single protein of length n and may be used in combination with the distribution of protein chain lengths (P(n)) of a cell available for various cell types from genomic/proteomic data in order to calculate temperature-dependent proteome stability.

3.14. The Model of Daniel (2010)

In 2010, Daniel et al. [91] proposed the equilibrium model to describe temperature-dependent catalytic activity of enzymes in non-limiting conditions. The group reported that the decrease of the catalytic rate constant (k_{cat}) of tested enzymatic reactions above the optimal temperature (T_{opt}) does not entirely correspond to thermal stability data and irreversible denaturation. Furthermore, they found that part of the activity loss above T_{opt} was reversible and probably associated with a "conformational, dynamic and solvent-based effect" altering the active site of the enzyme. To explain the higher than expected decrease of k_{cat} at a certain temperature, they suggested that an enzyme may be present in three states: (i) catalytically active (E_{act}); (ii) catalytically inactive, but not (significantly) unfolded (E_{inact}); and (iii) irreversibly denatured (X). They rapid changes of the Michaelis constant (K_m), describing an enzyme's affinity towards a substrate, that occur with temperature support their hypothesis of an "E_{inact} state", where the active site is altered. They argue that the active site may need a certain degree of flexibility to function properly and is therefore more prone to changes in temperature affecting conformation and dynamics.

$$E_{act} \underset{}{\overset{K_{eq}}{\rightleftharpoons}} E_{inact} \overset{k_{inact}}{\rightarrow} X \tag{40}$$

Daniel's group assumed a rapid equilibrium between E_{act} and E_{inact} and time-dependent denaturation at a certain temperature. The conversion rate of E_{act} to E_{inact} is thereby assumed to be faster than the rate of denaturation and catalytic reaction rate, respectively. The authors investigated the applicability of the model for >50 datasets of >30 enzymes of different reaction classes and structures (monomeric to hexameric) and concluded that their model is universally applicable and independent of the reaction type and enzymatic structure [91]. The model may therefore be suitable for describing a thermal growth curve, where a rate-controlling enzymatic reaction is often assumed to describe temperature-dependent growth (e.g., master reaction model, [60]).

$$u(T) = \frac{k_B}{h} \cdot T \cdot E_0 \cdot e^{-\left(\frac{\Delta G_{cat}^*}{R \cdot T}\right)} \cdot e^{\left[-\frac{k_B}{h} \cdot T \cdot e^{-\left(\frac{\Delta G_{inact}^*}{R \cdot T}\right)} \cdot \frac{e^{(\Delta H_{eq} \cdot (\frac{1}{T_{eq}} - \frac{1}{T})/R) \cdot t}}{1 + e^{(\Delta H_{eq} \cdot (\frac{1}{T_{eq}} - \frac{1}{T})/R)}} \right]} \cdot \left[1 + e^{(\Delta H_{eq} \cdot (\frac{1}{T_{eq}} - \frac{1}{T})/R)} \right]^{-1} \tag{41}$$

where k_B is the Boltzmann constant ~$1.381 \cdot 10^{-23}$, J K^{-1}; h is the Planck's constant ~$6.626 \cdot 10^{-34}$, J s^{-1}; T is the absolute temperature, K; E_0 is the total enzyme concentration composed of the sum of E$_{act}$, E$_{inact}$, and X, mol m^{-3}; ΔG^*_{cat} is the Gibbs free energy of activation of the enzymatic reaction; R is the universal gas constant ~8.314, J mol^{-1} K^{-1}; ΔG^*_{inact} is the Gibbs free energy of activation for the irreversible denaturation of the rate-controlling enzyme; ΔH_{eq} is the enthalpic change between E$_{act}$ and E$_{inact}$; T_{eq} is the equilibrium temperature where the rate-controlling enzyme is present as 50:50 E$_{act}$ and E$_{inact}$, K; and t is the time, s.

3.15. The Model of Kooijman (2010)

The DEB or Dynamic Energy Budget theory deals with the description of rates for physiological processes. Assimilation, growth, respiration, maintenance, or reproduction in individual, not further specified, "organisms" are analyzed and described by the generalized theory. The rates are described as a function of the environment, like the temperature or nutrient availability, and the state of the organism like size or age, for example. S.A.L.M. Kooijman, whose early concepts on energy budgets were published in 1986 [92], summarized his work in "DEB theory for Metabolic Organization" in 2010 [93]. In DEB theory, temperature-dependent growth is described by a reformulated Arrhenius equation in the numerator and complemented by a term inspired by Sharpe's model [63] for reduced rates at the high- and low-temperature end in the denominator. The equation therefore accounts for the amount of enzyme in its native state and considers a possible transition to an inactive state via hot and cold denaturation. Kooijman argues that Eyring's thermodynamic interpretation of the Arrhenius type of temperature dependence might only be understood as an approximation. It is an enormous step from Eyring's model, considering bimolecular reactions in the gas phase, to physiological rates with many compounds [94].

$$u(T) = \frac{k_1 \cdot e^{T_A/T_1 - T_A/T}}{1 + e^{T_{Al}/T - T_{Al}/T_l} + e^{T_{Ah}/T_h - T_{Ah}/T}} \tag{42}$$

where T_1 is a reference temperature, K, with the corresponding rate k_1, s^{-1}; T_A is the Arrhenius temperature (i.e., linear slope of the Arrhenius plot), K; T_l and T_h mark the cardinal temperatures flanking the thermal niche (low- and high-temperature denaturation, respectively), K; and T_{Al} and T_{Ah} account for Arrhenius temperatures at temperature boundaries of the thermal niche, K.

3.16. The Model of Huang (2011)

The group of Huang [95] developed a model by modifying and combining the Arrhenius equation with the theory and model of Eyring et al. [4,20,65].

$$u(T) = A \cdot T \cdot e^{-\left(\frac{\Delta G'}{R \cdot T}\right)^\alpha} \tag{43}$$

where $\Delta G'$, J mol^{-1}, accounts for an energy term; R is the universal gas constant ~8.314, J mol^{-1} K^{-1}; T is the absolute temperature, K; α is a Huang parameter, K^{-1}; and A describes the collision or frequency factor, s^{-1}. Huang reports that their model can only sufficiently describe growth behavior at a suboptimal temperature. To extend their model to the entire physiological temperature range, they used an expression known from Ratkowsky et al. [59]:

$$u(T) = A \cdot T \cdot e^{-\left(\frac{\Delta G'}{R \cdot T}\right)^\alpha} \cdot \left[1 - e^{c_2 \cdot (T - T_{max})}\right] \tag{44}$$

where c_2 is a Ratkowsky parameter, K^{-1}, and T_{max} is the maximum growth temperature, K. Huang's group reported good fits ($R^2 = 0.985$) for their model using data for thermal growth rates from five different bacteria.

3.17. The Model of Corkrey (2014)

In 2014, Corkrey et al. attempted to build a universal mechanistic model [58]. It was used to model the growth of 230 different strains of unicellular and multicellular organisms ranging from psychrophilic to hyperthermophilic, covering a temperature range of 124 °C, from −2 °C to +122 °C. They therefore argued that their findings might be used to model the dependence of the growth rate on temperature for all unicellular and multicellular life forms. Being able to find a good fit for their universal model to the thermal growth curves of various life forms, they concluded that there might be evidence for the presence of a single highly conserved reaction in the last universal common ancestor. Under all limiting conditions, a single-enzyme-catalyzed reaction rate, which controls the growth rate, is described in the numerator of the Corkrey model by an Arrhenius type of temperature dependency. The denominator accounts for the effects of temperature on protein conformation, which causes alterations in catalytic activity of the putative enzyme and therefore a change in the expected rate.

$$u(T) = \frac{T \cdot e^{\left(c - \frac{\Delta^{\ddagger} H^{\circ}}{R \cdot T}\right)}}{1 + e^{\left(-n \cdot \frac{\Delta H^{*} - T \cdot \Delta S^{*} + \Delta C_p \cdot (T - T_H^{*} - T \cdot \ln(T/T_S^{*}))}{R \cdot T}\right)}} \tag{45}$$

where $\Delta^{\ddagger} H^{\circ}$ is the enthalpy of activation, J/mol; R is the universal gas constant ~8.314, J mol^{-1} K^{-1}; c is a dimensionless scaling factor; T is the temperature, K; ΔC_p is the heat capacity change, J K^{-1} mol^{-1} -amino acid residue, upon denaturation of the putative enzyme; ΔH^{*} is the change of enthalpy, J mol^{-1} amino acid residue, at the convergence temperature T_H^{*}, K, for enthalpy of protein unfolding; and ΔS^{*} is the change of entropy, J K^{-1}, at the convergence temperature T_S^{*}, K, for the entropy of protein unfolding.

3.18. The Model of Hobbs (2014)

The heat capacity model or macromolecular rate theory was proposed by Hobbs et al. [96] and applied by Schipper et al. [97]. They state that enzymatic rates show Arrhenius behavior when increasing with temperature, until an optimum (T_{opt}), but argue that decreasing rates above T_{opt} cannot sufficiently be explained only by denaturation of the enzymes. They have reported the effect of a heat capacity change upon activation (between the ground state and transition state of a rate-limiting enzyme) that shapes the thermal growth curve. The change in heat capacity affects the temperature dependency of $\Delta^{\ddagger} G^{\circ}$ (Gibbs free energy difference, between a ground state and transition state) that is in turn responsible for determining the temperature dependency of enzymatic activity. They state that the change in heat capacity influencing enzymatic rates implicates temperature dependence for various biological rates, ranging from "enzymes to ecosystems". Hobbs et al. formulated their findings using Eying's Equation (46) as a scaffold and a term to describe the degree of temperature dependence of $\Delta^{\ddagger} G^{\circ}$ with $\Delta^{\ddagger} C_p^{\circ}$ Equation (47). Assuming $\Delta^{\ddagger} C_p^{\circ}$ to be zero, $\Delta^{\ddagger} G^{\circ}$ would be independent from temperature and the reaction behavior of the growth rate-limiting enzyme would follow an Arrhenius type of temperature dependence. Large negative values for $\Delta^{\ddagger} C_p^{\circ}$ would lead to a significant temperature dependence of $\Delta^{\ddagger} G^{\circ}$, leading to a non-Arrhenius behavior and explaining a decrease in reaction rate above T_{opt}, independent of denaturation. Compared to the master reaction model, heat capacity theory takes into account that enzymes are in fast equilibrium with the transition state and denaturation does not easily occur [61,96,97].

$$u(T) = \frac{k_B}{h} \cdot T \cdot e^{\left(-\Delta^{\ddagger} G^{\circ}(T)/R \cdot T\right)} \tag{46}$$

$$\Delta^{\ddagger} G^{\circ}(T) = \Delta^{\ddagger} H_{T_0}^{\circ} + \Delta^{\ddagger} C_p^{\circ} \cdot (T - T_0) + T \cdot \left(\Delta^{\ddagger} S_{T_0}^{\circ} + \Delta^{\ddagger} C_p^{\circ} \cdot \ln\left(\frac{T}{T_0}\right)\right) \tag{47}$$

where k_B is the Boltzmann constant ~$1.381 \cdot 10^{-23}$, J K^{-1}; h is the Planck's constant ~$6.626 \cdot 10^{-34}$, J s^{-1}; $\Delta^{\ddagger}G^{\circ}$ is the Gibbs free energy difference, between a ground state and transition state, J mol^{-1}; R is the universal gas constant ~8.314, J mol^{-1} K^{-1}; T is the absolute temperature, K; T_0 is a reference temperature, K; $\Delta^{\ddagger}H^{\circ}_{T_0}$ and $\Delta^{\ddagger}S^{\circ}_{T_0}$ are enthalpic and entropic change between reactants and the transition state at the reference temperature T_0; $\Delta^{\ddagger}C^{\circ}_p$ is the heat capacity change between reactants and the transition state, J mol^{-1} K^{-1}.

3.19. The Model of DeLong (2017)

In 2017, DeLong et al. [98] argued that models describing the thermal growth curve lack the assumption that the catalyzing enzyme lowers the activation energy of the rate-determining reaction. They therefore introduced a model that describes the reduction of required activation energy for the rate-determining reaction as a function of free energy (enzyme stability) of the catalyzing enzyme. This term was incorporated into the dividend of the exponential term of the Arrhenius function. The Arrhenius activation energy (E_a; see Equation (4)) was replaced with the difference of a baseline energy (E_b, J), describing kinetic requirements as if the reaction would take place outside an organism, lowered by the enzymatic contribution (E_c, J) inside an organism, yielding Equation (48):

$$u(T) = A \cdot e^{\frac{-(E_b - E_c)}{k_B \cdot T}} \tag{48}$$

where the authors used the Boltzmann constant k_B ~$1.381 \cdot 10^{-23}$, J K^{-1}, instead of the universal gas constant in the divisor of the exponential term and T is the absolute temperature, K. The dividend of the exponential term then accounts for the activation energy lowered by the enzymatic action. The extent of enzymatic contribution (E_c) depends on the activity status of the enzyme, which is given by probability terms of protein stability. The temperature-dependent protein stability is given by ΔG (Equation (49)), where ΔH, J, is the change of enthalpy and ΔC_p, J mol^{-1} K^{-1}, the change in heat capacity, both relative to the melting temperature T_m, K, between the folded and unfolded state. At T_m, ΔH is by definition zero and increases for temperatures below T_m. ΔC_p reflects the extent of free energy that can be kept by the enzyme without changing the temperature, which increases below T_m with a decreasing temperature. The authors assumed that the probability of the maximum contribution of the enzyme to reduce the activation energy by the amount E_L approaches 1 at $\Delta G_{\max}$. Hence, the probability function Equation (50) is composed of the maximum amount (E_L) by which an active enzyme can lower the activation energy and the probability of the enzyme being correctly folded and active, given by the ratio of ΔG to $\Delta G_{\max}$. This transformation yields probability terms for each parameter in Equation (49), as presented in Equations (51) and (52). Therefore, Equation (49) can be rewritten as Equation (53).

$$\Delta G = \Delta H \cdot \left(1 - \frac{T}{T_m}\right) + \Delta C_p \cdot \left(T - T_m - T \cdot ln\left(\frac{T}{T_m}\right)\right) \tag{49}$$

$$E_c = E_L \cdot \frac{\Delta G}{\Delta G_{\max}} \tag{50}$$

$$E_{\Delta H} = E_L \cdot \frac{\Delta H}{\Delta G_{\max}} \tag{51}$$

$$E_{\Delta C_p} = E_L \cdot \frac{\Delta C_p}{\Delta G_{\max}} \tag{52}$$

$$E_c = E_{\Delta H} \cdot \left(1 - \frac{T}{T_m}\right) + E_{\Delta C_p} \cdot \left(T - T_m - T \cdot ln\left(\frac{T}{T_m}\right)\right) \tag{53}$$

Combining the probability function Equation (53) for the degree of enzymatic contribution to lowering the baseline energy E_b of the rate-determining reaction with the Arrhenius-type Equation (48) yields the enzyme-assisted Arrhenius model in Equation (54).

$$u(T) = A \cdot e^{\frac{-(E_b - (E_{\Delta H} \cdot (1 - \frac{T}{T_m}) + E_{\Delta C_p} \cdot (T - T_m - T \cdot ln(\frac{T}{T_m}))))}{k_B \cdot T}} \tag{54}$$

3.20. Additional Temperature Models

In Table 1 further models describing a hunchback-shaped curve for temperature-dependent rates are summarized.

Table 1. Models describing a hunchback-shaped curve of temperature-dependent rates (not explained in the text).

Model	Equation	Source
Lehman et al.	$u(T) = \begin{cases} e^{-2.3 \cdot [\frac{(T - T_{opt})}{(T_{max} - T_{opt})}]^2}, & T > T_{opt} \\ e^{-2.3 \cdot [\frac{(T - T_{opt})}{(T_{min} - T_{opt})}]^2}, & T \leq T_{opt} \end{cases}$	[99]
Moison et al.	$u(T) = \begin{cases} log(2) \cdot 0.851 \cdot (1.066^T) \cdot e^{-|T - T_{opt}| \cdot \frac{3}{T_{low}}}, & T \leq T_{opt} \\ log(2) \cdot 0.851 \cdot (1.066^T) \cdot e^{-|T - T_{opt}| \cdot \frac{3}{T_{high}}}, & T > T_{opt} \end{cases}$	[100]
Bitaube Pérez et al.	$u(T) = A_1 \cdot e^{(\frac{E_1}{R \cdot T} \cdot \frac{T - T_{ref}}{T_{ref}})} - A_2 \cdot e^{(\frac{E_2}{R \cdot T} \cdot \frac{T - T_{ref}}{T_{ref}})}$	[101]
Alexandrov et al.	$u(T) = \dfrac{2 \cdot e^{(\frac{E_a}{R \cdot T_{opt}} - \frac{E_a}{R \cdot T})}}{\left(1 + \left(e^{(\frac{E_a}{R \cdot T_{opt}} - \frac{E_a}{R \cdot T})}\right)^2\right)}$	[102, 103]
Tevatia et al.	$u(T) = u_{opt} \cdot \dfrac{e^{(\frac{-E_d}{R \cdot T})}}{1 + K \cdot e^{(\frac{-E_d}{R \cdot T})}}$	[104]

T is the absolute temperature, K; T_{opt} is the optimal growth temperature, K; T_{max} and T_{min} are the upper and lower limit of the thermal curve, respectively; T_{low} and T_{high} are shaping parameters that determine asymmetry of the growth curve, -; A_i represents frequency factors, s^{-1}; E_i represents activation energies, J mol^{-1}; R is the universal gas constant ~8.314, J mol^{-1} K^{-1}; T_{ref} is a reference temperature, K; μ_{opt} is the optimal specific growth rate, s^{-1}; E_d is the activation energy for enzyme denaturation, J mol^{-1}; K is a dimensionless inactivation constant.

4. Biotechnological Applications for Targeted Temperature Variation Assisted by Temperature Models

4.1. Temperature with Potential for Bioprocess Design

Temperature is an easily measurable (continuously, in situ, and in real-time) and controllable process variable. In most cases, temperature is controlled at constant values to maintain suitable physiological conditions at a given optimum value. Besides temperature, these requirements are also common for other basic physico-chemical characteristics, including dissolved oxygen and the pH value, which are routinely controlled in most bioreactor cultivations. For bioprocess design and optimization, very few correcting variables (e.g., the addition of fresh feed media) are available to direct a bioprocess towards a desired outcome. There are two strategies for efficiently using the temperature to influence the process outcome. First, alterations in the process temperature may be used to target metabolism, deliberately trigger stress responses, modify enzymatic turnover, or activate existing regulatory mechanisms. Secondly, natural or synthetic regulation mechanisms may be introduced to engineer host microorganisms (see Section 2.2). As such, processes can be designed to include these regulatory mechanisms as an effective tool to influence the process. Even though several systems inducible by temperature have been discovered and made available to biotechnologists in the last decades, only very few have been applied for process design or optimization [12].

4.2. Application of Temperature Models and Temperature for Bioprocesses Design

Mukhtar et al. modeled the temperature-dependent soil nitrification potential rate with the square root model of Ratkowsky et al. and estimated the optimum temperature for the nitrification potential rate with the heat capacity model. The authors suggest that the knowledge of thermodynamic properties of the soil nitrification response may be used to improve the application of large-scale fertilization, while reducing eutrophication and connected negative environmental impacts [105]. For cleaning purposes of contaminated water with excess nitrogen, denitrifying fixed-bed bioreactors can be used for NO_3^- removal [106]. The group of Nordström et al. used the Eyring equation to simulate the temperature dependency of NO_3^- removal rates in a denitrifying fixed-bad wood chip bioreactor [107]. They used the heat capacity model of Hobbs et al. [96] to derive the optimum temperature for NO_3^- removal by their bioreactor. Nordström et al. reported that NO_3^- removal rates of reducing microbial consortium change over time, while the temperature optimum is shifted towards lower temperatures (from 24.2 to 16.0 °C) [107]. Downshifts in temperature may somewhat be beneficial for a bioprocess, depending on the intended outcome. Seel et al. were able to increase biomass yields to optimize nutrient assimilation at suboptimal temperatures (10 °C) for mesophilic isolates from chilled foods and refrigerators in defined medium compared to their reference strains [52]. They emphasized the importance of defining the "optimum temperature". Seel et al. distinguish between the optimum temperature for the maximum growth rate and the optimum for the maximum biomass yield. At 10 °C, which was around 15–25 °C below the optimum for the maximum growth rate of isolates, they reported an increase of 20%–110% of biomass formation. They argue that the generally assumed optimum growth temperature at μ_{max} may not be the optimum for all biological processes in the host. This has been described for protein production, membrane permeability, and cellular stress [108–112]. The works of Corkrey et al. [58,113] thermodynamically justified the applied findings of Seel et al. Corkrey's model described the connection between the temperature stability of proteins and the growth rate governed by an assumed essential enzymatic reaction, with the temperature for optimal enzyme stability being 10–15 °C below the temperature of μ_{max}. Seel et al. state that due to correct protein folding and protein turnover, energy can be conserved and the biomass yield improved. Perhaps the most popular example for altering temperature as a means of an optimized process outcome can be found during the production of recombinant protein, such as by using *E. coli* as a heterologous expression system [12]. During the process, typically at the stage of the induction of protein expression, a temperature shift is performed, lowering the process temperature by several °C. This temperature shift does not benefit overall transcription or translation activity, but instead, results in an increased amount of correctly folded protein. This is due to lower amounts of recombinant protein being less likely to form inactive, insoluble aggregates (inclusion bodies), and allowing more time for correct folding after translation due to lower protein production rates. In the case of enzyme production, correctly folded protein is a prerequisite of enzymatic activity, and therefore, in most cases, temperature shifts are applied [114]. Another possibility for using temperature as a method for optimizing protein production is the application of thermo-inducible promoter systems. Considering the perspective of bioprocess engineering, it is of high interest to combine strain engineering and process development approaches to maximize overall productivity and yields. Strong chemically inducible promoters and expression systems that are commonly used in laboratory-scale protein expression [115] cause a high degree of metabolic burden to the microorganisms, and as such, protein production results in a simultaneous reduction of growth. It is therefore beneficial for optimized process design to be able to uncouple biomass growth from protein biosynthesis. Furthermore, production of recombinant proteins at early stages of cultivation often results in reduced overall yields, as many proteins are sensitive to degradation by proteases [114,116]. To counteract this instability of recombinant proteins, inducible promoter systems are a possible tool. Thermo-inducible promoter systems have the advantage of not requiring intrusion into the process, as chemical inductors are not required. As such, the risk of contamination, as well as the costs of the process, may be reduced. Nalley et al. evaluated the effect of temperature on growth, fatty acid production, and the fatty acid profile for algae suitable for mass cultivation and biofuel production [117]. They assessed the effect of temperature on microalgae with the model of Norberg [77]. Nalley et al. report temperature-specific fatty acid production, which is mostly controlled by the temperature-dependent growth rate. Furthermore, they found that temperature dramatically influences the fatty acid profile, with an increase in polyunsaturated fatty acids and decrease in monounsaturated and short fatty acids when increasing the temperature. The thermostable phosphotriesterase-like lactonases (PLLs) from hyperthermophilic *Sulfolobus* genera present an industrially relevant molecule for bioremediation processes, such as in the degradation of highly toxic pesticides like organophosphates [118]. Thermostable PLLs are of particular interest due to their wide temperature and pH working range, as well as resistance to organic solvents. In contrast to mesophilic enzyme isolates, the application of extremozymes is not prone to low stability in solution or an elevated temperature (>30 °C) [119,120]. In the work of Restaino et al., a high-yield pre-industrial-scale process with an optimized purification method for PLLs was developed [11]. The authors exploited the thermostability of PLLs in their downstream and recombinant PLL production in fast-growing mesophilic *E. coli* for their upstream and bioproduction strategy. Impurities were removed by a thermo-precipitation step (65–75 °C), which was optimized using a statistical response surface method to compute optimal precipitation temperatures. The solubility of proteins can be altered by different variables, like the pH, protein concentration, ionic strength, or temperature. Ethanol may be used as a solvent for precipitation, but exhibits the tendency to denature proteins at temperatures above 0 °C. Therefore, cold EtOH is often used for protein fractionation [121]. The authors Cimini et al. investigated the influence of temperature on the industrially relevant capsular polysaccharide (CPS) of *E. coli*, K4 [122]. It exhibits a high similarity to the economically valuable but only expensively extractable chondroitin from animal tissue. Chondroitin is, for example, used in the pharmaceutical sector to prevent osteoarthritis [123]. Cimini et al. found a positive correlation between CPS production and temperature. As stated before, CPS production is thermoregulated and *E. coli* CPS are not expressed at temperatures <20 °C [124]. Another pharmaceutically relevant product and precursor for the commonly used anti-inflammatory drug desfluorotriamcinolone, is 16α-hydroxy hydrocortisone. Hydrocortisone is converted in a temperature-dependent manor by *Streptomyces roseochromogenes* to 16α-hydroxy hydrocortisone. The group of Restaino et al. was able to maximize the bioconversion of hydrocortisone to 16α-hydroxy hydrocortisone, while lowering side-product formation. By adjusting the process temperature to 26 °C and pH to 6, they were able to almost entirely (95%) convert hydrocortisone into the desired product 16α-hydroxy hydrocortisone [125]. Another example for lowering the temperature to obtain optimal expression, correctly folded, and working recombinant enzymes is the mammalian enzyme 6-*O*-sulfotransferase (6-OST). It can be recombinantly produced in *E. coli*. 6-OST is of particular interest as it is required for the industrial and biotechnological production of heparin. So far, the blood anticoagulant heparin has only been derived from animals. 6-OST side-specifically sulfonates a heparin precursor and marks a key step in heparin bioproduction. The group of Restaino et al. reported high cell density cultivation of *E. coli* in which recombinant mammalian 6-OST was produced using an induction strategy optimized for yield and productivity. The strategy involved lowering the temperature (37 to 25 °C) upon induction and using a combination of two inducer molecules to balance the metabolic burden. The combination of balanced biomass growth and the induction strategy resulted in an optimal recombinant enzyme expression and enhanced biomass productivity [126].

An interesting application involving measuring and controlling the temperature during bioprocesses is the estimation of metabolic activity by heat balancing. In a partially isolated bioreactor system, heat generation by metabolic processes can be calculated by measuring heat transfer from or to the bioreactor [5]. This calorimetric technique typically involves the calculation of a heat balance by calculating transfer from or to the heat exchanger, enthalpy balancing in exhaust gas, energy dissipation by stirring, and monitoring the temperature of added liquids [6]. A calorimetric control strategy for the growth rate of *Escherichia coli* [13] and *Saccharomyces cerevisiae* [14] by adjusting feed rates was developed. The authors report the successful establishment and control of a high-cell density cultivation, with the feed rate solely relying on heat balancing. Furthermore, besides applications in process control, recently, a calorimetric approach for the detection of prophage activation and release was proposed. The authors report that by evaluating differences in metabolic heat, reactivation of dormant infected bacterial cells can be detected [127]. In general, however, the development of calorimetric control strategies at a laboratory scale is difficult, as sufficient isolation and sensitive equipment to detect heat generated at a small-scale is required [128]. Even though, at a larger scale, increasing ratios of volume to surface favor the sensitivity of calorimetric approaches, calorimetric approaches are only scarcely applied in industrial biotechnology. Table 2 provides an overview of biological traits, and temperature models and/or temperature adjustments used to achieve a desired process outcome.

Table 2. Biological traits associated with modeling techniques and/or targeted temperature adjustments for the control, monitoring, and optimization of biotechnological processes.

Purpose	Process	Biological Basis	Applied Model/Temperature Adjustment	Outcome	Source
Control	Rec. protein production	Lower protein production rates	Downshift in T	Correctly folded proteins	[114,116,126]
Control	Rec. protein production	Thermo-inducible promotor	Upshift in T	Induced promoter	[12]
Monitoring	Antibiotic biosynthesis	Metabolic heat $\approx$ metabolic state	Calorimetry	Estimation of metabolic activity	[5]
Monitoring and control	Insecticidal crystal proteins production	Metabolic heat $\approx$ metabolic state	Calorimetry	Estimation of metabolic state, control of nutrient feed	[6]
Control	Biomass production	Metabolic heat $\approx$ metabolic state	Calorimetry	Calorimetric control of nutrient feed	[13,14]
Monitoring	Evaluating prophage activating chemicals	Metabolic heat difference $\approx$ activity state of prophage	Calorimetry	Detection of prophage activation + release	[127]
Optimization	Denitrification of wastewaters	Reducing microbial consortia	Eyring model (Equation (7))	Derive (shifts of) T_{opt} for NO_3^- removal rate	[106,107]
Control	Biomass production	Arguably more stable RNA + correctly folded protein + lower degradation rates at low T	Downshift in T, ~Assumptions of Corkrey's model (see 3.17)	Increased biomass yield, improved nutrient assimilation	[52,129]
Control	Fatty acid production for biofuel	Shorter and more unsaturated fatty acids at low T ~modulate membrane fluidity	Norberg model (Equation (31))	Temperature-specific fatty acid production (profile)	[117]
Optimization	Downstream processing	Thermostable extremozyme	RSM/Thermo precipitation	Purified phosphotriesterase-like lactonases	[11]

5. Summary and Conclusions

In bioprocess engineering, very few process variables are usually available online. Therefore, exploiting existing control variables to their full extent is a reasonable strategy for broadening existing toolsets of monitoring and control. Both underlying biological mechanisms of temperature sensing and adaptation and mathematical models for temperature effects have been well-described. However, temperature as a control variable is only scarcely applied in bioprocess engineering, so an exploitation strategy merging both in context has not yet been established.

This review presents and discusses the most important models for physiological, biochemical, and physical properties governed by temperature, along with application perspectives. As such, this review provides a toolset for the future exploitation of temperature as a control variable for optimization, monitoring, and control applications in bioprocess engineering.

Author Contributions: Conceptualization, P.N. and M.H.; Formal Analysis, P.N., L.L., R.H. and M.H.; Investigation P.N., L.L., R.H. and M.H.; Writing-Original Draft Preparation, P.N.; Writing-Review & Editing, P.N., L.L., R.H. and M.H.; Supervision, M.H. All authors have read and agreed to the published version of the manuscript.

Funding: P.N. is a member of the "BBW ForWerts" graduate program and receives a scholarship within the frame of the Baden-Wuerttemberg Landesgraduiertenfoerderung (LGF) awarded by the Ministry of Science, Research and the Arts (MWK) of Baden-Wuerttemberg, Germany.

Conflicts of Interest: The authors declare no conflicts of interest.

References

1. Hausmann, R.; Henkel, M.; Hecker, F.; Hitzmann, B. *Present Status of Automation for Industrial Bioprocesses*; Elsevier, B.V., Ed.; Elsevier: Amsterdam, The Netherlands, 2016; ISBN 9780444636744.

2. Klinkert, B.; Narberhaus, F. Microbial thermosensors. *Cell. Mol. Life Sci.* **2009**, *66*, 2661–2676. [CrossRef] [PubMed]

3. Noll, P.; Treinen, C.; Müller, S.; Senkalla, S.; Lilge, L.; Hausmann, R.; Henkel, M. Evaluating temperature-induced regulation of a ROSE-like RNA-thermometer for heterologous rhamnolipid production in *Pseudomonas putida* KT2440. *AMB Express* **2019**, *9*, 154. [CrossRef] [PubMed]

4. Arrhenius, S. Über die Reaktionsgeschwindigkeit bei der Inversion von Rohrzucker durch Säuren. *Z. Phys. Chem.* **1889**, *4*, 226–248. [CrossRef]

5. Von Stockar, U.; Duboc, P.; Menoud, L.; Marison, I.W. On-line calorimetry as a technique for process monitoring and control in biotechnology. *Thermochim. Acta* **1997**, *300*, 225–236. [CrossRef]

6. Voisard, D.; Von Stockar, U.; Marison, I.W. Quantitative calorimetric investigation of fed-batch cultures of *Bacillus sphaericus* 1593M. *Thermochim. Acta* **2002**, *394*, 99–111. [CrossRef]

7. Schaepe, S.; Kuprijanov, A.; Aehle, M.; Simutis, R.; Lübbert, A. Simple control of fed-batch processes for recombinant protein production with *E. coli*. *Biotechnol. Lett.* **2011**, *33*, 1781–1788. [CrossRef]

8. Zwietering, M.H.; De Koos, J.T.; Hasenack, B.E.; De Witt, J.C.; Van't Riet, K. Modeling of bacterial growth as a function of temperature. *Appl. Environ. Microbiol.* **1991**, *57*, 1094–1101. [CrossRef]

9. De Oliveira Filho, P.B.; Nascimento, M.L.F.; Pontes, K.V. *Optimal Design of a Dividing Wall Column for the Separation of Aromatic Mixtures Using the Response Surface Method*; Elsevier Masson SAS: Paris, France, 2018; Volume 43, ISBN 9780444642356.

10. MathWorks, Inc. Matlab-Documentation. 2019. Available online: https://de.mathworks.com/help/stats/response-surface-designs.html (accessed on 5 January 2020).

11. Restaino, O.F.; Borzacchiello, M.G.; Scognamiglio, I.; Fedele, L.; Alfano, A.; Porzio, E.; Manco, G.; De Rosa, M.; Schiraldi, C. High yield production and purification of two recombinant thermostable phosphotriesterase-like lactonases from *Sulfolobus acidocaldarius* and *Sulfolobus solfataricus* useful as bioremediation tools and bioscavengers. *BMC Biotechnol.* **2018**, *18*, 18. [CrossRef]

12. Aucoin, M.G.; McMurray-Beaulieu, V.; Poulin, F.; Boivin, E.B.; Chen, J.; Ardelean, F.M.; Cloutier, M.; Choi, Y.J.; Miguez, C.B.; Jolicoeur, M. Identifying conditions for inducible protein production in *E. coli*: Combining a fed-batch and multiple induction approach. *Microb. Cell Fact.* **2006**, *5*, 27. [CrossRef]

13. Biener, R.; Steinkämper, A.; Hofmann, J. Calorimetric control for high cell density cultivation of a recombinant *Escherichia coli* strain. *J. Biotechnol.* **2010**, *146*, 45–53. [CrossRef]

14. Biener, R.; Steinkämper, A.; Horn, T. Calorimetric control of the specific growth rate during fed-batch cultures of *Saccharomyces cerevisiae*. *J. Biotechnol.* **2012**, *160*, 195–201. [CrossRef] [PubMed]

15. Jensen, W.B.; Kuhlmann, J.; Brush, S.G.; Tyndall, J. Leopold Pfaundler and the origins of the kinetic theory of chemical reactions. *Bull. Hist. Chem.* **2012**, *37*, 29–41.

16. Bernoulli, D. *Hydrodynamica, Sive de Viribus et Motibus Fluidorum Commentarii*; Johannis Reinholdi Dulseckeri: Strasbourg, France, 1738.

17. Mikhailov, G.K. Daniel Bernoulli, Hydrodynamica (1738). In *Landmark Writings in Western Mathematics 1640–1940*; Grattan-Guinness, I., Corry, L., Guicciardini, N., Cooke, R., Crépel, P., Eds.; Elsevier: Amsterdam, The Netherlands, 2005; pp. 131–142. ISBN 9780444508713.

18. Clapeyron, P.E. Puissance motrice de la chaleur. *J. l'École Polytech.* **1834**, *14*, 153–190.

19. Jensen, W.B. The universal gas constant R. *J. Chem. Educ.* **2003**, *80*, 731–732. [CrossRef]

20. Eyring, H. The Activated Complex in Chemical Reactions. *J. Chem. Phys.* **1935**, *3*, 107–115. [CrossRef]

21. Laidler, K.J. A glossary of terms used in chemical kinetics, including reaction dynamics (IUPAC recommendations 1996). *Pure Appl. Chem.* **1996**, *68*, 149–192. [CrossRef]
22. Chalk, S.J. Enthalpy of activation, $\Delta\ddagger H°$. In *IUPAC Compendium of Chemical Terminology*; Nič, M., Jirát, J., Košata, B., Jenkins, A., McNaught, A., Eds.; IUPAC: Research Triangle Park, NC, USA, 2009; ISBN 0-9678550-9-8.
23. Johnson, F.H.; Eyring, H.; Stover, B.J. *The Theory of Rate Processes in Biology and Medicine*; John Wiley & Sons: New York, NY, USA, 1974; ISBN 9780471444855.
24. Mohr, P.W.; Krawiec, S. Temperature characteristics and Arrhenius plots for nominal psychrophiles, mesophiles and thermophiles. *J. Gen. Microbiol.* **1980**, *121*, 311–317. [CrossRef]
25. Smith, T.P.; Thomas, T.J.H.; Garcia-Carreras, B.; Sal, S.; Yvon-Durocher, G.; Bell, T.; Pawar, S. Metabolic rates of prokaryotic microbes may inevitably rise with global warming. *BioRxiv* **2019**, 524264. [CrossRef]
26. Ratkowsky, D.A.; Olley, J.; McMeekin, T.A.; Ball, A. Relationship between temperature and growth rate of bacterial cultures. *J. Bacteriol.* **1982**, *149*, 1–5. [CrossRef]
27. Shapiro, R.S.; Cowen, L.E. Thermal Control of Microbial Development and Virulence: Molecular Mechanisms of Microbial Temperature Sensing. *MBio* **2012**, *3*, e00238-12. [CrossRef]
28. Sengupta, P.; Garrity, P. Sensing temperature. *Curr. Biol.* **2013**, *23*, R304–R307. [CrossRef]
29. Pruss, G.J.; Drlica, K. DNA supercoiling and prokaryotic transcription. *Cell* **1989**, *56*, 521–523. [CrossRef]
30. López-García, P.; Forterre, P. DNA topology and the thermal stress response, a tale from mesophiles and hyperthermophiles. *BioEssays* **2000**, *22*, 738–746. [CrossRef]
31. Forterre, P.; Bergerat, A.; Lopez-Garcia, P. The unique DNA topology and DNA topoisomerases of hyperthermophilic archaea. *FEMS Microbiol. Rev.* **1996**, *18*, 237–248. [CrossRef]
32. López-García, P.; Forterre, P. DNA topology in hyperthermophilic archaea: Reference states and their variation with growth phase, growth temperature, and temperature stresses. *Mol. Microbiol.* **1997**, *23*, 1267–1279. [CrossRef] [PubMed]
33. Ono, S.; Goldberg, M.D.; Olsson, T.; Esposito, D.; Hinton, J.C.D.; Ladbury, J.E. H-NS is a part of a thermally controlled mechanism for bacterial gene regulation. *Biochem. J.* **2005**, *391*, 203–213. [CrossRef]
34. White-Ziegler, C.A.; Davis, T.R. Genome-wide identification of H-NS-controlled, temperature-regulated genes in *Escherichia coli* K-12. *J. Bacteriol.* **2009**, *191*, 1106–1110. [CrossRef] [PubMed]
35. Schröder, O.; Wagner, R. The bacterial DNA-binding protein H-NS represses ribosomal RNA transcription by trapping RNA polymerase in the initiation complex. *J. Mol. Biol.* **2000**, *298*, 737–748. [CrossRef] [PubMed]
36. Shin, M.; Song, M.; Joon, H.R.; Hong, Y.; Kim, Y.J.; Seok, Y.J.; Ha, K.S.; Jung, S.H.; Choy, H.E. DNA looping-mediated repression by histone-like protein H-NS: Specific requirement of Eσ70 as a cofactor for looping. *Genes Dev.* **2005**, *19*, 2388–2398. [CrossRef] [PubMed]
37. Lim, C.J.; Lee, S.Y.; Kenney, L.J.; Yan, J. Nucleoprotein filament formation is the structural basis for bacterial protein H-NS gene silencing. *Sci. Rep.* **2012**, *2*, 509. [CrossRef]
38. Kotlajich, M.V.; Hron, D.R.; Boudreau, B.A.; Sun, Z.; Lyubchenko, Y.L.; Landick, R. Bridged filaments of histone-like nucleoid structuring protein pause RNA polymerase and aid termination in bacteria. *Elife* **2015**, *4*, e04970. [CrossRef] [PubMed]
39. Chowdhury, S.; Maris, C.; Allain, F.H.T.; Narberhaus, F. Molecular basis for temperature sensing by an RNA thermometer. *EMBO J.* **2006**, *25*, 2487–2497. [CrossRef] [PubMed]
40. Wei, Y.; Murphy, E.R. Temperature-Dependent Regulation of Bacterial Gene Expression by RNA Thermometers. In *Nucleic Acids: From Basic Aspects to Laboratory Tools*; IntechOpen: London, UK, 2016; pp. 157–181.
41. Grosso-Becerra, M.V.; Croda-Garcia, G.; Merino, E.; Servin-Gonzalez, L.; Mojica-Espinosa, R.; Soberon-Chavez, G. Regulation of *Pseudomonas aeruginosa* virulence factors by two novel RNA thermometers. *Proc. Natl. Acad. Sci. USA* **2014**, *111*, 15562–15567. [CrossRef] [PubMed]
42. Waldminghaus, T.; Fippinger, A.; Alfsmann, J.; Narberhaus, F. RNA thermometers are common in α- and γ-proteobacteria. *Biol. Chem.* **2005**, *386*, 1279–1286. [CrossRef] [PubMed]
43. Narberhaus, F.; Waldminghaus, T.; Chowdhury, S. RNA thermometers. *FEMS Microbiol. Rev.* **2006**, *30*, 3–16. [CrossRef] [PubMed]
44. Waldminghaus, T.; Heidrich, N.; Brantl, S.; Narberhaus, F. FourU: A novel type of RNA thermometer in *Salmonella*. *Mol. Microbiol.* **2007**, *65*, 413–424. [CrossRef] [PubMed]
45. Sen, S.; Apurva, D.; Satija, R.; Siegal, D.; Murray, R.M. Design of a Toolbox of RNA Thermometers. *ACS Synth. Biol.* **2017**, *6*, 1461–1470. [CrossRef]

46. Elsholz, A.K.W.; Michalik, S.; Zühlke, D.; Hecker, M.; Gerth, U. CtsR, the Gram-positive master regulator of protein quality control, feels the heat. *EMBO J.* **2010**, *29*, 3621–3629. [CrossRef]

47. Krell, T.; Lacal, J.; Busch, A.; Silva-Jiménez, H.; Guazzaroni, M.-E.; Ramos, J.L. Bacterial Sensor Kinases: Diversity in the Recognition of Environmental Signals. *Annu. Rev. Microbiol.* **2010**, *64*, 539–559. [CrossRef]

48. Nishiyama, S.I.; Umemura, T.; Nara, T.; Homma, M.; Kawagishi, I. Conversion of a bacterial warm sensor to a cold sensor by methylation of a single residue in the presence of an attractant. *Mol. Microbiol.* **1999**, *32*, 357–365. [CrossRef]

49. Cedervall, T.; Johansson, M.U.; Åkerström, B. Coiled-coil structure of group A streptococcal M proteins. Different temperature stability of class A and C proteins by hydrophobic-nonhydrophobic amino acid substitutions at heptad positions a and d. *Biochemistry* **1997**, *36*, 4987–4994. [CrossRef] [PubMed]

50. Franzmann, T.M.; Menhorn, P.; Walter, S.; Buchner, J. Activation of the Chaperone Hsp26 Is Controlled by the Rearrangement of Its Thermosensor Domain. *Mol. Cell* **2008**, *29*, 207–216. [CrossRef] [PubMed]

51. Haslbeck, M.; Walke, S.; Stromer, T.; Ehrnsperger, M.; White, H.E.; Chen, S.; Saibil, H.R.; Buchner, J. Hsp26: A temperature-regulated chaperone. *EMBO J.* **1999**, *18*, 6744–6751. [CrossRef] [PubMed]

52. Seel, W.; Derichs, J.; Lipski, A. Increased biomass production by mesophilic food-associated bacteria through lowering the growth temperature from 30 °C to 10 °C. *Appl. Environ. Microbiol.* **2016**, *82*, 3754–3764. [CrossRef] [PubMed]

53. Huey, R.B.; Stevenson, R.D. Integrating Thermal Physiology and Ecology of Ectotherms: A Discussion of Approaches Department. *Am. Zool.* **1979**, *19*, 357–366. [CrossRef]

54. Bennett, A.F.; Lenski, R.E. Evolutionary Adaptation to Temperature II. Thermal Niches of Experimental Lines of *Escherichia coli*. *Evolution* **1993**, *47*, 1–12. [CrossRef] [PubMed]

55. Travisano, M.; Lenski, R.E. Long-term experimental evolution in *Escherichia coli*. IV. Targets of selection and the specificity of adaptation. *Genetics* **1996**, *143*, 15–26.

56. Cullum, A.J.; Bennett, A.F.; Lenski, R.E. Evolutionary adaptation to temperature. IX. Preadaptation to novel stressful environments of *Escherichia coli* adapted to high temperature. *Evolution* **2001**, *55*, 2194–2202. [CrossRef]

57. Takai, K.; Nakamura, K.; Toki, T.; Tsunogai, U.; Miyazaki, M.; Miyazaki, J.; Hirayama, H.; Nakagawa, S.; Nunoura, T.; Horikoshi, K. Cell proliferation at 122 °C and isotopically heavy CH_4 production by a hyperthermophilic methanogen under high-pressure cultivation. *Proc. Natl. Acad. Sci. USA* **2008**, *105*, 10949–10954. [CrossRef]

58. Corkrey, R.; McMeekin, T.A.; Bowman, J.P.; Ratkowsky, D.A.; Olley, J.; Ross, T. Protein thermodynamics can be predicted directly from biological growth rates. *PLoS ONE* **2014**, *9*, e96100. [CrossRef]

59. Ratkowsky, D.A.; Lowry, R.K.; McMeekin, T.A.; Stokes, A.N.; Chandler, R.E. Model for bacterial culture growth rate throughout the entire biokinetic temperature range. *J. Bacteriol.* **1983**, *154*, 1222–1226. [CrossRef] [PubMed]

60. Johnson, F.H.; Lewin, I. The growth rate of *E. coli* in relation to temperature, quinine and coenzyme. *J. Cell. Comp. Physiol.* **1946**, *28*, 47–75. [CrossRef] [PubMed]

61. Grimaud, G.M.; Mairet, F.; Sciandra, A.; Bernard, O. Modeling the temperature effect on the specific growth rate of phytoplankton: A review. *Rev. Environ. Sci. Biotechnol.* **2017**, *16*, 625–645. [CrossRef]

62. Hinshelwood, C.N. Influence of temperature on the growth of bacteria. In *The Chemical Kinetics of the Bacterial Cell*; Clarendon Press: Oxford, UK, 1946; pp. 254–257.

63. Sharpe, P.J.H.; Curry, G.L.; DeMichele, D.W.; Cole, C.L. Distribution model of organism development times. *J. Theor. Biol.* **1977**, *66*, 21–38. [CrossRef]

64. Sharpe, P.J.H.; DeMichele, D.W. Reaction kinetics of poikilotherm development. *J. Theor. Biol.* **1977**, *64*, 649–670. [CrossRef]

65. Eyring, H.; Stearn, A.E. The application of the theory of absolute reaction rates to proteins. *Chem. Rev.* **1939**, *24*, 253–270. [CrossRef]

66. Hultin, E. The influence of temperature on the rate of enzymic processes. *Acta Chem. Scand.* **1955**, *9*, 1700–1710. [CrossRef]

67. Lamanna, C.; Mallette, M.F.; Zimmermann, L.N. *Basic Bacteriology*, 4th ed.; Williams & Wilkins: Baltimore, MD, USA, 1973.

68. Schoolfield, R.M.; Sharpe, P.J.H.; Magnuson, C.E. Non-linear regression of biological temperature-dependent rate models based on absolute reaction-rate theory. *J. Theor. Biol.* **1981**, *88*, 719–731. [CrossRef]

69. Roels, J. *Energetics and Kinetics in Biotechnology*; Elsevier Biomedical Press: Amsterdam, The Netherlands, 1983; ISBN 0444804420.

70. Davey, K.R. A predictive model for combined temperature and water activity on microbial growth during the growth phase. *J. Appl. Bacteriol.* **1989**, *67*, 483–488. [CrossRef]

71. Davey, K.R. Modelling the combined effect of temperature and pH on the rate coefficient for bacterial growth. *Int. J. Food Microbiol.* **1994**, *23*, 295–303. [CrossRef]

72. Davey, K.R. Applicability of the Davey (linear Arrhenius) predictive model to the lag phase of microbial growth. *J. Appl. Bacteriol.* **1991**, *70*, 253–257. [CrossRef]

73. Lobry, J.R.; Rosso, L.; Flandrois, J.-P. A FORTRAN Subroutine for the Determination of Parameter Confidence Limits in non-linear models. *Binary* **1991**, *3*, 86–93.

74. Rosso, L.; Lobry, J.R.; Flandrois, J.P. An Unexpected Correlation between Cardinal Temperatures of Microbial Growth Highlighted by a New Model. *J. Theor. Biol.* **1993**, *162*, 447–463. [CrossRef] [PubMed]

75. Blanchard, G.F.; Guarini, J.M.; Richard, P.; Gros, P.; Mornet, F. Quantifying the short-term temperature effect on light- saturated photosynthesis of intertidal microphytobenthos. *Mar. Ecol. Prog. Ser.* **1996**, *134*, 309–313. [CrossRef]

76. Eppley, R.W. Temperature and phytoplankton growth in the sea. *Fish Bull.* **1972**, *70*, 1063–1085.

77. Norberg, J. Biodiversity and ecosystem functioning: A complex adaptive systems approach. *Limnol. Oceanogr.* **2004**, *49*, 1269–1277. [CrossRef]

78. Brandts, J.F. Heat effects on proteins and enzymes. In *Thermobiology*; Academic Press: London, UK, 1967; pp. 25–72.

79. Ratkowsky, D.A.; Olley, J.; Ross, T. Unifying temperature effects on the growth rate of bacteria and the stability of globular proteins. *J. Theor. Biol.* **2005**, *233*, 351–362. [CrossRef]

80. Privalov, P.L.; Khechinashvili, N.N. A thermodynamic approach to the problem of stabilization of globular protein structure: A calorimetric study. *J. Mol. Biol.* **1974**, *86*, 665–684. [CrossRef]

81. Murphy, K.P.; Privalov, P.L.; Gill, S.J. Common features of protein unfolding and dissolution of hydrophobic compounds. *Science* **1990**, *247*, 559–561. [CrossRef]

82. Murphy, K.P.; Gill, S.J. Solid model compounds and the thermodynamics of protein unfolding. *J. Mol. Biol.* **1991**, *222*, 699–709. [CrossRef]

83. Robertson, A.D.; Murphy, K.P. Protein structure and the energetics of protein stability. *Chem. Rev.* **1997**, *97*, 1251–1267. [CrossRef]

84. Baldwin, R.L. Temperature dependence of the hydrophobic interaction in protein folding. *Proc. Natl. Acad. Sci. USA* **1986**, *83*, 8069–8072. [CrossRef]

85. Ross, T. Assessment of a theoretical model for the effects of temperature on bacterial growth rate. In Proceedings of the Refrigeration Science and Technology Proceedings, Quimper, France, 16–18 June 1997; pp. 64–71.

86. Ross, T. A Philosophy for the Development of Kinetic Models in Predictive Microbiology. Ph.D. Thesis, University of Tasmania, Hobart, Australia, 1993.

87. Zeldovich, K.B.; Chen, P.; Shakhnovich, E.I. Protein stability imposes limits on organism complexity and speed of molecular evolution. *Proc. Natl. Acad. Sci. USA* **2007**, *104*, 16152–16157. [CrossRef] [PubMed]

88. Ghosh, K.; De Graff, A.M.R.; Sawle, L.; Dill, K.A. Role of Proteome Physical Chemistry in Cell Behavior. *J. Phys. Chem. B* **2016**, *120*, 9549–9563. [CrossRef] [PubMed]

89. Sawle, L.; Ghosh, K. How do thermophilic proteins and proteomes withstand high temperature? *Biophys. J.* **2011**, *101*, 217–227. [CrossRef] [PubMed]

90. Ghosh, K.; Dill, K.A. Computing protein stabilities from their chain lengths. *Proc. Natl. Acad. Sci. USA* **2009**, *106*, 10649–10654. [CrossRef] [PubMed]

91. Daniel, R.M.; Danson, M.J. A new understanding of how temperature affects the catalytic activity of enzymes. *Trends Biochem. Sci.* **2010**, *35*, 584–591. [CrossRef]

92. Kooijman, S.A.L.M. Energy budgets can explain body size relations. *J. Theor. Biol.* **1986**, *121*, 269–282. [CrossRef]

93. Kooijman, S.A.L.M. Dynamic Energy Budget theory for metabolic organisation: Summary of concepts of the third edition. *Water* **2010**, *365*, 68.

94. Kooijman, S.A.L.M. *Dynamic Energy Budgets in Biological Systems: Theory and Applications in Ecotoxicology*; Cambridge University Press: Cambridge, UK, 1993; ISBN 0-521-45223-6.

95. Huang, L.; Hwang, A.; Phillips, J. Effect of Temperature on Microbial Growth Rate-Mathematical Analysis: The Arrhenius and Eyring-Polanyi Connections. *J. Food Sci.* **2011**, *76*, 553–560. [CrossRef]

96. Hobbs, J.K.; Jiao, W.; Easter, A.D.; Parker, E.J.; Schipper, L.A.; Arcus, V.L. Change in heat capacity for enzyme catalysis determines temperature dependence of enzyme catalyzed rates. *ACS Chem. Biol.* **2013**, *8*, 2388–2393. [CrossRef] [PubMed]

97. Schipper, L.A.; Hobbs, J.K.; Rutledge, S.; Arcus, V.L. Thermodynamic theory explains the temperature optima of soil microbial processes and high Q10 values at low temperatures. *Glob. Chang. Biol.* **2014**, *20*, 3578–3586. [CrossRef] [PubMed]

98. DeLong, J.P.; Gibert, J.P.; Luhring, T.M.; Bachman, G.; Reed, B.; Neyer, A.; Montooth, K.L. The combined effects of reactant kinetics and enzyme stability explain the temperature dependence of metabolic rates. *Ecol. Evol.* **2017**, *7*, 3940–3950. [CrossRef] [PubMed]

99. Lehman, J.T.; Botkin, D.B.; Likens, G.E. The assumptions and rationales of a computer model of phytoplankton population dynamics. *Limnol. Oceanogr.* **1975**, *20*, 343–364. [CrossRef]

100. Moisan, J.R.; Moisan, T.A.; Abbott, M.R. Modelling the effect of temperature on the maximum growth rates of phytoplankton populations. *Ecol. Model.* **2002**, *153*, 197–215. [CrossRef]

101. Bitaubé Pérez, E.; Caro Pina, I.; Pérez Rodríguez, L. Kinetic model for growth of *Phaeodactylum tricornutum* in intensive culture photobioreactor. *Biochem. Eng. J.* **2008**, *40*, 520–525. [CrossRef]

102. Alexandrov, G.A.; Yamagata, Y. A peaked function for modeling temperature dependence of plant productivity. *Ecol. Model.* **2007**, *200*, 189–192. [CrossRef]

103. Quinn, J.; de Winter, L.; Bradley, T. Microalgae bulk growth model with application to industrial scale systems. *Bioresour. Technol.* **2011**, *102*, 5083–5092. [CrossRef]

104. Tevatia, R.; Demirel, Y.; Rudrappa, D.; Blum, P. Effects of thermodynamically coupled reaction diffusion in microalgae growth and lipid accumulation: Model development and stability analysis. *Comput. Chem. Eng.* **2015**, *75*, 28–39. [CrossRef]

105. Mukhtar, H.; Lin, Y.-P.; Lin, C.-M.; Petway, J.R. Assessing thermodynamic parameter sensitivity for simulating temperature responses of soil nitrification. *Environ. Sci. Process. Impacts* **2019**, *21*, 1596–1608. [CrossRef]

106. Schipper, L.A.; Robertson, W.D.; Gold, A.J.; Jaynes, D.B.; Cameron, S.C. Denitrifying bioreactors-An approach for reducing nitrate loads to receiving waters. *Ecol. Eng.* **2010**, *36*, 1532–1543. [CrossRef]

107. Nordström, A.; Herbert, R.B. Identification of the temporal control on nitrate removal rate variability in a denitrifying woodchip bioreactor. *Ecol. Eng.* **2019**, *127*, 88–95. [CrossRef]

108. Feller, G.; Gerday, C. Psychrophilic enzymes: Hot topics in cold adaptation. *Nat. Rev. Microbiol.* **2003**, *1*, 200–208. [CrossRef] [PubMed]

109. Feller, G. Life at low temperatures: Is disorder the driving force? *Extremophiles* **2007**, *11*, 211–216. [CrossRef] [PubMed]

110. Jaouen, T.; Dé, E.; Chevalier, S.; Orange, N. Pore size dependence on growth temperature is a common characteristic of the major outer membrane protein OprF in psychrotrophic and mesophilic *Pseudomonas* species. *Appl. Environ. Microbiol.* **2004**, *70*, 6665–6669. [CrossRef] [PubMed]

111. D'Amico, S.; Collins, T.; Marx, J.C.; Feller, G.; Gerday, C. Psychrophilic microorganisms: Challenges for life. *EMBO Rep.* **2006**, *7*, 385–389. [CrossRef]

112. Margesin, R.; Fonteyne, P.A.; Redl, B. Low-temperature biodegradation of high amounts of phenol by *Rhodococcus* spp. and basidiomycetous yeasts. *Res. Microbiol.* **2005**, *156*, 68–75. [CrossRef]

113. Corkrey, R.; Olley, J.; Ratkowsky, D.; McMeekin, T.; Ross, T. Universality of thermodynamic constants governing biological growth rates. *PLoS ONE* **2012**, *7*, e32003. [CrossRef]

114. Hannig, G.; Makrides, S.C. Strategies for optimizing heterologous protein expression in *Escherichia coli*. *Stud. Health Technol. Inform.* **1998**, *16*, 54–60. [CrossRef]

115. Gombert, A.K.; Kilikian, B.V. Recombinant gene expression in *Escherichia coli* cultivation using lactose as inducer. *J. Biotechnol.* **1998**, *60*, 47–54. [CrossRef]

116. Schmidt, M.; Babu, K.R.; Khanna, N.; Marten, S.; Rinas, U. Temperature-induced production of recombinant human insulin in high-cell density cultures of recombinant *Escherichia coli*. *J. Biotechnol.* **1999**, *68*, 71–83. [CrossRef]

117. Nalley, J.O.; O'Donnell, D.R.; Litchman, E. Temperature effects on growth rates and fatty acid content in freshwater algae and cyanobacteria. *Algal Res.* **2018**, *35*, 500–507. [CrossRef]

118. Jacquet, P.; Daudé, D.; Bzdrenga, J.; Masson, P.; Elias, M.; Chabrière, E. Current and emerging strategies for organophosphate decontamination: Special focus on hyperstable enzymes. *Environ. Sci. Pollut. Res.* **2016**, *23*, 8200–8218. [CrossRef] [PubMed]

119. Singh, B.K. Organophosphorus-degrading bacteria: Ecology and industrial applications. *Nat. Rev. Microbiol.* **2009**, *7*, 156–164. [CrossRef] [PubMed]

120. Horne, I.; Qiu, X.; Russell, R.J.; Oakeshott, J.G. The phosphotriesterase gene opdA in *Agrobacterium radiobacter* P230 is transposable. *FEMS Microbiol. Lett.* **2003**, *222*, 1–8. [CrossRef]

121. Zellner, M.; Winkler, W.; Hayden, H.; Diestinger, M.; Eliasen, M.; Gesslbauer, B.; Miller, I.; Chang, M.; Kungl, A.; Roth, E.; et al. Quantitative validation of different protein precipitation methods in proteome analysis of blood platelets. *Electrophoresis* **2005**, *26*, 2481–2489. [CrossRef]

122. Cimini, D.; Restaino, O.F.; Catapano, A.; De Rosa, M.; Schiraldi, C. Production of capsular polysaccharide from *Escherichia coli* K4 for biotechnological applications. *Appl. Microbiol. Biotechnol.* **2010**, *85*, 1779–1787. [CrossRef]

123. McAlindon, T.E.; LaValley, M.P.; Gulin, J.P.; Felson, D.T. Glucosamine and chondroitin for treatment of osteoarthritis: A systematic quality assessment and meta-analysis. *JAMA* **2000**, *283*, 1469–1475. [CrossRef]

124. Whitfield, C.; Roberts, I.S. Structure, assembly and regulation of expression of capsules in *Escherichia coli*. *Mol. Microbiol.* **1999**, *31*, 1307–1319. [CrossRef]

125. Restaino, O.F.; Marseglia, M.; Diana, P.; Borzacchiello, M.G.; Finamore, R.; Vitiello, M.; D'Agostino, A.; De Rosa, M.; Schiraldi, C. Advances in the 16α-hydroxy transformation of hydrocortisone by *Streptomyces roseochromogenes*. *Process Biochem.* **2016**, *51*, 1–8. [CrossRef]

126. Restaino, O.F.; Bhaskar, U.; Paul, P.; Li, L.; De Rosa, M.; Dordick, J.S.; Linhardt, R.J. High cell density cultivation of a recombinant *E. coli* strain expressing a key enzyme in bioengineered heparin production. *Appl. Microbiol. Biotechnol.* **2013**, *97*, 3893–3900. [CrossRef]

127. Xu, J.; Jiang, F.L.; Liu, Y.; Kiesel, B.; Maskow, T. An enhanced bioindicator for calorimetric monitoring of prophage-activating chemicals in the trace concentration range. *Eng. Life Sci.* **2018**, *18*, 475–483. [CrossRef]

128. Schubert, T.; Breuer, U.; Harms, H.; Maskow, T. Calorimetric bioprocess monitoring by small modifications to a standard bench-scale bioreactor. *J. Biotechnol.* **2007**, *130*, 24–31. [CrossRef] [PubMed]

129. Bakermans, C.; Nealson, K.H. Relationship of Critical Temperature to Macromolecular Synthesis and Growth Yield in *Psychrobacter cryopegellu*. *J. Bacteriol.* **2004**, *186*, 2340–2345. [CrossRef] [PubMed]

Review

Practical Solutions for Specific Growth Rate Control Systems in Industrial Bioreactors

Vytautas Galvanauskas *, Rimvydas Simutis, Donatas Levišauskas and Renaldas Urniežius

Department of Automation, Kaunas University of Technology, Kaunas 51367, Lithuania;
rimvydas.simutis@ktu.lt (R.S.); donatas.levisauskas@ktu.lt (D.L.); renaldas.urniezius@ktu.lt (R.U.)
* Correspondence: vytautas.galvanauskas@ktu.lt; Tel.: +370-37-300-291

Received: 27 August 2019; Accepted: 30 September 2019; Published: 2 October 2019

Abstract: This contribution discusses the main challenges related to successful application of automatic control systems used to control specific growth rate in industrial biotechnological processes. It is emphasized that, after the implementation of basic automatic control systems, primary attention shall be paid to the specific growth rate control systems because this process variable critically affects the physiological state of microbial cultures and the formation of the desired product. Therefore, control of the specific growth rate enables improvement of the quality and reproducibility of the biotechnological processes. The main requirements have been formulated that shall be met to successfully implement the specific growth rate control systems in industrial bioreactors. The relatively easy-to-implement schemes of specific growth rate control systems have been reviewed and discussed. The recommendations for selection of particular control systems for specific biotechnological processes have been provided.

Keywords: biotechnological processes; bioreactor control; specific growth rate control; batch-to-batch reproducibility

---

1. Introduction

Biotechnological processes play an increasingly important role in modern industry and health sectors. Many of the important active pharmaceutical ingredients are recombinant therapeutic proteins produced by the cultivation of genetically modified microorganisms or mammalian cells in bioreactors. These biotechnological processes are highly nonlinear and non-stationary. Therefore, modeling and control of the above bioprocesses are complicated control engineering tasks, especially in industrial recombinant protein production processes, in which high safety requirements and operational restrictions must be secured [1,2]. The goal of this contribution is to review and recommend practical and easily implementable control system schemes for biomass specific growth rate (further referred to as SGR or μ) control in industrial bioreactors. The recommendations are based on an analysis of the existing SGR control solutions and availability of the control schemes suitable for practical implementation in industrial bioreactors. The specific growth rate μ (1/h), is defined as the ratio of the cell's absolute growth rate and the amount of cells:

$$\mu = \frac{dX}{dt}\frac{1}{X} \tag{1}$$

where $X = xV$ (g) is the cell (biomass) amount; x (g/L) is the cell (biomass) concentration; and V (L) is the cultivation broth volume. The SGR is the most important variable in biotechnological processes, which influences the physiological state of microbial culture, production of cell biomass and desired products, and quantity and quality of products [3–8].

The development of relatively simple and reliable methods for SGR monitoring and control in industrial bioreactors is one of the most important control engineering tasks for successful implementation of the Process Analytical Technology (PAT) framework in bioengineering [9,10]. However, to properly exploit the benefits of SGR control systems in microbial and mammalian cell cultivation processes, basic bioprocess variables (temperature, pH, dissolved oxygen concentration, etc.) need to be controlled by commonly available and well-functioning control systems. Unfortunately, in many cases these systems do not ensure sufficient control quality [1,11,12] allowing to further proceed with SGR monitoring and control.

This paper is structured as follows: In Section 2, importance of the control quality of the basic control systems in the biotechnological processes is analyzed. Section 3 introduces important preconditions for implementation of SGR control systems in industrial bioreactors. Section 4 expands on strategies for SGR control suitable for industrial bioreactors. Finally, the authors give recommendations for application of the discussed SGR control solutions in various biotechnological processes.

2. Quality of Basic Control Systems in Industrial Bioreactors

The performance quality of automatic control systems for basic process variables is still low in most industrial microbial and mammalian cell cultivation processes [1,11]. Despite the fact that sophisticated control strategies for microbial cultivation processes are widely discussed in the academic community and research papers [2,13–15], the authors' experience shows that, at present, only simple, conventional automatic control systems are realized in the majority of industrial-scale (bio)reactors [11,16]. This situation is related to a common underestimation of the control systems' importance in improving the productivity and quality of the biotechnological processes. It is also related to the relatively high costs of implementation and maintenance of these advanced control systems and the resultant low acceptance of these systems by plant managers.

Bioreactors are the key operation units in biochemical and biopharmaceutical processes, in which the basic control systems attempt to control the cultivation environment outside of the cell. The commonly controlled variables of the cells' environment are temperature, pressure, pH, and dissolved oxygen concentration. The basic feedback control systems of industrial bioreactors for controlling bacterial cell cultures that produce biopharmaceutical products are presented in Figure 1. The temperature controller manipulates the flow rate of cooling water in the jacket. The pressure inside the bioreactor is controlled by manipulation of the off-gas flow rate. The pH controller manipulates the flow rate of ammonia solution (usually, the acid solution does not need to be added to bacterial cell culture cultivations, unless compensation of base excess is required). The dissolved oxygen controller output is split to manipulate the air flow rate and the agitation speed (at high cell density cultivation the air flow may be enriched by additional oxygen).

Today, the most important industrial cultivations of microbial and mammalian cells are carried out in the fed-batch mode. In fed-batch processes, one or more substrates are fed into the bioreactor during the process. The product remains in the bioreactor until the end of the cultivation cycle. Fed-batch processes overcome substrate inhibition and overflow effects. Such an operational mode allows a high cell density and product concentrations to be achieved [6]. By controlling the substrate feeding rate, the optimal conditions for the biotechnological process can be secured.

To realize the bacterial growth rate control systems, efficient glucose feeding algorithms need to be implemented, and in the mammalian cell cultivation processes, additionally, the feeding rate of glutamine needs to be controlled. It is important to note that modern industrial bioreactors are equipped with inexpensive and reliable devices to measure the composition of aeration gas in the inlet and outlet (fraction of O_2, CO_2) and the molar flow rate, Q. Hence, the oxygen uptake rate (OUR) and the carbon dioxide production rate (CPR) can be calculated from the online measurements as follows:

$$OUR = Q\left(O_2^{in} - O_2^{out}\right), \tag{2}$$

$$CPR = Q\left(CO_2^{out} - CO_2^{in}\right). \tag{3}$$

The above measurements allow an online estimation of the important process variables, OUR and CPR, during bacterial and mammalian cell cultivation processes [17,18]. Because of the lower cell density and respiratory intensity, OUR and CPR measurements based on the off-gas composition may cause larger measurement errors in mammalian cell cultivation processes. As an alternative technique, an OUR estimation using dissolved oxygen (DO) measurements may also be applied [19]. The OUR and CPR are the most important variables for indirect monitoring of the biomass growth rate in industrial bioreactors, as they comprehensively reflect the physiological state and metabolic activity of the aerobic biotechnological processes [1–3,11,20].

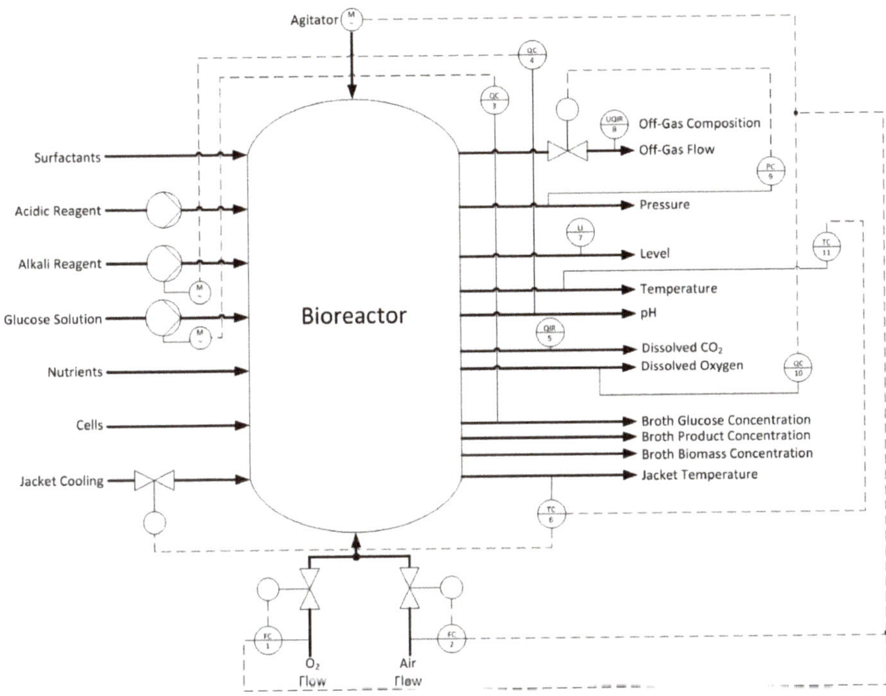

Figure 1. Basic control system loops for a typical microbial cultivation process.

Good performance of the basic control systems improves the batch-to-batch reproducibility/ repeatability of the processes [11,21]. The other advantage of a well-controlled bioreactor is the possibility to run the bioreactor at higher capacity or with better efficiency by operating the process closer to physical constraints. Good reproducibility is also an important condition for possible process improvements and modifications, as improvement in reproducibility by means of well-operating control systems allows a reduction in the number of expensive and time-consuming experiments required to compare the performance indices of the modified processes and to optimize the controlled technological regime [1].

The proportional-integral-derivative (PID) controllers are predominant controllers used in the basic control systems of microbial and mammalian cell cultivation processes. Quality of the bioprocess control depends on the complexity of the process dynamics, the process variable measurement noise and errors, tuning of the system controllers, and performance accuracy of the executive devices (valves and speed drives). Dynamics of the particular biotechnological parameter control process can be characterized by three resulting dynamic parameters: dead time, time constant and process gain. These parameters are commonly used to determine the tuning parameters of a PID controller. Control quality of the bioreactor operation mode critically depends on how well the controllers are set up and tuned to deal with the sources of the process variability [1,2,11,13]. Because of the nonlinearity and nonstationarity of the bioprocesses, proper tuning of the controllers requires appropriate efforts. The performance of PID controllers with fixed tuning parameters are not sufficiently accurate because of the significant variations in the process' dynamics. Consequently, various approaches have been proposed to tune the PID controller parameters in microbial cultivation processes under time-varying operating conditions, including gain-scheduling methods [12,22–24], first-principle models [25], tendency models [26], rule-based fuzzy systems [13], and other techniques [1,2,14,15]. The proposed approaches give a sound theoretical and practical basis to implement adaptive control schemes in bioreactor systems and show that implementation of the adaptive algorithms in basic control systems can significantly increase the performance of the systems. Gain-scheduling methods and tendency models are the most appropriate solutions for improving the quality of basic control systems in microbial cultivation processes because they are relatively simple to implement and pose low requirements on model complexity. The advantages have been extensively discussed and substantiated in previous studies [12,22–24,26]. An important task now is to broaden implementation of these algorithms in industrial bioreactors.

Well-functioning basic control systems create opportunities for further process improvements and also for implementation of the SGR control systems in bioreactors [1,11]. Development of relatively simple solutions to control the specific growth rate in fed-batch processes remains a timely and important task in view of implementing the PAT framework in industrial biotechnological processes.

3. Preconditions for Implementation of SGR Control Systems in Industrial Bioreactors

The basic requirements for SGR control systems designed to control microbial and mammalian cell cultivation processes can be formulated as follows:

- The systems should be as simple as possible and intuitive for the user. Process operators without special modeling/control knowledge should be able to supervise these systems.
- The systems must be based on measurement and control equipment that is currently used standard equipment in industrial bioreactors.
- Development time, cost, and benefits of the systems must be attractive to potential users.

According to the above requirements, most of the solutions for SGR control systems presented in scientific literature [3,6] are not attractive enough for industrial implementation. More complex monitoring and control systems, even if equipped with easy-to-use interfaces, may require retuning, model identification, and maintenance tasks in case the operational modes or microbial cultures have been changed. Often these tasks cannot be carried out by the biotechnology companies alone and may cause additional expenses for outsourcing and production delays. In the authors' opinion, this is the main reason SGR control systems have rarely been used in industrial bioreactors so far.

In this contribution, the authors provide an overview of those SGR control systems that meet the aforementioned requirements. In most widespread control systems, SGR is usually controlled by manipulating the substrate feeding rate [6,27]. In recombinant protein production processes, the temperature of the medium is also used to control cell growth [28]. Despite that the growth rate could be controlled (e.g., by manipulating the dissolved oxygen concentration in cultivation medium [29], temperature of the medium [28] or pH), to date these techniques have not been sufficiently investigated and have not been widely implemented in industrial practice [11].

When the growth rate is controlled by manipulating the substrate feeding rate, the substrate concentration in cultivation medium remains relatively low [30]. This allows avoiding production of the overflow metabolites in some of the most important microbial expression systems, so-called Crabtree-positive organisms, such as *S. cerevisiae* and *E. coli*. The presence of the overflow metabolites, such as acetate or ethanol, frequently leads to inhibition of both the biomass growth and formation of the proteins.

To ensure controllability and batch-to-batch reproducibility, the SGR needs to be controlled during the process at a level that is lower than the maximum SGR [11]. The maximum available SGR is observed during particular growth phases of the process, when the growth is not limited by substrate concentration [12,31], and depends on the specific culture, medium composition, concentrations of biomass and metabolites, as well as the oxygen transfer capabilities of the bioreactor. It is worth mentioning that direct control of the substrate concentration in bioreactor at the set-point does not guarantee that a constant SGR will be kept. This is because [11]:

- Cell growth at a limited rate occurs under low substrate concentrations. Because of this, online measurements, calibration of the measuring devices, and control of the substrate concentration are difficult to implement in industrial bioreactors.
- Sensor readings of the substrate concentration reflect only the local substrate concentration around the sensor, which may significantly differ from the average concentration in the bioreactor. Therefore, the substrate concentration control system is not able to control the SGR in the entire cultivation medium.

In the majority of recombinant protein production processes, the control objective is to maximize the amount of target protein at the end of the process while maintaining high batch-to-batch reproducibility. To achieve this goal, two steps most often are implemented for SGR control:

- During the first stage of the process, the SGR is kept at a trajectory that is 10–15% below the maximum available SGR.
- During the second stage, the SGR is kept at a trajectory that leads to the maximum specific production rate of the target product. Usually, the level of the SGR kept at this phase is significantly lower compared to that maintained at the first stage.

SGR control systems can be realized using open-loop and closed-loop control systems [6,11]. In the following sections, the authors analyze and evaluate SGR control systems that are best suited for industrial applications. The analyzed and evaluated control solutions are ordered in this review by their complexity (i.e., starting with the simplest open-loop systems and ending up with the control systems that employ cascade control schemes and SGR estimators).

4. Schemes for SGR Practical Control Systems

4.1. Open-Loop SGR Control Systems

The majority of industrial fed-batch microbial cultivation processes are operated using open-loop SGR control systems [11], in which the time profile of the substrate feeding rate is calculated using simple mass-balance models and a desired time profile of the SGR during the process. The desired SGR values, μ_{set}, can be described by the following equation:

$$\mu_{set} = \begin{cases} \mu_{set_1} = (0.85\ldots0.90)\mu_{max} \ for \ growth \ phase, \\ \mu_{set_2} = \mu_{opt} \ for \ production \ phase. \end{cases} \tag{4}$$

Based on the desired set values for SGR, the corresponding substrate feeding rate can be determined. Accumulation of the total biomass during cultivation and the substrate feeding rate for both stages of the process can be estimated from simple mass-balance equations:

$$\frac{dX}{dt} = \mu_{set_i} X, \ i = 1, 2. \tag{5}$$

The amount of biomass accumulated in the growth stage can be calculated from the equation

$$X(t) = X_0 e^{\mu_{set_1} t}, \tag{6}$$

and the amount of biomass accumulated in the production stage can be calculated from the equation

$$X(t) = X_0 e^{\mu_{set_1}\, t_g} \cdot e^{\mu_{set_2}\, (t-t_g)}, \tag{7}$$

where t_g (h) is the end time of the growth stage.

Using the predicted time trajectories of the biomass accumulation, $X(t)$, the substrate feeding rate $F_1(t)$ in the growth stage can be derived from the dynamic mass-balance equation for the substrate under steady-state conditions [2,6,8,20,30], which results in the following equation

$$F_1(t) = \frac{X_0 e^{\mu_{set_1}\, t}\mu_{set_1}}{Y_{xs1} S_F}, \tag{8}$$

and the substrate feeding rate $F_2(t)$ in the production stage can be estimated from the equation

$$F_2(t) = \frac{X_0 e^{\mu_{set_1}\, t_g} e^{\mu_{set_2}\, (t-t_g)}\mu_{set_2}}{Y_{xs2} S_F}, \tag{9}$$

where X_0 (g) is the total amount of biomass in the bioreactor at the beginning of cultivation process; $X(t)$ (g) is the time trajectory of the total biomass accumulated during the process; S_F (g/L) is the concentration of the substrate in the feeding solution; and Y_{xs1} and Y_{xs2} (g/g) are the yields of biomass on substrate in the growth and production phases, respectively. In substrate-limited processes, the substrate concentration in the bioreactor is low. Therefore, this concentration is not taken into account in Equations (8) and (9).

When the SGR control algorithm based on Equations (8) and (9) is developed, implementation of the control system is straightforward. For this purpose, only an actuator to dose the feeding substrate to the bioreactor is needed. For some recombinant protein production processes, the yield of biomass on substrate can be different in the growth (Y_{xs1}) and the production stages (Y_{xs2}). In such cases, the yields must be identified from experimental data for the particular process phase and must be taken into account when using Equations (8) and (9) to estimate the substrate feeding rates. Additionally, μ_{max} and μ_{opt} may vary during the process because of the increasing concentrations of metabolites, biomass, and other process variables. In this case, $\mu_{max}(t)$ and $\mu_{opt}(t)$ should be presented as time profiles, and a numerical integration procedure to predict the biomass growth and substrate feeding time profiles needs to be applied.

The substrate feeding time profiles estimated from Equations (8) and (9) can be directly used for implementing the open-loop SGR control systems in various biotechnological processes [6,7,22,30,32,33]. Certainly, more sophisticated bioprocess models and optimization procedures can be used to determine the feeding rate control algorithms in open-loop SGR control systems. These methods are widely reviewed and analyzed in many research and academic papers [3,6,8,27,34,35]. However, implementation of more sophisticated procedures in industrial bioprocesses requires specific knowledge in process modeling and efforts to develop more accurate models. Consequently, application of complex methods in industrial environment is not a commonplace.

The SGR open-loop control systems based on Equations (8) and (9) are easy to implement and do not require additional measurements. On the other hand, open-loop systems do not compensate for process disturbances. Consequently, possible variations in the substrate concentration of the feeding solution or deviations of the feeding flow rate are not compensated. These disturbances can decrease the performance of the biotechnological process. In the next sections, the authors analyze and provide relatively simple and already existing solutions to overcome these problems.

4.2. SGR Control Systems Based on CPR/OUR Estimations

Reliability and accuracy of SGR control can be increased by employing closed-loop control systems. One of the simplest closed-loop SGR control systems is proposed in Reference [27]. Here, based on a simplified assumption that the carbon dioxide production rate (CPR) during the process is in a linear relationship with the biomass growth rate

$$CPR(t) = \alpha\mu(t)X(t),\tag{10}$$

the specific growth rate μ can be estimated using the following equation

$$\mu(t) = \frac{CPR(t)}{\int_0^t CPR(\tau)d\tau},\tag{11}$$

where α is a model parameter, and τ is the integration time variable.

If real-time estimation of SGR is available, feedback control systems can be developed to automatically track a desired SGR time profile by manipulating the substrate feeding rate.

Results of the SGR control obtained in Reference [36] show that an acceptable control quality can be obtained by applying a typical control system based on PI controllers. To achieve better control quality, it is straightforward to adapt the controller parameters to the time-varying dynamics of the controlled process by applying gain-scheduling algorithms mentioned in Section 2 and using the CPR signal as a scheduling variable. The main drawback of the analyzed control approach is that, during SGR estimation, an assumption is made that the CPR during the process is proportional to the absolute biomass growth rate. In fact, it is known that more accurate results may be achieved if the Luedeking–Piret-type relationship is applied to correlate the CPR and the biomass growth rate [20,37]. This relationship additionally takes into account the CPR fraction that is related to the maintenance of the cell's vital functions and accounts for a significant part of the total CPR (for instance, in high-cell-density bacterial cultivation processes). Figure 2 shows the simulated trajectories of the biomass growth and the CPR of the recombinant *E. coli* cultivation process in a 1 m³ volume bioreactor as well as the comparison of the actual and the estimated SGRs. The latter is calculated from Equations (10) and (11). The actual CPR of the process is modeled using the equation

$$CPR(t) = \alpha\mu(t)X(t) + \beta X(t),\tag{12}$$

where parameter β determines the CPR fraction related to maintenance of the cell's vital functions.

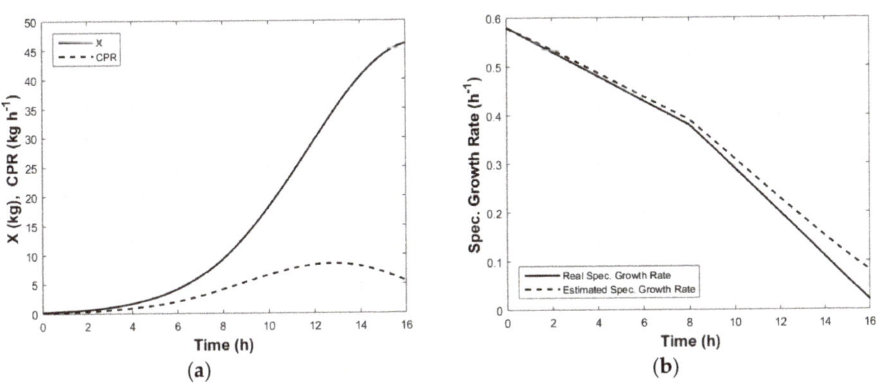

Figure 2. Simulated trajectories of the biomass growth and carbon dioxide production rate (CPR) (a), and the trajectories of the real specific growth rate (SGR) and that estimated from Equation (7) (b) in a typical recombinant *E. coli* cultivation process in a 1 m³ bioreactor.

In the simulation experiment, values of the parameters α and β were used that are typical for recombinant *E. coli* cultivation processes (α = 0.9 (gCO₂/gX), and β = 0.1 (gCO₂/(gX·h))), induction at *t = 8 h*) [38]. The simulation results, presented in Figure 2, show that the estimated SGR deviation from the real trajectory increases with an increasing amount of biomass (the estimated rate at the end of the process is *0.05 (1/h)* higher than the real one). Hence, it is advantageous to introduce empiric correlations correct the estimated SGR when applying this method in high-density cultivation processes. The magnitude of correction should be defined from earlier cultivation experiments.

To estimate SGR, *OUR* data can also be used. However, the measurements related to *OUR* estimation may be corrupted by the noise related to the off-gas composition, pressure, and the gas flow rate fluctuations if additional oxygen is used enrich the aeration air. Therefore, to control the high-density cultivation processes, it is recommended to use *CPR* data in SGR estimation relationships.

A more accurate SGR control system based on *OUR* or *CPR* measurements is proposed in References [25,39]. Realization of the proposed control system does not require a mathematical model and *a priori* knowledge of the culture of the microorganisms under control. It can be realized using standard programmable controllers/measurement devices and is well suited for control of industrial biotechnological processes. In the cited contributions, it is shown that if the substrate feeding rate is manipulated to control *OUR* during the process, in such a way that the *OUR* data-based ratio *R*

$$R = \frac{dOUR}{dt}\frac{1}{OUR},$$ (13)

is stabilized at the desired SGR set-point $R = \mu_{set}$, then the specific growth rate μ will asymptotically approach the set-point μ_{set} and will be controlled at that point. For control of the ratio *R*, the PI control algorithm was recommended, and controller gain was adapted to the time-varying dynamics of the controlled process using the gain-scheduling approach with the feeding rate as the scheduling variable. The block-scheme of the SGR control system and the simulation results of the system's performance are presented in Figure 3. The simulation and experimental investigation tests of the proposed SGR automatic control system have shown a stable performance and sufficiently accurate control of the SGR under stepwise changes to the process parameter values and high-level noise of the feedback signal measurements [25,39]. This control system can be efficiently applied in controlling biotechnological processes, in which the SGR set-point is constant or changes slowly. For realization of the system, either *OUR* or *CPR* online estimates can be used.

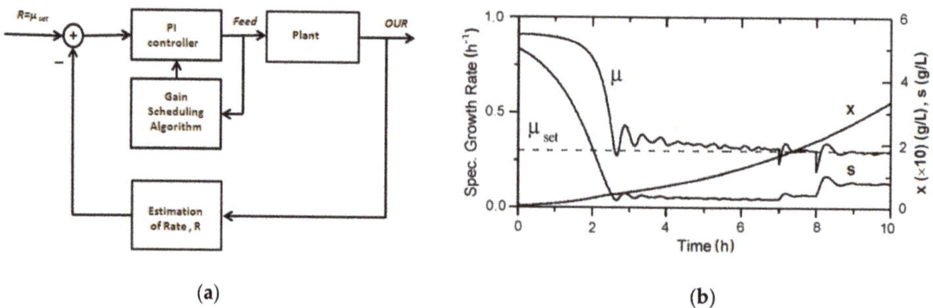

(a)   (b)

Figure 3. Block-scheme of the SGR control system (**a**) and the simulation results of the system performance (**b**). Reproduced with permission from D. Levišauskas, Biotechnology Letters; published by Springer Nature, 2001.

It should be stressed that the SGR control systems based on Equation (11), when applied in high-density cultivation processes, cause noticeable deviations at high cell concentrations. SGR control systems based on Equation (13) are less efficient when tracking time-varying SGR set-point profiles.

In the next sections, more complex control systems are discussed that overcome the above shortcomings.

4.3. SGR Control Systems Based on CPR/OUR Estimations and the Mass of CO2/O2 Produced/Consumed During Cultivation

Robust control of the SGR is a crucial problem when designing an efficient process, in which the SGR is to be controlled at the value $\mu_{set} < \mu_{max}$ in order to secure reproducibility of the processes. However, the already discussed SGR closed-loop control systems have two shortcomings: (a) for system implementation, an online estimation of the μ-values is required, and (b) high batch-to-batch reproducibility is not guaranteed. For example, if disturbances occur during a process (e.g., variation in the initial amount of biomass X_0) or in the instrumentation (e.g., if the substrate feeding is shortly interrupted), they cause slight deviations in the biomass growth trajectory from the desired trajectory, and such an offset cannot be eliminated later on, even if the controller exactly tracks the predefined μ-profile. An approach to cope with the disturbances that cause process reproducibility problems is proposed in Reference [40]. In this work, a desired SGR time profile $\mu_{set}(t)$, the initial amount of biomass X_0, and Equation (1) were used to estimate the biomass growth time profile $X(t)$ during the process. If the biomass growth profile $X(t)$ can be tightly controlled by manipulating the substrate feeding rate, the corresponding SGR profile will follow the desired $\mu_{set}(t)$ profile. This control system is more robust as compared to direct SGR control systems, as the short-term disturbances that occur in the control equipment and the process itself are compensated by controlling an integral variable—the amount of accumulated biomass $X(t)$. However, implementation of the above control system requires development of a reliable soft-sensor for the online estimation of the amount of accumulated biomass during the process. Therefore, the $X(t)$ online estimation problem complicates the practical realization of this control approach. To eliminate this shortcoming, simplified SGR control systems were developed and experimentally tested in bacterial and mammalian cell cultivation processes [41–43]. The main idea behind these control systems is to use the predetermined time profiles of $CPR_{set}(t)$ as the system's time-varying set-point, and the mass of $CO_2(t)$ produced during the process (mCO_{2set}) as an indirect metric for SGR control purposes. $CPR(t)$ is stoichiometrically related to the SGR and the biomass (Equation (9)), and the integral of this equation gives the mass $mCO_{2set}(t)$ produced during the process.

By manipulating the substrate feeding rate to control the predetermined set-point time profile $mCO_{2set}(t)$, the control system indirectly maintains the desired SGR during the process. The structure of the discussed cascade control system is depicted in Figure 4. The PI controller of the inner loop controls the $CPR_{set}(t)$ time profile, and the PI controller of the outer loop controls the set $mCO_{2set}(t)$ profile.

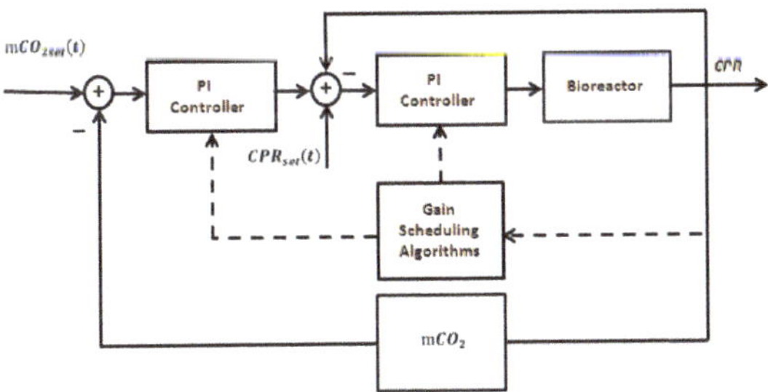

Figure 4. Cascade control system for indirect SGR control based on predetermined $CPR_{set}(t)$ and $mCO_{2set}(t)$ time profiles.

If the controlled process is tightly kept on the $CPR_{set}(t)$ and $mCO_{2set}(t)$ time profiles, the process will also follow the desired SGR time profile. The proposed control system ensures good quality of the SGR control, and, because of the cumulative nature of the set-point variable $mCO_2(t)$, random disturbances do not significantly distort the course and reproducibility of the process.

Implementation of the proposed control system can be realized in the following steps:

- Choose a rational $\mu_{set}(t)$ time profile for the process. A proper profile can be estimated from expert knowledge, mathematical model-based process optimization results, or from the analysis of a successful "golden batch" experiment.
- Choose an appropriate inoculum size (initial amount of the total biomass X_0) for the process and estimate the biomass growth time profile $X(t)$ using the $\mu_{set}(t)$ profile, Equation (4), and a numerical integration procedure.
- Estimate the $CPR_{set}(t)$ time profile using Equation (10) and the identified parameter values α and β. Note that the above parameter values may be different for the biomass growth and product formation stages.
- Integrate the $CPR_{set}(t)$ time profile to get the corresponding profile $mCO_{2set}(t)$ for the controlled process.
- Control the process by tracking the estimated profiles $CPR_{set}(t)$ and $mCO_{2set}(t)$. Control is realized using the cascade control system that manipulates the substrate feeding rate.

Various realizations of the above control system have been investigated by computer simulations of the system's performance and by controlling real processes of recombinant *E. coli* and mammalian cells (CHO) [18,41,42]. Typical results of the applied control system for controlling the recombinant *E. coli* fed-batch cultivation processes over six runs are presented in Figure 5. The laboratory-scale experimental results show that the proposed control approach leads to a stable and robust behavior of the controlled process. It should also be stressed that small variations in the initial amount of biomass X_0 and short instrumentation disturbances do not significantly affect the reproducibility of the process.

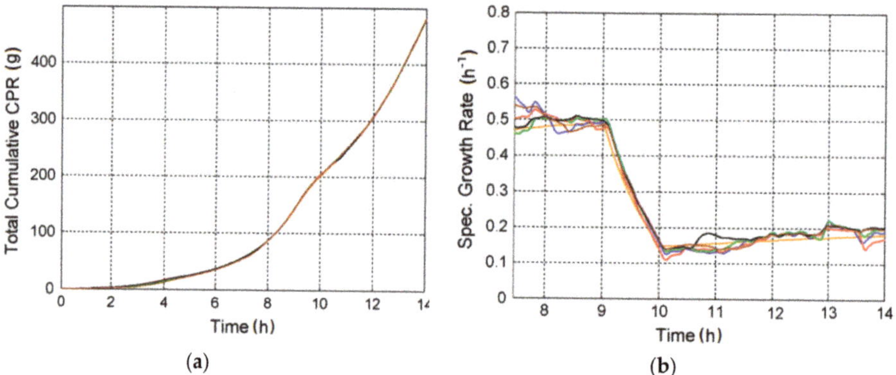

Figure 5. Typical experimental results of the total cumulative CPR (**a**) and SGR indirect control ($\mu_{set}(t)$ = 0.5 1/h at the first process stage and $\mu_{set}(t)$ = 0.175 1/h at the second stage) (**b**) during the recombinant *E. coli* cultivation process. Reproduced with permission from M. Jenzsch et al. J. of Biotechnology; published by Elsevier, 2007.

Because of the significant changes in the process dynamics during cultivation, it is possible to improve control quality of the cascade control system by adapting controller parameters. Tuning parameters of the PI controllers can be adapted to time-varying dynamics of the controlled process using the gain-scheduling approach. The controller adaptation scheme using the gain-scheduling algorithm is shown in Figure 4 by the dashed lines. In the gain-scheduling algorithms, one can use CPR or OUR measurements as the gain-scheduling variables.

Instead of using the $OUR(t)$ or $CPR(t)$ time profiles, the performance of the inner control loop of the cascade control system can also be improved by implementing the SGR estimator, developed from Equations (12) and (13) [44]. Investigation results presented in Reference [44] have shown that the control system with the SGR estimator outperforms the control system depicted in Figure 4 when the controlled process is affected by disturbances to the substrate feeding rate. On the other hand, implementation of the modified control system requires additional calculations related to online estimation of the SGR.

The structure of the SGR control system presented in Figure 4 may be used as a basis for development of closed-loop control systems for controlling the processes of various microbial cultures in industrial bioreactors. Because it is technically simple to implement and possible to improve batch-to-batch reproducibility, this system could be used as a benchmark to compare the control quality of various SGR control systems and to evaluate their potential implementation in industrial bioreactors.

5. Concluding Remarks and Recommendations

In recent years, numerous research papers have been published, in which original solutions and sophisticated control techniques were developed for the automatic control of microbial and mammalian cell cultivations processes. However, the majority of the proposed control systems are too complicated to be attractive for robust control of industrial biotechnological processes. Therefore, the well-known statement of Luyben [16], "Complex elegant control systems look great on paper but soon end up on 'manual' in an industrial environment", is also valid for the majority of the control systems developed for biotechnological processes.

In this paper, relatively simple control approaches that can be applied in microbial and mammalian cell cultivation processes are discussed and recommended for practical application. The reviewed algorithms and systems designed for indirect control of the specific growth rate can significantly increase robustness and batch-to-batch reproducibility of industrial-scale biotechnological processes. The recommended control algorithms and systems are based on CPR or OUR online measurements and on the total mass of oxygen consumed or the total mass of carbon dioxide produced during the process. In the case when additional oxygen is used during the processes, it is recommended to use the CPR and mCO_2 signals in the control system algorithms because of their lower estimation errors compared to those when using OUR and mO_2 signals. To estimate oxygen uptake and carbon dioxide production rates, several industrially well-established gas analyzers and mass flow meters are available. Basic instrumentation for installation of the SGR control systems, the online gas analyzer, combines parallel measurement of CO_2 and O_2 concentrations in the off-gas using two space-saving sensors. The analyzer can be used both for lab- and industrial-scale bioreactors. In the industrial gas analyzers, compensators for gas pressure and humidity are incorporated. Consequently, these analyzers ensure good precision and reliability of the measurements.

The available instrumentation and discussed control methods and systems provide a possibility for wider application of SGR control in biotechnological processes. At the very beginning of the process, accuracy of the indirect measurements is usually low and insufficient to track exactly the SGR set-point time profile in closed-loop control systems. Consequently, it is recommended to start the process using the feeding rate open-loop control strategies determined by Equations (8) and (9) and, after three to four hours, to switch the SGR control to the closed-loop control system. For the low-density cell cultivation processes, the SGR control system based on Equation (11) is recommended. For processes, in which the SGR set-point is kept constant, the control system based on Equation (13) (Figure 3) is well suited. For more advanced applications, the SGR control system presented in Figure 4 is recommended. The above system can be applied as a benchmark to compare the control quality of various SGR control systems.

Author Contributions: Conceptualization, V.G., R.S., and D.L.; methodology, V.G., R.S., and D.L.; software, V.G. and R.S.; validation, V.G., R.S., and D.L.; formal analysis, V.G., R.S. and D.L.; investigation, V.G., R.S., D.L., and R.U.; resources, V.G., R.S. and D.L.; data curation, V.G.; writing—original draft preparation, V.G. and R.S.; writing—review and editing, V.G., R.S., D.L., and R.U.; visualization, V.G.; supervision, V.G.; project administration, V.G.; funding acquisition, V.G., R.S., D.L., and R.U.

Funding: This research was funded by the European Regional Development Fund according to the supported activity "Research Projects Implemented by World-class Researcher Groups" under the Measure No. 01.2.2-LMT-K-718.

Conflicts of Interest: The authors declare no conflicts of interest.

References

1. Boudreau, M.A.; McMillan, G.K. *New Directions in Bioprocess Modelling and Control: Maximizing Process Analytical Technology Benefits*; ISA: Eindhoven, The Netherlands, 2007; ISBN 978-1-556-17905-1.
2. Dochain, D. *Bioprocess Control*; ISTE: London, UK, 2008; ISBN 978-0-470-61112-8.
3. Schuler, M.M.; Marison, I.W. Real-time monitoring and control of microbial bioprocesses with focus on the specific growth rate: Current state and perspectives. *Appl. Microbiol. Biotechnol.* **2012**, *94*, 1469–1482. [CrossRef] [PubMed]
4. Rocha, I.; Veloso, A.; Carneiro, S.; Costa, R.; Ferreira, E. Implementation of a specific rate controller in a fed-batch, *E. coli* fermentation. *IFAC Proc. Vol.* **2008**, *41*, 15565–15570. [CrossRef]
5. Puertas, J.; Ruiz, J.; Vega, M.R.; Lorenzo, J.; Caminal, G.; González, G. Influence of specific growth rate over the secretory expression of recombinant potato carboxypeptidase inhibitor in fed-batch cultures of *Escherichia coli*. *Process Biochem.* **2010**, *45*, 1334–1341. [CrossRef]
6. Mears, L.; Stocks, S.M.; Sin, G.; Gernaey, K.V. A review of control strategies for manipulating the feed rate in fed-batch fermentation processes. *J. Biotechnol.* **2017**, *245*, 34–46. [CrossRef]
7. Chenikher, S.; Guez, J.S.; Coutte, F.; Pekpe, M.; Jacques, P.; Cassar, J.P. Control of the specific growth rate of *Bacillus subtilis* for the production of biosurfactant lipopeptides in bioreactors with foam overflow. *Process Biochem.* **2010**, *45*, 1800–1807. [CrossRef]
8. Jenzsch, M.; Simutis, R.; Luebbert, A. Generic model control of the specific growth rate in recombinant *Escherichia coli* cultivations. *J. Biotechnol.* **2006**, *122*, 483–493. [CrossRef]
9. Food and Drug Administration. Guidance for Industry. PAT–A Framework for Innovative Pharmaceutical Development, Manufacturing, and Quality Assurance. 2004. Available online: https://www.fda.gov/regulatory-information/search-fda-guidance-documents/pat-framework-innovative-pharmaceutical-development-manufacturing-and-quality-assurance (accessed on 4 August 2019).
10. Gnoth, S.; Jenzsch, M.; Simutis, R.; Lübbert, A. Process Analytical Technology (PAT): Batch-to-batch reproducibility of fermentation processes by robust process operational design and control. *J. Biotechnol.* **2007**, *132*, 180–186. [CrossRef]
11. Simutis, R.; Lübbert, A. Bioreactor control improves bioprocess performance. *Biotechnol. J.* **2015**, *10*, 1115–1130. [CrossRef]
12. Gnoth, S.; Kuprijanov, A.; Simutis, R.; Lübbert, A. Simple adaptive pH control in bioreactors using gain-scheduling methods. *Appl. Microbiol. Biotechnol.* **2010**, *85*, 955–964. [CrossRef]
13. Babuška, R.; Damen, M.R.; Hellinga, C.; Maarleveld, H. Intelligent adaptive control of bioreactors. *J. Intell. Manuf.* **2003**, *14*, 255–265. [CrossRef]
14. Smets, I.Y.; Claes, J.E.; November, E.J.; Bastin, G.P.; Impe, J.F. Optimal adaptive control of (bio)chemical reactors: Past, present and future. *J. Process Contr.* **2004**, *14*, 795–805. [CrossRef]
15. Bastin, G.; Impe, J.F. Nonlinear and adaptive control in biotechnology: A tutorial. *Eur. J. Contr.* **1995**, *1*, 37–53. [CrossRef]
16. Luyben, W.L. *Process Modeling, Simulation, and Control for Chemical Engineers*, 2nd ed.; McGraw-Hill: New York, NY, USA, 1990; ISBN 978-0-07-039159-8.
17. Aehle, M.; Kuprijanov, A.; Schaepe, S.; Simutis, R.; Lübbert, A. Simplified off-gas analyses in animal cell cultures for process monitoring and control purposes. *Biotechnol. Lett.* **2011**, *33*, 2103–2110. [CrossRef] [PubMed]
18. Aehle, M.; Schaepe, S.; Kuprijanov, A.; Simutis, R.; Lübbert, A. Simple and efficient control of CHO cell cultures. *J. Biotechnol.* **2011**, *153*, 56–61. [CrossRef] [PubMed]
19. Lin, J.; Takagi, M.; Qu, Y.; Yoshida, T. Possible strategy for on-line monitoring and control of hybridoma cell culture. *Biochem. Eng. J.* **2002**, *11*, 205–209. [CrossRef]
20. Pirt, S.J. *Principles of Microbe and Cell Cultivation*; Blackwell Scientific Publications: Oxford, UK, 1985; ISBN 978-0-632-01455-2.
21. Urniezius, R.; Galvanauskas, V.; Survyla, A.; Simutis, R.; Levisauskas, D. From Physics to Bioengineering: Microbial Cultivation Process Design and Feeding Rate Control Based on Relative Entropy Using Nuisance Time. *Entropy* **2018**, *20*, 779. [CrossRef]

22. Åkesson, M.; Hagander, P. A Gain-Scheduling Approach for Control of Dissolved Oxygen in Stirred Bioreactors. *IFAC Proc. Vol.* **1999**, *32*, 7608–7613. [CrossRef]
23. Galvanauskas, V. Adaptive pH control system for fed-batch biochemical processes. *Inf. Technol. Control* **2009**, *38*, 225–231.
24. Gnoth, S.; Simutis, R.; Lübbert, A. Selective expression of the soluble product fraction in *Escherichia coli* cultures employed in recombinant protein production processes. *Appl. Microbiol. Biotechnol.* **2010**, *87*, 2047–2058. [CrossRef]
25. Levišauskas, D. An algorithm for adaptive control of dissolved oxygen concentration in batch culture. *Biotechnol. Tech.* **1995**, *9*, 85–90. [CrossRef]
26. Levišauskas, D.; Simutis, R.; Galvanauskas, V. Adaptive set-point control system for microbial cultivation processes. *Nonlinear Anal. Model. Control* **2016**, *21*, 153–165. [CrossRef]
27. Lee, J.; Lee, S.Y.; Park, S.; Middelberg, A.P.J. Control of fed-batch fermentations. *Biotechnol. Adv.* **1999**, *17*, 29–48. [CrossRef]
28. Velut, S.; de Maré, L.; Hagander, P. Bioreactor control using a probing feeding strategy and mid-ranging control. *Contr. Eng. Pract.* **2007**, *15*, 135–147. [CrossRef]
29. Galvanauskas, V.; Volk, N.; Simutis, R.; Lübbert, A. Design of Recombinant Protein Production Processes. *Chem. Eng. Comm.* **2004**, *191*, 732–748. [CrossRef]
30. Lim, H.C.; Shin, H.S. *Fed-Batch Cultures: Principles and Applications of Semi-Batch Bioreactors*; Cambridge University Press: Cambridge, UK, 2013; ISBN 978-0-521-51336-4.
31. Jenzsch, M.; Gnoth, S.; Beck, M.; Kleinschmidt, M.; Simutis, R.; Lübbert, A. Open-loop control of the biomass concentration within the growth phase of recombinant protein production processes. *J. Biotechnol.* **2006**, *127*, 84–94. [CrossRef]
32. Shiloach, J.; Fass, R. Growing *E. coli* to high cell density–A historical perspective on method development. *Biotechnol. Adv.* **2005**, *23*, 345–357. [CrossRef]
33. Sun, Z.; Ramsay, J.A.; Guay, M.; Ramsay, B.A. Automated feeding strategies for high-cell-density fed-batch cultivation of *Pseudomonas putida* KT2440. *Appl. Microbiol. Biotechnol.* **2006**, *71*, 423–431. [CrossRef]
34. Wechselberger, P.; Sagmeister, P.; Engelking, H.; Schmidt, T.; Wenger, J.; Herwig, C. Efficient feeding profile optimization for recombinant protein production using physiological information. *Bioprocess Biosyst. Eng.* **2012**, *35*, 1637–1649. [CrossRef]
35. Yüzgeç, U.; Türker, M.; Hocalar, A. On-line evolutionary optimization of an industrial fed-batch yeast fermentation process. *ISA Trans.* **2009**, *48*, 79–92. [CrossRef]
36. Chul Lee, S.; Gyeom Kim, C.; Keun Chang, Y.; Nam Chang, H. Dissolved Oxygen Concentration and Growth Rate Control in Fed-Batch Fermentation Process. *IFAC Proc. Vol.* **1992**, *25*, 267–270. [CrossRef]
37. Luedeking, R.; Piret, E.L. A kinetic study of the lactic acid fermentation. Batch process at controlled pH. *Biotechnol. Bioeng.* **2000**, *67*, 636–644. [CrossRef]
38. Jenzsch, M.; Simutis, R.; Eisbrenner, G.; Stückrath, I.; Lübbert, A. Estimation of biomass concentrations in fermentation processes for recombinant protein production. *Bioprocess Biosyst. Eng.* **2006**, *29*, 19–27. [CrossRef] [PubMed]
39. Levisauskas, D. Inferential control of the specific growth rate in fed-batch cultivation process. *Biotechnol. Lett.* **2001**, *23*, 1189–1195. [CrossRef]
40. Jenzsch, M.; Gnoth, S.; Kleinschmidt, M.; Simutis, R.; Lübbert, A. Improving the batch-to-batch reproducibility in microbial cultures during recombinant protein production by guiding the process along a predefined total biomass profile. *Bioprocess Biosyst. Eng.* **2006**, *29*, 315–321. [CrossRef] [PubMed]
41. Jenzsch, M.; Gnoth, S.; Kleinschmidt, M.; Simutis, R.; Lübbert, A. Improving the batch-to-batch reproducibility of microbial cultures during recombinant protein production by regulation of the total carbon dioxide production. *J. Biotechnol.* **2007**, *128*, 858–867. [CrossRef]
42. Aehle, M.; Bork, K.; Schaepe, S.; Kuprijanov, A.; Horstkorte, R.; Simutis, R.; Lübbert, A. Increasing batch-to-batch reproducibility of CHO-cell cultures using a model predictive control approach. *Cytotechnology* **2012**, *64*, 623–634. [CrossRef]

43. Kuprijanov, A.; Schaepe, S.; Aehle, M.; Simutis, R.; Lübbert, A. Improving cultivation processes for recombinant protein production. *Bioprocess Biosyst. Eng.* **2012**, *35*, 333–340. [CrossRef]

44. Levisauskas, D.; Galvanauskas, V.; Simutis, R.; Repsyte, J. Inferential control system for tracking desired trajectory of biomass growth in fed-batch cultivation process. In *Recent Advances in Circuits, Systems and Automatic Control, Proceedings of the 14th International Conference on Circuits, Systems, Electronics, Control & Signal Processing (CSECS '15), Konya, Turkey, 20–22 May 2015*; WSEAS Press: Sofia, Bulgaria, 2015; pp. 57–62.

MDPI
St. Alban-Anlage 66
4052 Basel
Switzerland
Tel. +41 61 683 77 34
Fax +41 61 302 89 18
www.mdpi.com

Processes Editorial Office
E-mail: processes@mdpi.com
www.mdpi.com/journal/processes